SUPERCONDUCTIVITY

SUPERCONDUCTIVITY

K.N. SHRIVASTAVA

(EDITOR)

NOVA SCIENCE PUBLISHERS, INC.
NEW YORK

Art Director: Maria Ester Hawrys
Assistant Director: Elenor Kallberg
Graphics: Denise Dieterich, Kerri Pfister,
          Erika Cassutti and Barbara Minerd
Manuscript Coordinator: Gloria H. Piza
Book Production: Tammy Sauter, Gavin Aghamore
Circulation: Irene Kwartiroff and Annette Hellinger

Library of Congress Cataloging–in–Publication Data
available upon request

ISBN 1-56072-360-2

© *1996 Nova Science Publishers, Inc.*
  6080 Jericho Turnpike, Suite 207
  Commack, New York 11725
  Tele. 516-499-3103 Fax 516-499-3146
  E Mail Novasci1@aol.com

Printed in the United States of America

Contents

PREFACE

Superconductivity was discovered by Kammerlingh Onnes in Leiden in 1911. At that time the transition temperature below which this phenomenon occurs were very low. Somehow the entire sample never became superconducting so that it took many years before the Meissner effect, which is the expulsion of magnetic field from the sample, could be discovered in 1933. Bardeen, Cooper and Schrieffer found that the phenomenon is caused by the spin singlet zero momemtum pairs in 1957. Later in 1960, Josephson showed that the pairs of electrons can tunnel through a thin layer of an oxide. Thus there is a phase coherence and the unit charge of twice the electron charge can be measured. In 1975, Shrivastava found that the zero-field splitting goes through a minimum value at a characteristic temperature. This temperature was high compared with the transition temperatures of the superconductors but the system becomes repulsive above this temperature and attractive below this temperature. Thus the temperature at which there was a minimum in the energy level splitting is a transition temperature. In 1984, the energy minima shown by Shrivastava were at 38, 42 and 51 K in the Ampere Congress in Zürich in 1984. In 1986, Bednorz and Müller found that the transition temperature of a superconductor may be 34 K which is about 12 degrees higher than the highest value of 22.5 K known at that time. The Nobel prizes were awarded to Kammerlingh Onnes;John Bardeen, Leon Cooper and Robert Schrieffer; Brian Josephson; and Jorg Bednorz and and Alex Müller. It should be noted that Landau, Giever and Esaki were also awarded the Nobel prizes for their work closely related to the area of superconductivity.

The number of persons, about 10, who were awarded the Nobel prize for their work related to superconductivity shows that it is surely amongst the greatest of the discoveries of this century. Other than fundamental studies, it has applications in industry. Therefore many industries and governments are interested in this subject.

We have been actively working in superconductivity for the last many years during which several discoveries have been made by us. Therefore it was thought to have a meeting in Hyderabad. The present book is generated from this meeting. Its several chapters compare in standard with the best ones in the world.

K. N. Shrivastava
Editor

Combined Phononic and Lochonic Mechanism and the Isotope Effect in Cuprate Superconductors

K.P. Sinha and A. Rastogi

Department of Physics
Indian Institute of Science
Bangalore 560 012

Abstract

The combined phonon and lochon (local charged boson) mediated pairing mechanism developed by us for cuprate superconductors has been used to calculate the oxygen isotope exponent as a function of critical temperature T_c) which, in turn, depends on doping. The expression for the oxygen isotope exponent α_o has the form

$$\alpha_o = (\tfrac{1}{2}) \frac{\lambda_p}{\lambda_p + \bar{\lambda}_e} + \tfrac{1}{2} g^2 \ln \left(\frac{\omega_e}{\omega_p}\right) \frac{\lambda_p \, \bar{\lambda}_e}{(\lambda_p + \bar{\lambda}_e)^2}$$

$$+ \tfrac{1}{2} g^2 \frac{(1+\lambda_b) \bar{\lambda}_e}{(\lambda_p + \bar{\lambda}_e)^2}$$

Where λ_p and $\bar{\lambda}_e$ are the phononic and electronic coupling constants, ω_p and ω_e are the corresponding cut off and g^2 is a parameter connected with the hybridisation between wide band and localised states. Further λ_e depends on doping and decreases for decreasing carrier concentration from the maximum value.

The expression is capable of giving positive (and even $\alpha_o > \frac{1}{2}$)or negative values of the exponent depending on the values of the parameters involved for perticular system. As $\bar{\lambda}_e$ decreases with decreasing hole concentration (ie. decreasing temperature), the values of α_o increase with decreasing T_c. The theoretical graph gives the right trend of the variation of α_o as a function of T_c (or doping) observed in some cuprates.

1.   <u>Introduction</u>

More than twenty five years ago an interaction mechanism involving fermions and lochons (local charged bosons, local pairs etc.) was suggested by us for the enhancement of superconducting transition temperature of some systems [1]. The model implicitly contained the concept of negative U centers (or complexes) harbouring local pairs or bipolarons. Later several groups elsewhere also developed similar models [2-4]. This mechanism was again invoked by us and others to explain the superconducting and normal state properties of high T_c oxide systems (from 1987 onwards) [5-12]. However, in our approach a combined lochon and phonon mediated processes are taken into account for the pairing of fermions belonging to a wide band [5-9]. In this model, any deviation from ideal Fermi liquid behaviour arises from the interaction of fermions with lochons in the system [9]. The fermion-lochon interaction mechanism involves the splitting of a lochon into a pair of fermions and the confluence of a fermion pair to give a lochon. Opinion is veering towards the view that the high T_c oxide systems contain both the Fermi and Bose fluid features; the predominance of one or the other depends on the degree of doping [14].

The recent measurements of oxygen isotope effects in cuprates present very interesting features [15-19]. For small doping range in which the T_c is considerably lower the isotope exponent tends towards larger values. Alexandrov has attempted to explain this on the basis of polaronic theory [20]. The purpose of the present paper is to give an alternative explanation based on the combined phononic and lochonic mechanism. Carbotte and Akis have also phenomenologically discussed the role of a combined phononic and electronic mechanism without specifying the nature of the latter interaction [21]. Similarly, Newns et al. invoke the role of electronic mechanism arising out of van Hove singularity in correlated quasi-two dimensional Fermi gas and try to explain the isotope shift as a function of doping [22]. However, the cuprates are entropy (disorder) stabilized metastable systems, and the critical van Hove effects may be smeared out. Also, it cannot explain the observed value of the ratio $2\Delta(o)/T_c \sim 6$ [23].

2.   <u>Combined Phononic and Lochonic Mechanism</u>

The Hamiltonian for the combined phonon and lochon mechanism in the Nambu spinor representation is given by [9]

$$H = H_o + H_{cp} + H_{cl},\qquad(1)$$

$$H_o = \sum_{\underline{k}} \epsilon_{\underline{k}} \psi_{\underline{k}}^{+} \sigma_z \psi_{\underline{k}} + \sum_{\underline{q}} \omega_{\underline{q}} a_{\underline{q}}^{+} a_{\underline{q}} + \sum_{1} E_{1} b_{1}^{+} b_{1}\qquad(2)$$

$$H_{cp} = \sum_{\underline{q}} P_{\underline{q}} \phi_{\underline{q}} \psi_{\underline{k}}^{+} \sigma_z \psi_{\underline{k}'},\qquad(3)$$

$\phi_q = (a_q + a^+_{-q})$, phonon field operator, a^+_q and a_q being phonon creation, annihilation operator, P_q is the coupling constant involving carriers and phonons.

$$H_{cl} = \Sigma \, B_l \, \psi^+_k \, [\sigma_+ b_l + \sigma_- b^+_l] \psi_k, \tag{4}$$

$$b^+_l = C^+_{l\uparrow} \, C^+_{l\downarrow}; \quad b_l = C_{l\downarrow} \, C_{l\uparrow} \tag{5}$$

are the lochon creation, annihilation operators, lochons being the composite of fermions localized at site l, and B_l is the carrier-lochon coupling constant. The carrier field operators are expressed as

$$\psi_k = \begin{pmatrix} C_{k\uparrow} \\ C^+_{-k\downarrow} \end{pmatrix}, \quad \psi_k = \begin{pmatrix} C^+_{k\uparrow} & C_{-k\downarrow} \end{pmatrix} \tag{6}$$

$C_{k\sigma}(C^+_{k\sigma})$ being the carrier annihilation (creation) operator in the state $|k\sigma\rangle$, k = wave vector, σ spin index and ε_k the corresponding single particle energy; ω_q and E_l are the phonon and lochon energies; and $\sigma_{\pm} = 1/2(\sigma_x \pm i\sigma_y)$, σ_x, σ_y, σ_z being the Pauli matrices. Although we have not written the term for the Coulomb repulsion, its effect will be incorporated in the expressions to follow.

The superconducting order parameter in the temperature Green function formalism is given by:

$$Z\Delta(\omega_n,k) = \frac{T}{(2\pi)^d} \sum_{\omega_{n'}} dk' \Gamma(\omega_n - \omega_{n'}, k - k') F^+(\omega_{n'}, k'), \tag{7}$$

where $\omega_n = (2n + 1)\pi T$ are the Matsubara frequencies, $T =$ temperature. The anamolous Green's function

$$F^+(\omega_n, \underline{k}) = \frac{\Delta(\omega_n, \underline{k})}{[\omega_n^2 + \varepsilon^2 + \Delta^2(\omega_n, \underline{k})]} \qquad (8)$$

The explicit form of the renormalization function Z is not written here. The vertex function is a sum of the phononic and electronic parts

$$\Gamma = \Gamma_{ph} + \Gamma_{1o}, \qquad (9)$$

where

$$\Gamma_{ph} = \lambda_p \, D_{Ph}(\omega_n - \omega_{n'} ; \underline{k} - \underline{k}') \qquad (10)$$

λ_p being the carrier - phonon coupling constant, and the phonon Green's function is given by

$$D_{ph} = \frac{\omega^2}{(\omega_n - \omega_{n'})^2 + \omega^2} \qquad (11)$$

and

$$\Gamma_{1o} = \lambda_e D_{1o}(\omega_n - \omega_{n'}) + V_c \qquad (12)$$

λ_e being the carrier-lochon coupling constant and D_{1o} is the lochon Green function

$$D_{1o} = \frac{1}{(\omega_n - \omega_{n'}) - E_1} \qquad (13)$$

and V_c represents the usual Coulomb repulsion. As the electron-phonon interaction is not taken to be the dominant mechanism in the present formulation, we shall treat Z as a renormalization parameter with $Z = 1 + \lambda_p$.

3.   <u>Critical temperature and the oxygen isotope exponent</u>

Thus in the modified weak-coupling approxiamtion, the solution of equation (7) near $T = T_c$, after the usual contour integration and using two-cut offs corresponding to phonon and electronic modes, gives the critical temperature T_c as

$$1 + \lambda_p = \lambda_p \ln(1.14 \frac{\omega_p}{T_c}) + \overline{\lambda}_e (1.14 \frac{\omega_e}{T_c}) , \qquad (14)$$

where ω_p and ω_e are the Lorentzian cut offs and T_c the critical temperature in dimensionless units for the phonon and electronic mechanisms respectively. This yeilds an expression for the transition temperature T_c in the combined mechanism as

$$T_c = 1.14(\omega_p)^{\nu_p}(\omega_e)^{\nu_e} \exp\left[- \left(\frac{\lambda_p + \overline{\lambda}_e}{1 + \lambda_p}\right)^{-1}\right] \qquad (15)$$

where,

$$\nu_p = \frac{\lambda_p}{\lambda_p + \overline{\lambda}_e} , \quad \nu_e = \frac{\overline{\lambda}_e}{\lambda_p + \overline{\lambda}_e} , \qquad (16)$$

with the present form of $\overline{\lambda}_e$ extended as

$$\overline{\lambda}_e = \lambda_e(\xi_e) \exp(-g^2) - \mu, \qquad (17)$$

μ being the coupling constant connected with Coulomb repulsion. Here $\exp(-g^2)$ arises from the polaronic type reduction of the hybridisation between wide band and localised states and is temperature independent for $K_B T \ll \omega_o$, ω_o being the frequency of the boson mode involved [24]. This will involve phononic mode other than those considered in Eqn. (3) and (4). Further, as shown earlier, the attractive interaction, induced by lochons $\lambda_e(\xi_c)$ will depend on carrier concentration [5-9].

$$\lambda_e(\xi_c) = \lambda_e^o f(\xi_c) \simeq \lambda_e^o \, \xi_c(1 - \xi_c), \qquad (18)$$

where $f(\xi_c)$ is a modulating factor and depends on the carrier concentration with ξ_c expressed in normalized dimensionless form (ξ_c ranges from 0 to 1). The reason is that the numbers of free carriers and local charged bosons (lochons) are not independently conserved. They are changing into one another (pair wise) owing to the mutual interaction. This dependence gives the variation of T_c as a function of hole concentration (doping); it will have a maximum for a certain level of doping and will decrease for other ranges. It also explains why in the overdoped region T_c becomes very small and the system just becomes a better metal. This effect may be further modulated if some remnants of van Hove singularity in the density of states $N(\varepsilon)$ remain. Also the polaronic reduction factor is likely to be influenced by the degree of doping; g^2 will increase with decreasing ξ_c.

The oxygen isotope exponent α_o for the average reference mass M_o (with $\omega_p = $ constant $M_o^{-1/2}$) and noting that

$$g^2 = \text{constant } M_o^{1/2} \quad [20] \qquad (19)$$

is given by the expression

$$\alpha_o = - \frac{\partial \ln T_c}{\partial \ln M_o} \qquad (20)$$

where T_c and M_o are in dimensionless units.

Its explicit form using the relation (15) and the above dependence of ω_p and g^2 on M_o turns out to be:

$$\alpha_o = (1/2) \frac{\overline{\lambda}_p}{\lambda_p + \overline{\lambda}_e} + (1/2)g^2 \ln\left(\frac{\omega_e}{\omega_p}\right)[\lambda_p \overline{\lambda}_e/(\lambda_p + \overline{\lambda}_e)^2]$$

$$+ (1/2)g^2 \frac{(1 + \lambda_p)\overline{\lambda}_e}{(\lambda_p + \overline{\lambda}_e)^2} \qquad (21)$$

the parameters ω_p and ω_e are expressed in dimensionless units. It is clear from the above expression that when the electronic part vanishes i.e. $\overline{\lambda}_e = 0$, the BCS values $\alpha_o = 1/2$ is restored. Qualitatively, the full expression is capable of giving positive (and even $\alpha_o > 1/2$) or negative values of the isotope exponent depending on the values of the parameters involved. In what follows we give an estimate for some typical choice of parameters.

4.   Discussion of results

In Fig. 1, we have given a plot of the computed values of α_o versus T_c to display the general behaviour. It shows a clear trend of the increase of oxygen isotope exponent as the transition temperature T_c decreases for some cuprate compounds. When the lochonic contribution is minimal for very low hole concentration, the isotope exponent tends towards the phononic value 0.5 and can even exceed. In this graph, the experimental points for a few systems have been displayed in the same figure. The values of various parameters involved are taken as follows: $\lambda_p = 0.15$, $\lambda_e^o =$

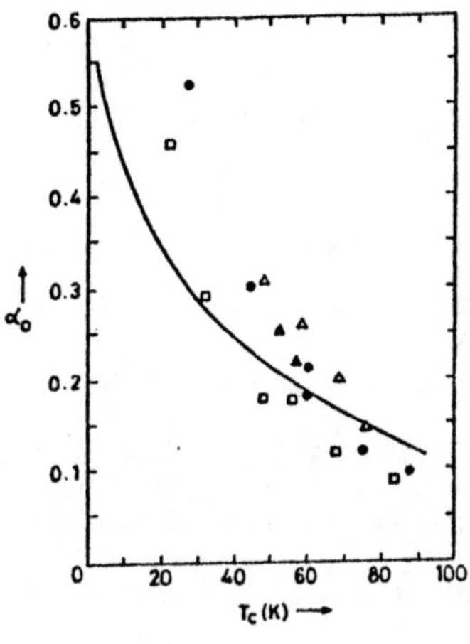

Fig.1 *The oxygen isotope exponent α_o as a function of T_c. Solid line represents the theoretical graph; various symbols are experimental points for systems as shown:*

($\bullet$, $\blacktriangle$) $Y_{1-x}Pr_xBa_2Cu_3O_{6.92}$

($\square$) $YBa_2(Cu_{1-x}Co_x)O_{7-\delta}$

($\triangle$) $YBa_{2-x}La_xCu_3O_7$

3.0, μ = 0.1, ω_p = 100, ω_e = 500, $g^2(\xi_c)$ = 0.03, 0.0375, 0.05, 0.075, 0.15 for ξ_c varying from ξ_c = 0.5, 0.4, 0.3, 0.2, 0.1 respectively.

In the case of $La_{2-x}Sr_xCuO_4$ and $La_{2-x}Ba_xCuO_4$, the low temperature orthorhombic (LTO) to low temperature tetragonal (LTT) structural phase transition which occurs at x $\simeq$ 0.12 [16], some parameters, for example the local phonon mode and the associated coupling constants λ_p and $\overline{\lambda}_e$ and in particular g^2 will undergo sudden change. This may cause a dip in T_c and peak in the α_o value. The tilting instability of the CuO_6 octahedra is perhaps driven at some critical doping. Although, X-ray diffraction measurements show little change in the bulk lattice parameters, there will be local rearrangement of atoms. This, in turn, is bound to affect the parameters in question. Thus deviations in the α_o values for $La_{2-x}M_xCuO_4$ is due to local changes in atomic configurations which produce concomitant changes in the corresponding parameters. Hence, the experimental data for this system are not compared with a theoretical plot. In the case of YBCO system Pr doping in $(Y_{1-x}Pr_x)Ba_2Cu_3O_7$ changes the hole concentration and hence T_c. Similarly, in $YBa_2(Cu_{1-x}Co_x)_3O_{7-\partial}$, cobalt substitution for copper leads to partial hole filling in the planes; thus decreasing mobile hole concentration. Experimental values for system $YBa_{2-x}La_xCu_3O_7$ are also shown [19]. The increase of α_o with decreasing T_c value shows that the decline in the lochonic mechanism leads to an increase in the value of α_o. The increase in α_o, however, is not a universal function of T_c [25]. The combined mechanism does explain the general trend. We have not attempted a precise quantitative calculation for each system as the various parameters involved cannot be computed at this stage of the development of the field. Changing the values of the parameters to get better agreement will amount to curve fitting. That doping affects the electronic component of the combined mechanism considerably can be seen from the figure.

5.   Concluding remarks

Based on our combined phononic and electronic (lochonic) mechanism, we have shown the trend of α_o as a function of T_c which depends on doping. The decrease in the electronic component leads to an increase in α_o. This confirms that although the mechanism involving local charged bosons dominates in the high T_c region, phonons are also involved to a lesser degree in the pairing mechanism. In contradistinction to the bipolaronic mechanism leading to Bose condensation, the combined mechanism discussed here involves Cooper pairing of the fermions through the mediation of phonons and lochons. In agreement with Carbotte and Akis [21], we find that a pure phonon mechanism alone will not be able to explain the isotope effect and other properties of the cuptrates systems [26,27].

References

[1]   B.N. GANGULY, U.N. UPADHYAYA and K.P. SINHA, Phys. Rev. 146, 317 (1966).

[2]   D.M. EAGLES, Phys. Rev. 178, 668 (1969)

[3]   S.P. IONOV, Izv. Acad. Nauk SSSR Ser Fiz, 49, 310 (1985).

[4]   J. RANNINGER and S. ROBASZKIEWICZ, Physica B135, 468 (1985).

[5]   K.P. SINHA and M. SINGH, J. Phys. C21, L 231 (1988).

[6]   M. SINGH and K.P. SINHA, Solid State Commun, 70, 149 (1989).

[7]   K.P. SINHA, Physica B163, 664 (1990).

[8]   K.P. SINHA and M. SINGH, Phys. Status Solidi(b) 159, 787 (1990).

[9]   K.P. SINHA, Solid State Commun. 79, 735 (1991).

[10] R. MICNAS, J. RANNINGER and S. ROBASZKIEWICZ, Revs. Mod. Phys. 62, 113 (1990) and references therein.

[11] K.P. SINHA, Ind. J. Phys. 66A, 1 (1992) and references threrein.

[12] D.M. EAGLES, Physica C211, 319 (1993) and references therein.

[13] C.I. VENTURA, B.R. ALASCIO and R. ALLUB in Proc. XV Int. Workshop in Condensed Matter Theories (Argentinia, 1991).

[14] A.S. ALEXANDROV and N.F. MOTT, Supercond. Sci. Technol, 6, 215 (1993).

[15] M.K. CRAWFORD, M.N. KUNCHUR, W.E. FARNETH, E.M. McCARRON III, S.J. POON, Phys. Rev. B41, 282 (1990).

[16] M.K. CRAWFORD, W.E. FARNETH, E.M. McCARRON III, R.L. HARLOW and A.H. MONDDEN, Science 250, 1390 (1990).

[17] J.P. FRANCK, J. JUNG, M.A.K. MOHAMED, S. GYGAX AND G.I. SPROULE, Physica, B169, 697 (1991).

[18] J.P. FRANCK, S. GYGAX, G. SOERENSEN, E. ALTSHULTER, A. HNATIW, J. JUNG, M.A.K. MOMAMED, M.K. YU, G.I. SPROULE, J. CHRZANOWSKI and J.C. IRWIN, Physica. C185-189, 1379 (1991).

[19] H.J. BORNEMANN, D.E. MORRIS, H.B. LIU, A.P.B. SINHA, P. NARWANKAR AND M. CHANDRACHOOD, Physica, C185-189, 1359 (1991).

[20] A.S. ALEXANDROV, Phys. Rev. B46, 4932 (1992).

[21] J.P. CARBOTTE and R. AKIS, Solid State Commun. 82, 613 (1992).

[22] D.M. NEWNS, H.R. KRISHNAMURTHY, P.C. PATTNAIK, C.C. TSUEI and C.L. KANE, Phys. Rev. Lett. 69, 1264 (1992).

[23] A.A. ABRIKOSOV, J.C. CAMPUZANO and K. GOFRON, Physica C214, 73 (1993).

[24] S. ROBASZKIEWICZ, R. MICNAS and J. RANNINGER, Phys. Rev. B36, 180 (1987).

[25] J.P. FRANCK, S. HARKER and MING YU, Physica, B194-196, 2087 (1994).

[26] J.C. PHILLIPS, Phys. Rev. B43, 6257 (1991).

[27] V.H. CRESPI AND M.L. COHEN, Phys. Rev. B48, 398 (1993).

Symmetry Considerations in an Electronic Mechanism of Superconductivity

Partha Bhattacharyya

Theoretical Physics Group
Tata Institute of Fundamental Research
Homi Bhabha Road, Colaba
Colaba, Bombay 400 005

ABSTRACT

We show that Josephson transitions as a means of restoring gauge symmetry by a zero mode in an electronic mechanism of superconductivity implies scale invariance having magnetic monopole as the quantum. Quantum interference of time reversed paths across an energy barrier results in a unitary transformation such that a finite displacement is achieved without change of the dual variable. A possible physical interpretation is the coexistence of a Peierl's spin density wave at $Q/2 = mV_s$ that shows as a Fermi surface that is nested.

One of the most intriguing possibilities in the discovery of superconductivity in cuprate superconductors is the possibility of new symmetries of the order parameter that are different from the conventional S-wave pairing in phonon mediated superconductivity and superfluidity in He^3 where the pairing is p-wave. Among the ideas that are interesting are recent experiments $(1-3)$ where the authors have presented measurements of phase coherence in dc SQUIDS and tunnel junctions made from single crystals of $YBa_2Cu_3O_{6.8}$ and thin films of Pb, a conventional S-wave superconductor. Although, the authors claim that the symmetry of the pairing state can be determined unambiguously by the experiment, their results are not conclusive strongly suggesting, however, an order parameter having a symmetry $d_{x^2-y^2}$. Excellent accounts and comparisons are given in review by Dynes (4) and Schrieffer (5). An important distinction between conventional isotropic S-symmetry superconductor and one of a $d_{x^2-y^2}$ symmetry has zero crossings along the diagonals. This results in lines on the Fermi surface where the gap is zero and hence the gaplessness in the density of excitations

$$N(E) \propto E, \ |E| \leq \Delta$$

$$d_{x^2-y^2} \tag{1a}$$

$$\propto Re\left\{\frac{E}{\sqrt{E^2-\Delta^2}}\right\} |E| > \Delta$$

Compared to

$$N(E) \propto Re\left\{\frac{E}{\sqrt{E^2 - \Delta^2}}\right\} \tag{1b}$$

where in Eq. (1a), Δ is the maximum value of the gap. It is important to remark, however, that higher symmetry is not necessarily the observation of points or lines of $\Delta = 0$ on the Fermi surface. For instance it is, in principle, possible to retain S-symmetry and have zero crossings that retain the square rotational symmetry, however, rather than the four zero crossings in a 2π rotation one expect a minimum of eight. It is worth pointing out that zero crossings are possible in conventional superconductors, in a CDW superconductor, for instance.

We remark and will show that these remarks on d-symmetry are also obtained when Josephson transitions across lines of disclinations is the mode of restoration of gauge symmetry. Indeed, when one starts with a singlet configuration, in a charge (or color) conjugate ground state, we are immediately convinced of the need to think of electronic states excitations is from an equilibrium $S\bar{S}$ configuration, where S represents a half quantized SG soliton or, equivalently, half charged lines of disclination when the phase ϕ is in $[\pi/2, -\pi/2]$. Evidently, the quantum transition is best described by a Hamiltonian having normal state electronic energy states on the diagonal and the transition amplitude Δ on the off-diagonal. This is similar to the usual BCS Hamiltonian having $T_c = J \exp(-1/9)$, where J is an electronic energy. (The quantum nature of the duality between number and phase implies some features that are not seen in the usual BCS-theory).

In the usual description of a Josephson transition, space is no longer simply connected, since there are regions where the phase slips through π. This results in the creation of an fermionic excitation, the magnetic monopole (6) or a bosonic excitation

that is a mode of scale symmetry restoration. Schrieffer has given a nice summary of some recent theoretical implication. The quantum number of the electron describing a quantum transition can now be thought of as a quantum number describing the electron as an elementary excitation from a zero momentum condensate Quantum interference between time reversed paths implies a scaling in units of a localization length that is manifested as an umklapp transition on a quasicrystalline lattice. A finite displacement operator $\exp(iaP)$ where a is a unit of length gives a phase factor $\exp(iKa) = \exp(i\phi)$ having a representation as generator of infinitesimal rotations, the angular momentum operator L_z with eigenvalue $1/2$. By duality, the angular momentum operator $\exp(i\theta L_z)$ is the generator of scaling $K \rightarrow K + \Delta K$, where $\Delta K = ((\theta)(\pi)K$. A little reflection will convince one that this statement is merely the conservation of angular momentum complying that scale symmetry is the manifestation of gauge symmetry.

The step operators $L_{\pm}$ create an elementary excitation with energy Δ that is dispersionless implying that the zero mode is also the Goldstone mode of the quasicrystalline order. Similarly, the step operators $\exp(i\theta L_{\pm})$ induce a phase gradient $\nabla\theta \propto Q$ where Q is a Brillouin zone vector consistent with Z_2 boundary conditions and the dual operator $P_{\pm}$ requires propagation at $x \pm v_f t$. A physical manifestation of this duality is then the representation of the superconducting order parameter as acoexisting spin density wave (SDW) Peierls' state having an intrinsic supercurrent $Q/2 = mv_S$. This manifests in the observation of a nested Fermi surface in the normal state in $YbCo$. The idea that pairing in the $d_{x^2-y^2}$ state results in apparently disconnected regions of the Fermi surface finds a nice interpretation.

Quasicrystallography (7)

Consider a trajectory of a charged particle in a compact space R^N having a representation in a gauge group $G(h_1, \cdots, h_r)$ where $\{h_i\}$ are the Cartan generators. We can associate a vector with each point P_r corresponding to the time evolutions of $\mathbf{h} \equiv (\exp(2i\pi x_1), \cdots, \exp(2i\pi x_n))$ that is a vector conjugate to an $SU(N)$ matrix. This is also the central element x in the Lie algebra such that $\hat{\tau}(x) = x$, where $\hat{\tau}$ is the group of outer automorphisms. In $SU(2)$, $\mathbf{h}$ is the Pauli matrix τ_3 corresponding to the rotation by a phase around the z axis. The generic class in $SU(N)$ corresponding to N distinct $\exp(2i\pi x_\mu)$ is of the form $SU(N)/U(1) \times \cdots N$ times and intersects in $N!$ points the Cartan torus $T_1(\mathbb{Z}_2)$ for a singlet configuration where $\mathbb{Z}_2$ boundary conditions are required.

An invariant trajectory is maintained by having a sequence of quantum transitions corresponding to permutations in the group of inner automorphisms of the generators $\{h_1, \cdots, h_n\}$ symmetry considerations alone have shown that when the free particle motion of an electron in a gauge group G/H, where G is simply connected induces a Goldstone mode in the dual space, the rotation in the phase of the wave function $\exp(i\theta L_z) \sim P_z$ where $\mathbf{P}$ is the generator of infinitesimal translations. Similarly, we think of $L_{\pm}$ as the creation operator for an Einstein mode where $P_{\pm}$ is the generator of reflections on a quasicrystalline lattice. This is most easily achieved when superconducting order is in coexistence with a spin density wave (SDW) order parameter modulated at a wave vector $\mathbf{Q}/2 = m\mathbf{v}_S$. A prediction of symmetry considerations is the observation of nested Fermi surfaces seen in some high temperature superconductors.

It is evident, now, that permutations of $\{x_\mu\}$ are generated by transpositions. These may be realized as (improper) orthogonal transformations corresponding to reflections in the $(n-1)$-dimensional hyperplane orthogonal to the vectors $e_{\mu\nu} = e_\mu - e_\nu$ lying in V of the form

$$\mathbf{x} \to S_{\mu\nu} \ \ \mathbf{x} = \mathbf{x} - (\mathbf{e}_{\mu\nu} \cdot \mathbf{x})\mathbf{e}_{\mu\nu}^{\star}$$

$$e_{\mu\nu} = e_\mu - e_\nu, \quad \mathbf{e}_\mu = (\mathbf{e}_1 + i\mathbf{e}_2)_\mu, \quad \underset{\sim}{e}_\alpha \cdot e_\beta = \delta_{\alpha\beta} \tag{2}$$

where the unit vectors parametrize the quasicrystalline lattice sites and are defined on $SO(3)_2$ i.e. $SO(3)$ with opposite sites ($\phi \to \pi + \phi$) equivalenced. Topological excitations of this equivalence class is valued in $\mathbb{Z} \oplus \mathbb{Z}_2$ implying the existence of an elementary excitation quantized in half the usual units. A possible manifestation is the monopole quantization condition

$$1 = e^{\star} \cdot m^{\star} \tag{3}$$

where $(e^{\star}, m^{\star})$ are the (electric, magnetic) charges. Thus if the magnetic charge is $1/2$, the electric charge is that of a Cooper pair and the elementary flux quantum is $\Phi_0 = h/2e$. The equivalence classes of quasicrystalline lattice sites also makes possible the non symmetry related equivalence of states. In other words, just as an infinitesimal transition induces a reflection $\psi \to \psi + \pi$, reflection symmetry of the dual order parameter, namely $\underset{\sim}{e}_{\mu\nu} \to e_{\mu\nu}$ implies that the excitations are defined from a Peierls' state, such that

$$\exp(i\Phi/\Phi_0) H(\Phi_0) \exp(-i\Phi/\Phi_0) = H(\Phi_0 + \Phi) \tag{4}$$

This is one of the important observations of this paper.

Microwave attenuation and d-wave superconductivity.

One of the anomalies in the electronic properties of cuprate superconductors is revealed in the microwave modulation characteristics in small magnetic fields. It is given by the virgin curve

$$\Delta P(H) = P(H) - P(0) = (H/H_0)/(1 + (H/H_0)) \tag{5}$$

that seems to be a feature of samples prepared by many different techniques. Regarding H_0, the following points are notable: (i) $H_0 \sim a + b(1-t)$ where $t = T/T_c$; (ii) there is no systematics between H_0 and the grain size, ($\sim$ microns) and (iii) H_0 is a weak function of frequency. Consiering that the microwavelengths are longer than the London length, a dynamical length, we are interested (8) in the restoration of gauge symmetry by a $k = 0$ mode. Translational invariance requires that a linearly polarized mode is thought of as left and right polarized waves where the plane of polarizaation rotates. In a colour singlet configuration, this is achieved by the action of the magnetization operator L_+

that creates a spin flop in the singlet configuration. This is a scale transformation $k \to k \pm Q/2$ on a quasicrystalline lattice in the usual propagation of a spin wave mode (a phason). An approximate description can be obtained by considering a Hamiltonian that has a dulaity between the mangetization variable and the co-ordinate having rotational symmetry and a duality where the zero mode is an eigenmode

$$H\alpha \int \sqrt{1 + |\nabla \underset{\sim}{h}(y)^2} \, dy + \int \left(h^2(y)/2\lambda_{sq}^2 \right) d^2 y$$

where $\underset{\sim}{h}(y)$ is a magnetization variable. The spin wave contribution to the correlation function is obtained by recognizing that $(Q = 1/\lambda_{sq})$

$$
\begin{aligned}
\langle |h_Q|^2\rangle &\alpha \int_q \langle \exp(i\phi(z))\exp(-i\phi(z))\rangle_q \\
&= \int^Q \langle \exp\left(-\langle \phi(z)\phi(z')\rangle\right)\rangle|_q \\
&= \int^Q (1/z - z')|_q = \int^Q \frac{dqq}{q^2} \\
&= \tfrac{1}{2}\ln\left(Q^2/Q_c^2\right),
\end{aligned}
$$

where Q_c is a spin fluctuation wave vector of moduloating the SDW order parameter consistent with the S invariance $(a \rightarrow a_D, a_D \rightarrow a)$ $\underset{\sim}{a}$, $\underset{\sim}{a}_D$ being the usual and dual gauge fields). A magnetic field has the effect of modulating the CDW order parameter $k \rightarrow k + Q_H, Q_H\alpha\sqrt{H}$ when implying that scale invariance is maintained in a magnetic field when $(q \rightarrow q + Q_H)$ when symmetrized gives $\langle |h_Q|^2\rangle_H \rightarrow \langle h_{Q+}h_{-Q-}\rangle = \tfrac{1}{2}\ln((Q^2 + Q_H^2)/Q^2)$ renormalized that correlations disappear when $H \rightarrow 0$. We have shown that correlations calculated by this method are consistent with the requirements of a disappearance of interfacial tension and that the microwave attentuation given by the change in the free energy is seen experimentally. An interesting possibility is that the reflected wave from a SNS interface is $\exp(-i(kx + Q_N + \delta_N))$ where $\theta_N = 2\pi/N$ and $\delta_N = \delta - \theta_N$ where $\delta = \pi$ fermions, pseudoscalar bosons and $\delta = 0$ bosons. For the $k = 0$ mode when $k \rightarrow k + Q_H$ in a magnetic field, the difference in the magnetization curve between the directions is proportional to $\delta_H\alpha Q_H$ implying magnetization $M \propto Q_H \propto \sqrt{H}$.

One of the interesting features of a single electron transition is that a quantum transition is that there are density of states in the gap. If the energy $|E| < \Delta$, the transition is at $\omega = 0$ and is along the great circle of $SO(3)$.

$$
N(E) = Im\ G(k,\omega = 0)|_{\text{phys.},E=E_k}
$$

where $|_{\text{phys.}}$ denotes that the states are connected by $\exp(iaL_\mu)$ that is just the step operator P_μ, the momentum operator. Then,

$$
\begin{aligned}
N(E) &= \langle |P_\mu T(\quad)P_\mu|\rangle|_{E=E_k} \\
&= (\mathbf{P}^2 - E^2)\frac{(n_{k_+} - n_{-k_-})}{\omega - E}\Big|_{\omega=0} k_\pm = k \pm \Delta k \qquad (8) \\
&= (E/\Delta)(-\partial n/\partial E) \\
&= (E/4\Delta)\,\mathrm{Sech}^2(E/2\Delta)
\end{aligned}
$$

that is the result of a time reversal operator, the minus sign though for $E \ll 2\Delta$, $N(E) \propto E$, the being $\phi \rightarrow \phi + \pi$ strict linearity ought to allow some deviations. Eq. (6) is approximately the same as Eq. (1) adduced as evidence for d-wave. For $|E| > 2\Delta$, the physical states are connected by a unit charge transition corresponding to a Cooper pair state being either occupied or unoccupied and gives the usual result

$$N(E) \propto E/\sqrt{E^2 - \Delta^2}$$

after taking into account that the physical states each contribute a factor E/ξ.

We have seen, similarly, that the step operator $L_\pm$ create an Einstein mode having an energy Δ that is the energy of a single electron excitation in the dispersion $E^2 = \xi^2 + \Delta^2$ where $\xi = (p - p_f)v_f$ is the energy of an electron along the light cone. This operator is the generator of translations without rotations implying a creation of q where $\mathbf{G}$ is one of the Brillouin zone vectors and $\Delta\mathbf{G}$ is an infinitesimal scale change. The transition rate $1/\tau_\Delta$ for a single electron transition is proportional to the Bose occupation factor $n_B(\omega_q, q \to 0)$ implying $\tau_\Delta \propto 1/T$. While in the superconducting state, this is probably the reason for the electronic Raman line (9) seen both in the 90 k and the 60 k phase, the normal state susceptibility is given as is usual in spin diffusion in normal fermi liquids as

$$\chi'(\omega) \propto (\omega^2\tau_\Delta^2/(1 + \omega^2\tau_\Delta^2))/\coth(\omega\tau_\Delta) \tag{9}$$

in approximate agreement with the predictions of a marginal fermi liquid.

SUMMARY

A possibility giving an electronic mechanism of superconductivity having transition temperatures $T_c \sim 100$ K is the possibility of a Josephson mode restoring symmetry in a singlet state of a colour conjugate configuration. The equilibrium configuration is an equal number of lines of disclinations quantized in half the usual units. A singlet electron transition across induces a monopole configuration that induces logarithmic terms in the free energy. This is believed to be responsible for the some of the anomalous features of microwave attenuation. Another feature of such a quantum transition is that the electron moves without acceleration implying that the density of states is given approximately as $N(E)\alpha 1/n_B(E/\Delta) \sim (E/\Delta)$. An important possibility is the coexistence of SDW order. This result, as also the feature of intrinsically equivalent configurations arising as a result of the connectedness of space are intuitively appealing as a manifestation of symmetry in the context of d-wave superconductivity.

References

1. D.A. Wohlman et al., Phys. Rev. Lett. $\underline{71}$, 2134 (1993)

2. T. Yokoya et al., Preprint. See Ref. 4

3. P. Chaudhari and S.-Y. Lin, Phys. Rev. Lett. $\underline{72}$, 1084 (1994)

4. J.R. Schrieffer, Solid State Commun. $\underline{92}$, 129 (1994)

5. R.C. Dynes, Solid State Commun. $\underline{92}$, 53 (1994)

6. N. Seiberg and E. Witten, Nucl. Phys. $\underline{B426}$, 19 (1994)

7. C. Itzykson, Int'l. Journal of Mod. Phys. $\underline{A1}$, 65 (1986)

8. P. Bhattacharyya, Physica B$\underline{194\text{-}196}$, 1361 (1994), S.M. Bhagat et al, Physica C$\underline{202}$, (1992)

9. T.P. Deveraux et al, Phys. Rev. Lett. $\underline{72}$, 396 (1994)

Vortex Overlapping: Magnetization and Specific Heat of High Temperature Superconductors

K.K. Nanda [†]

Institute of Physics
Sachivalaya Marg
Bhubaneswar – 751005, India

Abstract

We study the temperature and field dependence of magnetization and specific heat of high temperature superconductors by incorporating the vortex overlapping in the London theory [Ref. 9]. At low fields, this vortex overlapping mechanism gives results which are in between that obtained from the usual London theory and the variational model proposed by Hao *et al.* The value of the upper critical field (B_{c2}), in-plane penetration depth (λ_{ab}) and the anisotropy of effective mass (γ) obtained from the specific heat data are reasonable aggreement with earlier reported values.

1. INTRODUCTION

Since the discovery of high temperature superconductors (HTSCs), numerous authors have solved the 3-dimensional (3D) anisotropic London equations to calculate the thermodynamical quantities when the applied field B_a are arbitrarily oriented. The HTSCs are extremely type-II superconductors with the Ginzburg-Landau (GL) parameter κ of the order of 10^2. For $B_a \sim 1$ T, these superconductors remain in the mixed state. In this field range, the superconducting coherence length (ξ) of HTSCs is much smaller than the interspacing between vortex cores. Hence, the core energy is believed to be insignificant compared to the total energy of the superconductors which justifies the use of the London theory. In this theory, the order parameter $\mid \psi \mid^2$ is considered to be constant throughout the sample [1]. Due to the strong overlap of the vortex cores at higher fields, $\mid \psi \mid^2$ is no more constant in space. If f be the normalised order parameter defined as the ratio of the order parameter in a field to that in zero field, the spatial average of square of the normalised order parameter $<\mid f^2 \mid>$ will be $(1 - b)$ with $b = B/B_{c2}$, where B is the magnetic flux density and B_{c2} is the upper critical field [2]. The overlapping between vortices can be accounted for approximately by introducing a field-dependent penetration depth $\lambda_{eff} = \frac{\lambda}{(1-b)^{1/2}}$ in the London theory.

The behavior of magnetization (M) and specific heat (C_p) of the HTSCs in magnetic fields are very unusual. The reversible magnetization below the transition temperature T_c in case of HTSCs is understood by using 3D description of the London theory [3]. It has also been applied to understand the temperature dependence of specific heat in magnetic field [4]. In the range of intermediate magnetic fields B_a with $\phi_0/\lambda^2 << B_a << \phi_0/\xi^2$, where λ is the penetration depth and ϕ_0 is the flux quantum, the reversible magnetization of a superconductor is a linear function of $ln B_a$ which follows directly from the London theory. The accuracy of this theory becomes insufficient for determination of superconducting parameters because of the additional suppression of the order parameter in the vortex core by other vortices [5]. It has also been pointed out by Koshelev [6] that the variational approach used in Ref. [5] exaggerates the effect of suppression of the order parameter in the vortex cores at small fields. In this work, we study the temperature and field dependence of magnetization and specific heat by the vortex overlapping mechanism. From the specific heat data, the value of B_{c20}, in-plane penetration depth $(\lambda_{ab}(0))$ and anisotropy of effective mass (γ) are obtained by least square fitting procedure

2. MAGNETIZATION

Magnetization measurements represent an important tool for studying the magnetic behavior in superconductors. The HTSCs exhibit the magnetic properties of extreme type-II superconductors characterized by large κ.

2.2 Magnetization of Y(123)

It is found experimentally that in the vicinity of the transition temperature T_c, rounding in the $M(T)$ dependence of Y(123) is observed [7,8]. In this temperature range the magnetization is proportional to $(T - T^*)^2$, where T^* is almost field independent and close to T_c. Also, the slopes of the $M - T$ curves decrease as the applied magnetic field is increased.

According to GL-theory, the penetration depth and the upper critical field varies with temperature according to

$$\lambda(T) = \frac{\lambda_0}{\sqrt{2(1 - t)}} \quad and \quad B_{c2} = B_{c20}(1 - t)^{2\nu}, \tag{1}$$

where $t = T/T_c$. The value of ν is 1/2 in the mean field region and 2/3 in the critical region. In this context, we note that for large κ and low applied fields, the order parameter is almost constant outside the core and the GL-theory reduces to London theory.

Incorporating the overlapping of vortices in the London theory the magnetization for 3D superconductors can be written as [9],

$$M = M_L = B - H = -\frac{\phi_0}{32\pi^2\lambda_{eff}^2}ln(\frac{B_{c2}}{B_a}) = -\frac{\phi_0 f^2}{32\pi^2\lambda^2}ln(\frac{B_{c2}}{B_a}). \tag{2}$$

In the case of HTSCs, a transition from 2D to 3D superconductivity is expected when the coherence length along $c-$axis (ξ_c) is equal to the interlayer distance (s). For Y(123), it occurs at 75 K and it is very close to T_c for Bi(2212). Thus, in case of Y(123), the Josephson interlayer coupling regime is realised at all temperatures below 75 K, and above this temperature, there is no thermal or quantum fluctuations of the vortices [10,11]. Also, 3D description should be sufficient to explain the physical properties of Y(123) in the mixed state. The field dependence of magnetization are shown in Fig 1(a). It is noted that the vortex overlapping mechanism at low fields gives results which are in between that obtained from the usual London theory and the variational approach proposed by Hao et al. [5] and the magnetization is approximately logarithmic in B_a. The M vs T

curves at different fields according to Eq. (2) are shown in Fig. 1(b). There are vortex overlapping that give rounding in the magnetization near the transition temperature.

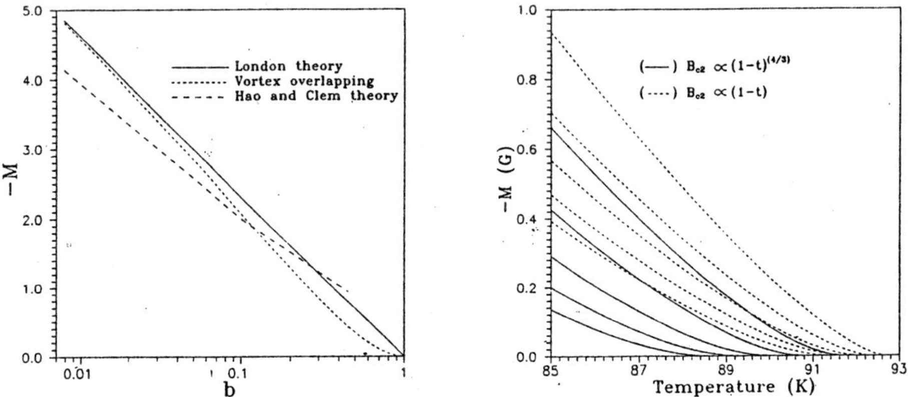

Fig.1: (a) The magnetization (in units of $\phi_0/32\pi^2\lambda^2$) vs b. (b)The temperature dependence of magnetization in different fields (from top to bottom) 1, 2, 3, 4, 5 T for 3D superconductors according to Eq. (3).

2.2 Magnetization of Bi(2212)

Experimental data on Josephson-coupled layered superconductors show that $\frac{dM}{dB}$ is positive at $T < T^*$ and negative at $T > T^*$, i. e. all $M(T)$ curves at different B_a cross at $T = T^*$, where $M(T^*)$ is field independent [12-15]. Thus, $T_c(B_a)$ increases with increasing B_a, opposite to the usual behavior. Recently, Grover et al. [16] have defined (T_{2D}, B_{2D}) line as $\frac{dM}{dB_a}=0$ line. The overall shape of the B_{2D} line, however, is not a sharp phase transition but a gradual crossover spread over a narrow temperature interval as determined from the intersection of $M-T$ curves at different fields [16]. It was shown by Bulaevskii Ledvij and Kogan (BLK) [17] that T^* is independent of the field. If T_{2D} and T^* are the same, it would imply that $M-T$ curves for all B_a would not intersect at the same T^*. As B_a values increase, the temperature-band of intersection of $M-T$ curves would expand towards the lower temperature opposite to that found by Ref. [18] where the curves at higher fields cross each other at slightly higher temperatures.

Moreover, in the intermediate field range, the $-4\pi M$ vs lnB_a curve of Bi(2212) is not linear as predicted from the London theory. The magnitude of the slope of this curve decreases slowly with increasing magnetic field parallel to $c-$axis [19-21]. Bulaevskii and co-workers [10,21] have explained that the observed variation of the slopes with magnetic field for Bi(2212) is due to the thermal and quantum fluctuations of vortices. Whereas, at high temperatures and close to T_c, the magnetization data can be understood by BLK approach [17]. Eq. (2) gives the magnetization for 3D superconductors. For B_a parallel to $c-$axis, in Josephson-coupled layered superconductors, the thermal fluctuations of vortices contribute to the free energy. The change in magnetization due to fluctuations is given by [17],

$$M_{th} = \frac{k_B T}{\phi_0 s} ln \frac{16\pi k_B T \kappa^2}{\alpha \phi_0 s B_a \sqrt{e}}, \tag{3}$$

where s is the interlayer spacing , α is a dimensionless parameter of order unity and k_B is Boltzmann constant. Eq. (3) is obtained by assuming that the vortex lines collapse into vortex pancakes in each plane which are thought of as interacting two-dimensional particles of area $\alpha\pi\xi^2$. Thus, for Josephson-coupled superconductors in magnetic field parallel to $c-$axis, the total magnetization becomes

$$M = M_L + M_{th} = -\frac{\phi_0(1 - \frac{B_a}{B_{c2}^{\|c}})}{32\pi^2\lambda_{ab}^2} ln(\frac{B_{c2}^{\|c}}{B_a}) + \frac{k_B T}{\phi_0 s} ln \frac{16\pi k_B T \kappa^2}{\alpha \phi_0 s B_a \sqrt{e}}, \tag{4}$$

where λ_{ab} is the in-plane penetration depth and $B_{c2}^{\|c}$ is the upper critical field along $c-$axis. This expression fails to explain the low temperature behavior at fields below 7 T because of quantum fluctuations [21]. Whereas, it describes the experimental data in the vicinity of T_c.

The temperature dependence of magnetization of Josephson-coupled layered superconductors for different fields (1T−5T) is shown in Fig. 2(a) using parameters $\lambda(0){=}1500$ Å, $B_{c2}(0){=}113.5$ T, $S{=}15$ Å, $\kappa{=}100$, $\alpha{=}3.2$ and $T_c{=}95$ K taken from Ref. 17. The dashed lines represent the BLK results and the solid lines represent the vortex overlapping results of Eq. (4). From Fig. 2(a), we see that T^* is not a rigorous crossing point, but that the higher field curves cross at low temperatures $i.e$ T^* is field dependent and the temperature-band of intersection of $M - T$ curves expands towards the lower temperature as B_a values increase. The width of the temperature-band depends on the variation of upper critical field as shown in Fig. 2(b). In this Fig., we have plotted the magnetization versus temperature for different fields (1T−5T) considering different variation of B_{c2}.

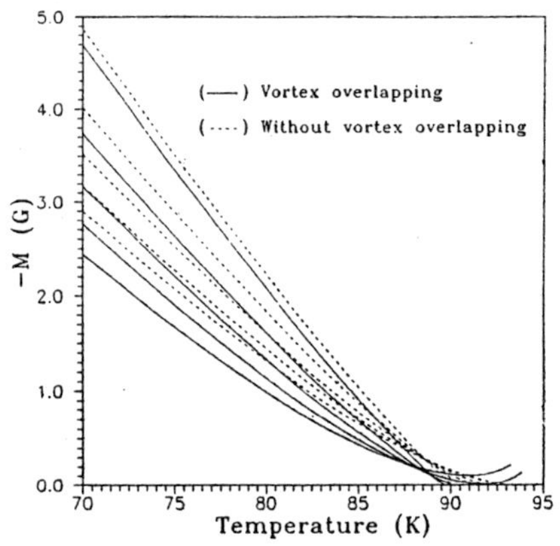

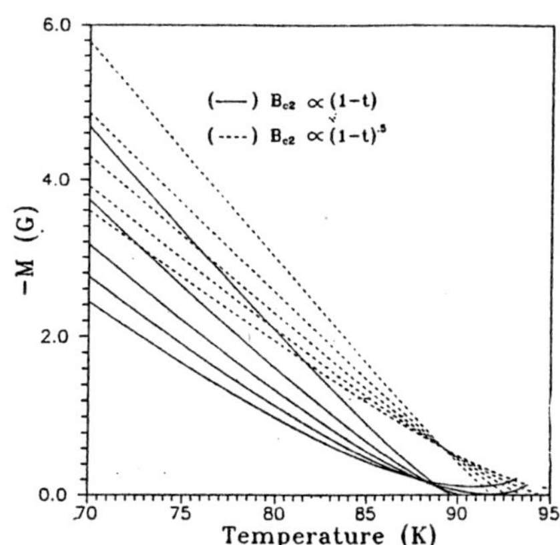

Fig.2: The magnetization vs temperature curves for a Josephson-coupled layered super-conductors in different fields (from top to bottom) 1, 2, 3, 4, 5 T parallel to c−axis. (a) The dashed lines represent the BLK result and the solid lines represent the result according to Eq. (4). (b): The dashed are for $B_{c2} \sim (1 - t)^{0.5}$ and the solid lines are for $B_{c2} \sim (1 - t)$. This shows that the width of the temperature-band depends on the variation of the upper critical field. The parameters are given in the text.
The dashed lines are for $B_{c2} \sim (1 - t)^{0.5}$ and the solid lines are for $B_{c2} \sim (1 - t)$. It is clear that the width of the temperature-band is narrow in the former case compared to the latter. For $\nu > 1/2$, the width of the temperature-band increases further. Thus, it is predicted that the upper critical field varies as $(1 - t)^{2\nu}$ with ν value in between 0 and 1/2.

3. SPECIFIC HEAT

The specific heat is a key signature so far as the superconducting transition is concerned. The influence of a magnetic field on HTSC does not show the familiar aspect known for conventional superconductors. To explore the field dependence of specific heat, we examine the variation $\Delta C = C_p(0, T) - C_p(B_a, T)$ with respect to magnetic field.

3.1 Specific heat of Y(123)

Any theory that gives a proper description of the reversible magnetization, should also give proper description of the variation of specific heat as a function of the magnetic field. Experiments on HTSCs show that the applied field B_a suppresses the magnitude of the jump in the specific heat rather than suppressing the temperature of the onset of the jump. Athreya *et al.* [7] and Salamon *et al.* [22] have found that the change in specific heat by the application of magnetic fields ΔC is not linear in B_a. It has been observed that the temperature dependence of the peak position in ΔC lowers with increasing magnetic field [22].

The HTSCs are highly anisotropic from conventional superconductors. In case of 3D anisotropic superconductors, the magnetic field dependence of specific heat depends on the orientation of B_a with respect to the crystallographic axes. For any general orientation of B_a, the change in specific heat is [9],

$$\Delta C = \frac{(2\nu)\phi_0 B_a f^2 \gamma^{-1/3}\epsilon(\theta)}{16\pi^2 T_c \lambda_m^2(0)} \frac{t}{1-t} \tag{5}$$

with $f^2 = (1 - \frac{B_a \epsilon(\theta)}{\gamma B_{c2}^{\|c}})$, where $\lambda_m = (\lambda_{ab}^2 \lambda_c)^{1/3}$, $\epsilon(\theta) = (sin^2\theta + \gamma^2 cos^2\theta)^{1/2}$, θ is the angle between B_a and crystallographic $c-$axis, γ is the anisotropy of the effective mass and $B_{c2}^{\|c}$ is the upper critical field for B_a parallel to $c-$axis.

Eq. (5) shows that ΔC is not linear in B_a but it varies with $B_a - (B_a^2/B_{c2})$ and the maximum value of t is limited by $B_a = B_{c2}$. We analyse the specific heat data of Athreya *et al.* [7] and obtain the value of $B_{c2}^{\|c}$, $\lambda_{ab}(0)$ and γ as follows. In the mean field region $\nu=1/2$. The least square fitted value of $B_{c2}(0)$ and $\lambda_{ab}(0)$ are found to be 296 T and 1936 Å from the specific heat data for $B_a \parallel c-$axis. We take $T_c=93$ K. The value of $\lambda_{ab}(0)$ is comparable with $\lambda_{ab}(0)=1400\pm500$Å obtained from magnetization measurement [23] and the value of B_{c2} is also in excellent aggreement with that obtain from magnetization measurement [24]. Using this value of $\lambda_{ab}(0)$, $B_{c2}(0)$ and the data for $B_a \perp c-$axis, the least squares fitted value of γ is estimated to be 6.1. This value of γ is reasonable agreement with earlier reported value [25]. Fig. 3 shows the comparison between the data of Athreya *et al.* [7] and the theoretical curves for a grain-aligned Y(123) crystal.

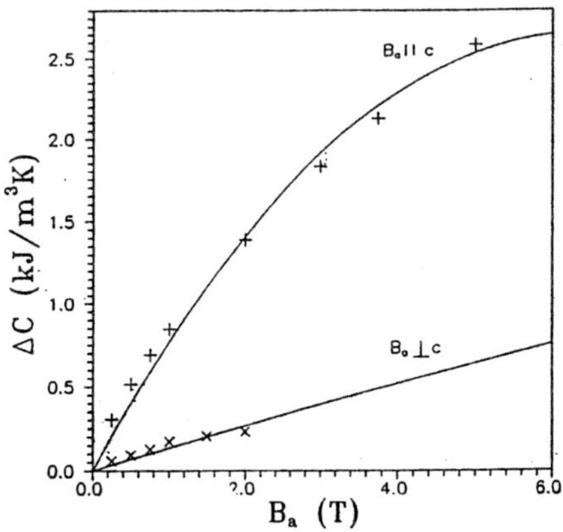

Fig.3: The field dependence of ΔC of a grain-aligned Y(123) crystal at $T=89$ K. The data is obtained from Fig. 5 of Ref. 7. The solid lines are the theoretical curves according to Eq. (5) with parameters $T_c=93$ K, $\lambda_{ab}(0)=1936$ Å, $B_{c20}^{\parallel c}(89)=296$ T, $\nu=1/2$ and $\gamma=7$.

3.2 Specific heat of Bi(2212)

Recent measurement on Bi(2212) single crystal with $B_a \parallel c-$axis [18] shows that there is no mean field specific heat jump at T_c. It is seen that the variation of specific heat with B_a is symmetrical about T_c. Like magnetization all the $C_p(B_a,T)$ curves pass through a new point $C_p(B_a,T^+) = C_p^+$ which is 10 K below T_c. The crossing point T^* of the magnetization $M(B_a,T^*)=M^*$, coincides with the point at which the specific heat peak occurs. The field dependence of the specific heat anomaly is observed at $T < T_c$. The effect of magnetic field in $B_a \perp c-$axis is very very less compared to $B_a \parallel c$. The measurement of Kierspel *et al.* [26] on single crystal with $B_a \perp c-$axis shows that the anomaly is not suppressed. Also, there is no evidence for any broadening of the anomaly. Whereas, the specific heat anomaly in case of polycrystalline Bi(2212) sample shows strong fluctuation and a small mean field step in zero magnetic field. The anomaly is suppressed and broadened by the application of magnetic field. The downward shift of the anomaly is very small. Both scaling theory [27] and the mean field theory cannot explain the experimental data [18]. Certainly, more measurements on specific heat anomaly of Josephson-coupled superconductors are needed, in addition to further theoretical investigations.

4. CONCLUSION

In conclusion, we have shown that the observed non-linear behavior of ΔC with magnetic field B_a of $YBa_2Cu_3O_{7-\delta}$ can be understood quantitatively by the vortex overlapping mechanism. This mechanism can be used in the intermediate field region for the determination of the superconducting parameters. The upper critical field, in-plane penetration depth and the anisotropy of effective mass obtained are reasonable aggreement with earlier reported values. It is also predicted that the upper critical field varies as $(1-t)^\nu$ with the value of ν lying in between 0 and 1/2 for Bi(2212). Whereas, the value of ν is 1/2 in .the mean field region and 2/3 in the critical region for Y(123).

REFERENCES

[1] A. L. Fetter and P. C. Hohenberg, in *Superconductivity*, edited by R. D. Parks (Marcel, Dekker, New York, 1969), Vol. **2**, Chap. 14, p. 817.

[2] R. P. Huebner, *Magnetic Flux Structures in Superconductors*, Vol. **6** of Springer Series in Solid State Sciences (Springer-Verlag, Berlin, 1979); E. H. Brandt, J. Superconductivity **6** (1993) 201.

[3] S. Mitra *et al.*, Phys. Rev. B **40** (1989) 2674; G. Triscone *et al.*, Physica C **224** (1994) 263.

[4] S. B. Ota, Phys. Rev. B **43** (1991) 1237; S. B. Ota, K. K. Nanda, and S. N. Behera, Physica B **194 − 196** (1994) 1387.

[5] Z. Hao *et al.* Phys. Rev. B **43** (1991) 2844; Z. Hao and J. Clem Phys. Rev. Lett. **67** (1991) 2371.

[6] A. E. Koshelev, Phys. Rev. B **50** (1994) 506.

[7] K. S. Athreya *et al.*, Phys. Rev. B **38** (1988) 11846.

[8] U. Welp *et al.*, Phys. Rev. Lett. **60** (1989) 1908.

[9] K. K. Nanda, preprint.

[10] L. N. Bulaevskii, M. P. Maley, and I. F. Schegolev, Physica B **197** (1994) 506.

[11] D. A. Brawner *et al.*, Phys. Rev. Lett. **71** (1993) 785.

[12] P. H. Kes *et al.* Phys. Rev. Lett. **67** (1991) 2383.

[13] Q. Li *et al.*, Phys. Rev. B **46** (1992) 3195.

[14] Qiang Li *et al.*, Phys. Rev. B **47** (1993) 2854.

[15] J. R. Thompson *et al.*, Phys. Rev. B **48** (1993) 14031.

[16] A. Grover *et al.*, Physica C **220** (1994) 353; A. Grover *et al.*, Pramana - Journ. of Physics, **42** (1994) 193; A. K. Grover *et al.*, Proc. of ETL conference to be published (1994).

[17] L. N. Bulaevskii, M. Ledvij, and V. G. Kogan, Phys. Rev. Lett. **68** (1992) 3773.

[18] A. Junod *et al.*, Physica C **229** (1994) 209.

[19] M. Tuominen, A. M. Goldmann, Y. Z. Chang, and P. Z. Jiang Phys. Rev. B **42** (1990) 412.

[20] V. G. Kogan *et al.*, Phys. Rev. Lett. **70** (1993) 1870.

[21] J. C. Martinez *et al.*, Phys. Rev. Lett. **72** (1994) 3614.

[22] M. B. Salamon *et al.*, Phys. Rev. B **38** (1988) 885.

[23] L. Krusin-Elbaum, R. L. Greene, F. Hotzberg, A. P. Malozemoff, and Y. Yeshurun, Phys. Rev. Lett. **62** (1989) 217.

[24] A. Bezinge, J. L. Jorda, A. Junod, and J. Muller, Solid State Commun. **64** (1987) 79.

[25] A. Junod *et al.*, Physica C **211** (1994) 304.

[26] H. Kierspel *et al.*, Physica C **235** − **240** (1994) 1765.

[27] Z. Tēsanovic *et al.*, Phys. Rev. Lett. **69** (1992) 3563.

Strongly Correlated Electron Systems

Indrani Bose

Department of Physics
Bose Institute
93/1 A.P.C. Road
Calcutta – 700009, India

Abstract. In this article, some of the interesting problems associated with strongly correlated electron systems are discussed. These include the motion of a hole in an antiferromagnetically-interacting spin background , the breakdown of Fermi Liquid theory , possible phases of the correlated systems and the binding of holes. Recent results on a specially-constructed coupled-chain model are also described.

1. Introduction

The discovery of high-T_c superconductivity in the cuprates by Bednorz and Müller in 1986[1] has given rise to unprecedented research activity. Experiments show that the cuprate systems have a rich phase diagram. Consider the parent compound La_2CuO_4 . The compound is doped by replacing some of the La ions by Ba or Sr ions. The phase diagram is the T-x diagram where T is the temperature and x is the dopant concentration. There are various phases like antiferromagnetic insulating , paramagnetic insulating , strange metallic , conventional metallic and superconducting phases. Also, there are two possible structures : tetragonal and orthorhombic. The insulating phase is believed to be caused by strong Coulomb correlation between electrons[2]. The electrons avoid each other by staying localized at respective lattice sites. This type of insulator is known as a Mott insulator and is different from conventional band insulators. The strange metallic phase is characterized by anomalous properties. It is believed that an explanation of these properties may offer clues as to the mechanism of superconductivity.

The cuprate systems because of a layered structure are highly anisotropic materials. A common structural ingredient is the copper-oxide (CuO_2) plane. The dominant electronic and magnetic properties are associated with this plane. For example, the resistivity ρ_c for current flow perpendicular to the plane is many times larger than the in-plane resistivity ρ_{ab}[3]. The temperature dependence of the resistivity shows that the cuprate system is metallic in the ab plane and semiconducting-like perpendicular to it. The resistivity ρ_{ab} increases linearly with temperature over a wide range. This property can not be explained by the conventional theory of metals. Other anomalous properties include a Hall coefficient which is dependent on temperature , the different temperature dependences of the NMR spin-lattice relaxation rates of copper and oxygen ions, the presence of a strong background continuum in the optical conductivity and Raman scattering measurements[3] etc. The anomalous properties can not be explained by the conventional Fermi liquid theory of metals. The non-Fermi liquid (NFL) behaviour is believed to be caused by strong Coulomb correlation. This has provided the motivation for a large number of studies on strongly correlated electron systems.

The simplest model that takes correlation into account is the Hubbard model :

$$H_U = -t \sum_{i,j,\sigma} C_{i\sigma}^{+} C_{j\sigma} + h.c. + U/2 \sum_{i,\sigma} n_{i\sigma} n_{i-\sigma} \qquad (1)$$

$$n_{i\sigma} = C_{i\sigma}^{+} C_{i\sigma}$$

The first term describes the motion of an electron from the j-th to the i-th site of the lattice with spin σ. The last term is the on-site Coulomb repulsion term. The energy of the system is raised by the amount U if the site i is doubly occupied by an up-spin and a down spin electron. In the limit of strong correlation (U/t >> 1) , the Hubbard model maps onto the t-J model[4] :

$$H_{t-J} = -t \sum_{i,j,\sigma} [C_{i\sigma}^{+}(1 - n_{i-\sigma}) (1 - n_{j-\sigma}) C_{j\sigma} + h.c.$$
$$+ J \sum_{<ij>} (\vec{S}_i . \vec{S}_j - 1/4\, n_i\, n_j)] \qquad (2)$$

where $\vec{S}_i$ is a spin-1/2 operator at the site i and J is the antiferromagnetic coupling. The hopping term allows an electron with spin σ to move to the site i with the same spin provided the i-th site does not contain an electron with spin $-\sigma$. Also, the j-th site is singly occupied originally. In other words, double occupancy of a site by electrons is strictly prohibited at any stage. The model has only three possible states per site ,an electron with spin up or down or empty ,i.e.,occupied by a hole. The double occupancy prohibition is the origin of correlation in the system. Electrons correlate their positions in order to avoid double occupancy of a site. In the cuprate systems the CuO_2 plane looks like a square lattice with copper ions sitting at the lattice sites and the oxygen ions sitting on the bonds in between. Thus each unit cell contains one copper ion and two oxygen ions. The relevant electronic orbitals are copper $3d_{x^2-y^2}$ and oxygen $2p_x$ and $2p_y$,giving rise to three bands in the system. The t-J Hamiltonian in (2) should be appropriately generalised to give a realistic three-band description to the CuO_2 plane. On the other hand ,Zhang and Rice[5] have argued that in the low energy sector , the three band model reduces to a one band model described by the Hamiltonian (2). There is a single electronic orbital at each lattice site. In the half-filled limit,each site is occupied by a single electron.The first term of the t-J model in this case has no effect and the Hamiltonian reduces to the antiferromagnetic (AFM) Heisenberg Hamiltonian describing the exchange interaction between spins.Doping in this scenario is equal to the removal of some electrons giving rise to holes.It is these holes which are responsible for charge transport in the metallic and the superconducting phases of the cuprates.The t-J model provides lots of physical insight on strongly correlated electron systems.In Section 2 ,we cosider the case of one hole and give a brief overview of some of the results obtained so far.In Section 3,the NFL behaviour of 1d systems is discussed along with a possible extension to 2d systems.In Section 4,we discuss the possible phases of strongly correlated systems and also describe some recent work on coupled-chain models.

2.Dynamics of a single hole

We first consider the case of a single hole.The hole moves in the background of antiferromagnetically interacting spins described by the t-J Hamiltonian in (2).The model is defined on the square lattice corresponding to the CuO_2 plane.The t term in the t-J model favours the delocalization of the hole but the propagation of the hole disturbs the AFM spin arrangement favoured by the J term.Thus there is a competition between the t and J terms.This is easy to understand if one considers the $t-J_z$ model in which the interaction between the spins is of the Ising type.The Néel state with an antiparallel arrangement of spins is the exact ground state of the AFM Ising Hamiltonian.As the hole moves in the Néel background it leaves behind a trail (string) of wrongly-oriented (parallel) spin pairs (Fig.1) which raises the exchange energy of the system.The cost in energy is proportional to the length of the path that the hole has traversed.The hole thus finds itself in a linear confining potential which tends to

localize it.The corresponding eigenvalue problem can be solved[6].
Complicated paths for hole motion have been suggested[7] which
restore the Néel background thus giving mobility to the hole.When

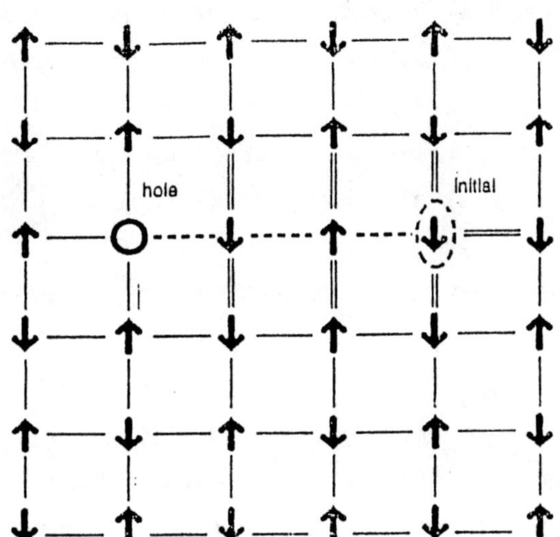

Fig.1 The motion of a hole in a Néel background :   "initial"
denotes the initial location of the hole ;the circle denotes the
hole position after three hops and the dashed line the path of the
hole.The double lines indicate wrongly-oriented spin pairs which
cost in exchange energy.The number of such links grows with the
link of the path.

the full t-J model is considered ,the XY term $1/2 (S_i^+ S_j^- + S_i^- S_j^+$
) of the Heisenberg Hamiltonian (i,j are nearest-neighbour (n.n.)
sites),gives rise to spin flips which constitute the so-called
quantum fluctuation.The effect of this is to repair the damage
caused by the strings to the Néel order allowing the hole to move
coherently.The Néel state is ,however,not the exact ground state
of the Heisenberg Hamiltonian.The validity of the string picture
for the full t-J model is thus in question.However,numerical

calculations on small clusters show that the ground and excited state energies of the hole are consistent with those obtained in the string picture[8].

In the general dynamical problem the pertinent question to ask is whether a quasi-particle (QP) description can be given to the hole. The hole as it moves is surrounded by a cloud of spin excitations. The composite object can be termed as a QP. The quantity of interest is the spectral function $A(\vec{k},\omega)$ given by

$$A(\vec{k},\omega) = 1/\pi \; \mathrm{Im}\; G(\vec{k},\omega) \tag{3}$$

where $G(\vec{k},\omega)$ is the hole Green's function or propagator describing the spectrum of energies of the hole excitations. The spectral function is the quantity that is measured in photoemission experiments. For a particular value of the momentum wave vector $\vec{k}$, if $A(\vec{k},\omega)$ has a sharp peak as a function of ω, then we may think of this peak as describing a coherent excitation or QP. The positions of these OP peaks as a function of momentum wave vector determine the band dispersion and hence the QP mass. If no peak can be observed in the spectral function then the hole is incoherent and a particle-like description of its motion can not be given.

For a noninteracting system of electrons, the free particle Green's function in momentum space is given by

$$G_0(\vec{k},\omega) = 1/(\omega - \varepsilon_k + i\delta) \tag{4}$$

where δ is an infinitesimal quantity and ε_k is the single particle excitation energy. The corresponding spectral function $A(\vec{k},\omega)$ is a delta function $\delta(\omega - \varepsilon_k)$. The Fermi liquid theory describes an interacting system in terms of QPs rather than free particles. The QPs have a renormalized energy ε'_k and a lifetime τ_k.

$$G_{QP}(\vec{k},\omega) = Z_k/(\omega - \varepsilon'_k + i\tau_k^{-1}) + D(\vec{k},\omega) \tag{5}$$

where the QP weight Z_k has value less than 1. The missing intensity is spread over energies away from the QP energy in the incoherent part $D(\vec{k},\omega)$. The spectral function corresponding to the QP part of the propagator is a Lorentzian with the δ-function peak broadened out. The rest of the spectral function is a featureless background. The single particle density of states (D.O.S.) and the momentum distribution are given in terms of $A(\vec{k},\omega)$ by

$$N(\omega) = \sum_k A(\vec{k},\omega)$$

and
$$n(\vec{k}) = \int d\omega\, A(\vec{k},\omega) \qquad (6)$$

respectively. The QP weight Z_k is a measure of the discontinuity in the momentum distribution at the Fermi wave vector k_F. For a non-interacting system the discontinuity has the magnitude 1. If $Z_k = 0$, the QP picture fails and the FL theory is no longer valid. For the cuprate systems we consider hole propagators instead of electron propagators. The ground state with a static hole is a linear combination of various spin configurations. The hole can have a Bloch-like propagation if there is a finite overlap of the spin configuration after the hole hops with the spin configuration prior to the hop. The QP weight Z_k is proportional to this overlap. As mentioned in Section 1, the anomalous normal state properties of the cuprates are believed to be caused by strong correlation leading to a breakdown of the FL theory. Two scenarios have been proposed in this context. In the Marginal Fermi Liquid (MFL) theory proposed by Varma et al[9] the QP weight Z_k goes to zero logarithmically at the Fermi surface. In the Luttinger liquid theory of Anderson[10,11], Z_k goes to zero as a power law. In the light of these theories, the question as to whether the hole can move as a QP in a background of interacting spins becomes relevant. Kane et al[12] have developed a QP theory for the motion of a single hole in an AFM background in which the ground state of the spins is described either by a quantum Néel state or a d-wave resonating-valence-bond (RVB) state. By considering the $J \ll t$ limit, they have shown in a self-consistent diagrammatic perturbation theory that interactions with spin excitations strongly renormalize the hole spectrum. They found the existence of a QP peak in the spectral function at low energy. The characteristic energy describing the coherent motion of the hole is J and not t (t > J). This is because the hole motion is strongly coupled to the spin dynamics. The spin correlations destroyed by the string created due to hole motion can be recovered by spin fluctuations after a characteristic time 1/J. The effective mass of the QP is quite large since the mobility of the QP is retarded by the cloud of spin excitations. Besides the QP peak the rest of the spectral function is incoherent. As already referred to ,Dagotto et al[8] have carried out exact diagonalization study of the t-J model on a 4 × 4 lattice. The spectral function of the hole has three distinct features : (i) a low energy QP peak describing the simultaneous hole hopping and spin flip processes ,(ii) levels of the Ising string picture that survive the inclusion of the spin flip term and (iii) an incoherent background with a total width of order t. Other issues which havve been studied in the context of single hole motion are the possibility of ferromagnetism in the ground state, the quantum numbers of the hole in the ground state and the nature of the spin distortion caused by the moving hole. Several studies have been undertaken to clarify these issues ,employing a variety of techniques[6]. Two toy models have been constructed for which some exact, analytical results can be derived. One model consists of a chain of octahedra[13]. The exact ground state in the undoped state can be determined. In this state, the basal spins in each octahedron are in a RVB state which is a linear combination of two valence bond states (a valence bond describes a spin singlet). In one state, the bonds are horizontal and in the other the bonds are vertical. A single hole introduced into the system causes a distortion in the spin background of the octahedron in which the hole is located.The hole along with the spin distortion is called a spin polaron which has the size of a single octahedron.Exact eigenstates have been constructed which describe coherent Bloch-type motion of the spin polaron.There are also eigenstates in which the polaron is localized,i.e.,can not propagate.The second toy model consists of two chains coupled by both vertical and diagonal links[14].The exact ground state in the undoped state consists of singlets along the vertical links of the coupled-chain(CC) model.Exact analytic solution shows that a QP consisting of a hole and a free spin-1/2 can propagate coherently in a Bloch type of state in the singlet background of spins (details to be given in Section 4).Brandt and Gieskus[15] have obtained exact solutions of the Hubbard model in the limit of infinite Coulomb repulsion.The model is studied on perovskite-like lattices for any dimension d $\geq$ 2.The exact ground state and an exact eigenstate have been obtained.Tasaki[16] generalizes this model to show that the exact ground state can be described as a RVB state.It is also argued that the state may exhibit superconductivity.These works,however,do not address the question of coherence of hole motion in an AFM spin background.

3.Non-Fermi liquid (NFL) behaviour

　　The unconventional properties of the metallic state of the cuprate systems signify that the conventional FL theory has to be either modified or discarded in favour of a NFL theory.Let us first discuss the major features of the FL theory.Consider a Ferni gas which is a system of non-interacting fermions,say,electrons.At T = 0 ,the energy levels are filled up to a maximum energy level called the Fermi level.Correspondingly,there exists a Fermi surface (FS) in momentum space which separates the occupied electronic states from the unoccupied electronic states.Excitations are created in the system by taking electrons from states below the FS to empty states above the FS. Now consider a Fermi liquid which is a fermionic system with interactions. Landau formulated the FL theory which gives a good account of the low-lying single particle excitations of the system of interacting electrons. He showed that the interacting Fermi system can effectively be described as a system of weakly interacting QPs. There is a one-to-one correspondence between the excitations in the non-interacting system and the QP excitations in the interacting system. The QPs have an effective mass and lifetime τ. The corresponding Green's function has already been written down in Eq. (5). The fermionic QPs carry both spin and charge and have a sufficiently long lifetime so that $1/\tau \sim \omega^2$ where ω the excitation energy of the QP is measured w.r.t. that of the FS. In the cuprate systems, angle-resolved photoemission spectroscopy measurements show[3] that a FS exists but the decay rate $1/\tau$ of the QP varies linearly with ω rather than ω^2 as in the FL theory. The QP is thus not well-defined.

A class of systems for which the breakdown of FL theory can be explicitly demonstrated is known as the Luttinger liquids (LLs). In analogy with the correspondence between the Fermi gas and the Fermi liquid, the low energy excitations of the LL have features similar to those of an exactly-solvable model of interacting fermions known as the Luttinger model. The model was solved exactly by Mattis and Lieb[17] using the so-called Bosonization technique. The model describes a fermionic system in 1d. The important physics takes place near the two Fermi points $+k_F$ and $-k_F$. The fermionic spectrum has a linear dispersion relation and two branches 'a' and 'b'. The 'a' branch corresponds to excitations around the Fermi point $+k_F$ and the 'b' branch corresponds to excitations around the Fermi point $-k_F$. The non-interacting part of the Hamiltonian is given by

$$H_0 = v_F \sum_{k,\sigma} \{ (k - k_F) \, a^+_{k\sigma} \, a^+_{k\sigma} + (- k - k_F) \, b^+_{k\sigma} \, b_{k\sigma} \} \qquad (7)$$

The branch energies extend to $-\infty$ giving rise to unphysical states. The inclusion of these states makes the eigenvalue problem mathematically tractable. The interaction terms of a Luttinger-type model , in general, describe four types of scattering processes: backward, forward, Umpklapp and forward scattering on the same branch (Fig.2). The corresponding coupling constants are g_1 , g_2, g_3 and g_4. The original Luttinger (L) model does not contain backscattering processes , i.e., g_1 = 0 which makes it exactly solable. If $g_1 \neq 0$ and is positive , one can show through a renormalization group (RG) analysis that the system has behaviour characteristic of the L model (fixed point $g^* = 0$). The feature of

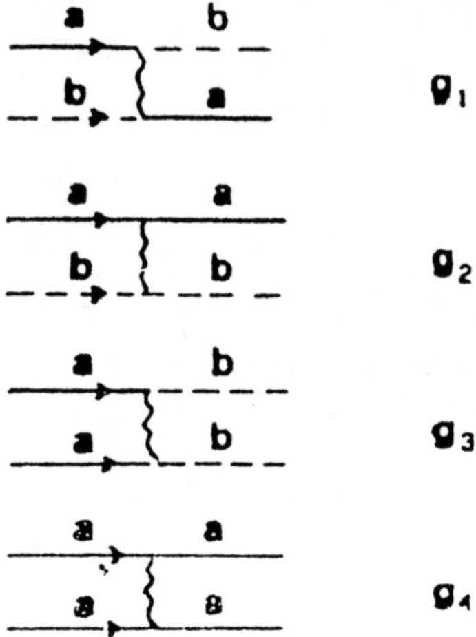

Fig.2 Diagrams corresponding to the four types of scattering processes in a Luttinger-type model. Solid lines correspond to branch "a" and the dashed lines to branch "b". The g_i's are the coupling constants.

the L model that allows its solution is its exactly linear fermion dispersion.The low energy excitation spectrum has three parts[18] :bosonic excitations describing collective density fluctuations ,charge and current excitations.There are three velocities v_S ,v_N and v_J associated with the three excitations.The velocities obey the relation

$$v_N = v_S \exp (- 2\phi)$$
$$v_J = v_S \exp (2\phi) \tag{9}$$

The parameter ϕ determines the exponents in the power -law decays of the correlation functions of the system.Haldane in 1981[18] considered interacting models in 1d characterised by non-linear dispersion.He showed that all such models can be grouped under a single class known as Luttinger liquids (LLs).Due to the non-linearity of the fermion dispersion relation,the models can not be exactly diagonalized by bosonization.However,the same types of excitations as in the L model occur though with interactions between the elementary excitations.This resembles the relation between the Fermi liquid and the free Fermi gas.In the former the QP excitations are weakly interacting.The relation (8) between the velocities is preserved in the case of LLs.Also,the relations in (9) hold true.Support for the universality of the L model features is provided by the class of models which can be solved exactly by the Bethe Ansatz (BA)[19].For these models,v_S ,v_N and v_J can be ccalculated exactly and the relation in (8) verified[20].The BA-solvable models thus belong to the class of LLs.The original L model and other fermionic models describe spinless fermions.Later,1d models with internal symmetries such as spin were found to have the LL character.Examples of LLs are interacting spinless fermions,interacting spin-1/2 fermions,uniaxially anisotropic AFM spin systems,ferromagnets (easy plane only),1d Hubbard model etc.

LLs exhibit non-Fermi liquid behaviour in two ways: (i) the correlation functions show power-law behaviour with coupling-dependent anomalous exponents ,(ii) spin and charge excitations have different velocities leading to spin-charge decoupling.The two effects are independent of each other.The basic correlation functions that have been considered are momentum distribution function,spin ,charge and superconducting (both singlet and triplet) correlation functions[21].The correlation functions and the associated exponents have been calculated using various methods: (i)direct calculation based on the exact wave function in the strong correlation limit,(ii)bosonization theory,(iii) conformal field theory combined with finite-size corrections to the BA results and (iv) quantum Monte Carlo simulation.The exponents in turn determine a number of physical properties like temperature dependence of the NMR relaxation rate,x-ray scattering intensities,and effect of impurities on possible low-temperature ordered states in systems of coupled chains[22].As an example of anomalous behaviour,consider the momentum distribution function n_k of the 1d Hubbard model; n_k exhibits a power-law singularity at $k = k_F$ and a weaker singularity at $k = 3k_F$.

$$n_k = n (k_F) - C \mid k - k_F \mid^\alpha Sgn (k - k_F) \qquad (10)$$

where C is a constant and the exponent α depends on the model parameters.Typically α varies between 0 and 1/8.A FL would have a jump in n_k at $k = k_F$ corresponding to $\alpha = 0$.The exponents of the correlation functions are anomalous because of their dependence on the model parameters.In ordinary critical phenomena ,the correlation function exponents are universal,i.e.,independent of the strengths of the coupling constants.

Spin-charge decoupling is a signature of NFL behaviour.In FL theory,QPs carry both spin and charge and there is no decoupling of the spin and charge degrees of freedom.For the 1d Hubbard model,spin-charge decoupling can be demonstrated in a rigorous manner from the BA exact results.One way of seeing this is to consider the limit $U/t \to \infty$ .In this limit,the wave function has a simple structure.Let f $(x_1 ,\ldots, x_N ; y_1 ,\ldots, y_M)$ be the amplitude of the wave function describing N electrons located at $x_1,\ldots,x_N$ and M down spins at the positions $y_1 ,\ldots,y_M$ .Then in the limit $U/t \to \infty$,one can show that[23]

$$f(x_1 ,\ldots, x_N ; y_1 ,\ldots, y_M)$$
$$= \det [\exp (i k_i x_j)] \; \Phi (y_1 ,\ldots,y_M) \qquad (11)$$

The determinant corresponds to spinless fermions with momenta k_i's.$\Phi (y_1 ,\ldots, y_M)$ is just the BA exact wave function of the 1d 'sqeezed' Heisenberg spin system which is obtained by eliminating holes from the original system.One can prove (11) from the BA results or much more easily by standard degenerate perturbation theory in the $U/t \to \infty$ limit.

The breakdown of FL theory is well-established for 1d interacting systems.A crucial question is whether such breakdown can be demonstrated for 2d or higher dimensional systems.Anderson in a series of papers[11,24-28] has argued that the ground state of a strongly correlated electron system in 2d descrbed by the Hubbard or t-J-type models is a LL.The low energy excitation spectrum shows evidence of spin-charge decoupling.There are neutral spin-1/2 excitations called spinons and spinless charge +e excitations called holons.According to Anderson the Fermi surface experimentally observed is the spinon Fermi surface.The anomalous normal state properties of the cuprates in the LL theory are a consequence of spin-charge decoupling[29].There are ,however,no rigorous theories as yet which demonstrate spin-charge decoupling or breakdown of the FL theory in 2d.Fabrizio and Parola[30] have considered a model of interacting electrons defined on two chains connected by a transverese hopping $t_\perp$ .They have solved the model exactly using the bosonization technique and shown that spin-charge separation occurs.Valenti and Gros[31] have shown using a variational wave function first suggested by Hellberg and Mele[32] for the 1d t-J model that (i) it is possible to define LLs in 2d variationally and (ii) that the projected kinetic energy favours LL-type correlations in 2d ,as it does in 1d.Chen and Lee[33] have shown that the static spin correlation function $S(\vec{q}$) of the LL variational wave function agrees well with the $S(\vec{q}$) of the exact ground state of a 10 × 10 cluster.Putikka et al[34] have studied the single spin momentum distribution function n_k ,and the static spin and charge (density) correlation functions by the high temperature series expansion method.By comparison to known results for 1d they find the evidence of spin-charge separation in the 2d t-J model.Chen et al[35] have studied the spin and density correlation functions of the 2d Hubbard and t-J models at low electronic density using the power method,quantum Monte Carlo and a perturbation expansion.The results obtained are consistent with the high temperature expansion results.They have ,however,given arguments that the examination of the density correlation function is not sufficient to obtain evidence for spin-charge separation at low electronic density.To summarise,the issue of spin-charge separation in a 2d strongly correlated system is still controversial though a large body of work on the cuprate systems is based on this concept.

4. Coupled-chain models of strong correlation

In this Section,we first discuss the possible phases of strongly correlated electron systems and then review the recent work done on the coupled-chain (CC) models .The phases of 1d interacting electron systems can be classified into four types: (i) Both spin and charge excitations are gapless.

(ii) The spin excitation has a gap while the charge excitations are gapless.

(iii)The spin excitations are gapless but the charge excitations have a gap.

(iv) Both the spin and charge excitations have gaps.

The third type of phase corresponds to the Mott insulator.The fourth type describes the Mott insulator with an Ising-like anisotropy or a dimer state caused by frustration or spin-Peierls transition.The first and second types of phases describe metallic electrons.The Hubbard model in 1d has three of the phases described.The repulsive and attractive Hubbard models away from the half-filled limit belong to type 1 and type 2 respectively.The repulsive Hubbard model in the half-filled limit is an example of type 3.

Next,consider the extended Hubbard model in 1d.

$$H_{ex} = - t \sum_{i,j,\sigma} (C^{+}_{i\sigma} C_{j\sigma} + h.c.) + U/2 \sum_{i} n_{i\sigma}n_{i-\sigma} + V \sum n_{i} n_{j} \qquad (12)$$

The phase diagram of the model has been obtained by both exact solution for some special parameters and also the weak -coupling RG analysis known as g-ology[36,37].The phase diagram of the model has two regions : the Luttinger region (equivalently called the Tomonaga-Luttinger (TL) region) and the Luther-Emery (LE) region (Fig.3).The TL region corresponds to positive backward scattering $(g_{1} > 0)$ and belongs to type 1 of the phases described above.The LE region has $g_{1} < o$ and belongs to type 2.Away from half-filling,the TL (LE) region corresponds to U > 2 V (U < 2V).In the TL region ,there is a phase in which triplet pairing is the most dominant correlation accompanied by logarithmically weaker singlet pairing correlation.This region is sandwiched between the

spin density wave (SDW) dominating region and a region of phase separation.In the LE region,the charge density wave (CDW) or the

singlet pairing correlation is the most dominant correlation
depending on the values of the parameters.The attractive Hubbard

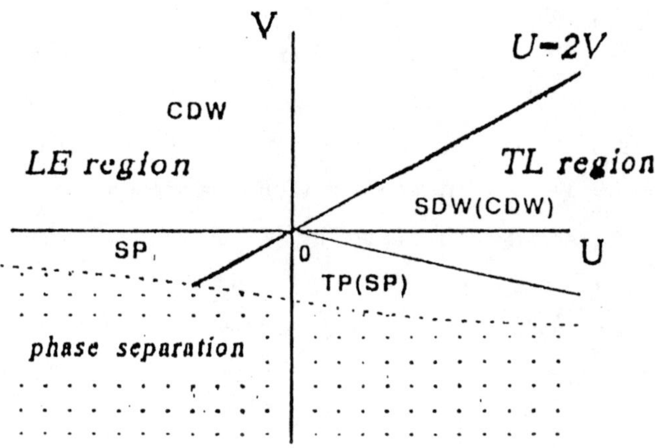

Fig.3 Schematic phase diagram of the extended Hubbard model in 1d
away fom half-filling.The terms SP ,TP ,CDW ,SDW stand for singlet
pairing ,triplet pairing,charge density wave and spin density wave
respectiely.Correlations within brackets have logarithmically
faster decay than the dominant ones.

model belongs to the LE region.One interesting fact which emerges
from the study of the phase diagram is that both in the TL and LE
regions,the superconducting pairing correlation occurs in the
vicinity of an area of phase separation.It is thus of interest to
gain more insight about this region.There is experimental
evidence[6] that $La_2CuO_{4+\delta}$ has a regime where phase separation into
oxygen-rich and oxygen-poor phases occur.The dopant oxygen atoms
may cluster together independent of the physics of electrons in
the CuO_2 planes.The opposite point of view is that the oxygen ions
follow the holes to minimize energy and the oxygen separation is
due to phase separation into hole-rich and hole-poor regions.In

the t-J model at large J/t ,phase separation is known to occur.The holes cluster together in order to minimize the loss of exchange energy associated with broken (missing) AFM links around the holes.Large J/t is not realized in the cuprate systems and the issue whether phase separation occurs in physically interesting regimes is controversial.Long range Coulomb interaction effects, which are neglected in the usual Hubbard model,may be unfavourable to the phase-separated state.

We next consider the possibility of a superconducting phase in models of strongly correlated electrons.Various mean field (MF) theories of the models show the existence of such a phase but the theories being approximate in nature uncertainty exists about the results.Numerical studies which include exact diagonalization on small clusters and quantum MC simulation[6] ,do not show the existence of a superconducting phase in the usual one and three band Hubbard models.Such studies,considered to be more reliable than MF theories,are ,however,limited by the small size of the lattice and the temperature accessible to the MC simulation.Riera and Dagotto[38,39] have considered a multi-band Hubbard model which includes long -range repulsive Coulomb interactions.Numerical analysis of pairing correlations on small clusters shows that a superfluid condensate is formed.The results appear to be true both in 1d and 2d.There is ,however,a non-trivial competition with a CDW order.The CDW states become stable as the strength of the long-range repulsion term $1/r$ is increased.Nevertheless, superconductivity exists in a wide range of parameter space.The long range interaction is found to suppress phase separation.Using exact diagonalization and variational MC techniques ,Dagotto et al[40] have found indications that the 2d t-J model has a superconducting phase near phase separation in the regime of quarter -filling, $\langle n \rangle \sim 1/2$.At this density,the dominant pairing channel is $d_{x^2-y^2}$ but transition to s-wave superconuctivity is observed as the electronic density is decreased.They have also studied the one-band t-U-V model on the square lattice and a two-band Hubbard model on a chain.The results obtained lead to the conjecture that electronic models tend to have superconducting phases in the vicinity of phase separation.The attractive forces which give rise to phase separation are also responsible for strong pairing correlations in its neighbourhood.In this regime,it is energetically favourable for the holes to form pairs rather than cluster together.The holes gain in kinetic energy by moving in pairs.Though a superconducting phase has been identified in the phase diagrams of models of correlated electrons,the possibility of such a phase in the realistic regimes (small J/t and densities close to half-filling,as is appropriate for the cuprate systems) is still open.

We now turn to the coupled-chain (CC) models described by t-J-type models.The CC model consists of two chains coupled by links.A site of the model is either occupied by an electron or empty.The system of electrons is described by an appropriate Hamiltonian.The model has been studied to address issues like 'quantum confinement',spin gap and superconductivity.As already mentioned in Section 3 ,an important issue of current interest is whether the anomalous properties of the LL in 1d survive in 2d and 3d.A CC model which interpolates between 1d and 2d is an ideal candidate to explore the possibility.The RG studies[41] show an instability of the LL behaviour when the interchain single-electron hopping (SEH) $t_\perp$ is taken into account.Anderson[24] has ,however,suggested that the intrachain correlations including spin-charge separation should be exactly treated before switching on the SEH.According to him,no coherent single electron motion is possible between the chains if the hopping parameter $t_\perp$ has a value less than a critical value.This leads to the idea of quantum confinement .The concept has been invoked to explain the temperature dependence of the resistivity ρ_c in a direction perpendicular to the copper-oxide plane.Coherent single electron transport between successive CuO_2 planes is blocked due to confinement in the planes and this gives rise to semiconducting-like ($\rho_c \sim 1/T$) behaviour.On the other hand ,pair hopping processes between the chains is possible.

The cuprate syatem $La_{2-x}Sr_xCuO_4$ is a single-layer material ,i.e.,there is one CuO_2 plane in one unit cell.In bilayer materials like $YBa_2Cu_3O_{7-\delta}$ there are two CuO_2 planes per unit cell.Experiments indicate that the physical properties of single-layer materials are different from those of multilayer materials.The superconducting transition temperature T_c is generally higher for compounds which have more CuO_2 planes per unit cell.Experiments also show the existence of a spin gap in the bilayer material $YBa_2Cu_3O_6{}^6$.The spin gap phase is observed at low doping and exists well above T_c.Recent analysis of the experimental data [44] suggests that the spin gap phase is only observed in multi-layer materials but not in single-layer materials.Ubbens and Lee[45] have argued that the spin gap is related to the pairing of electrons in nearby CuO_2 planes.The CC model which consists of two coupled chains instead of coupled planes as in the bilayer materials exhibits a spin gap in its excitation spectrum which survives small amounts of doping.The existence of the spin gap is reminiscent of the Luther-Emery type of models which have been discussed earlier in this Section.Pairing of electrons occurs naturally in the CC model in the form of singlets along the vertical links connecting the chains.Also,superconducting pairing correlations are greatly enhanced in the CC model and this is connected to the fact that the spin gap survives finite doping. For a review of other models of correlated electrons which have a spin gap in their spectrum see Ref. 46.

The compounds $Sr_2Cu_4O_6$[47] and $(VO)_2P_2O_7$[48] provide good examples of lattices consisting of weakly-coupled double chains. The first compound is a member of the homologous series $Sr_{n-1}Cu_{n+1}O_{2n}$ which differ from known high-T_c cuprates because of the presence of a parallel array of line defects in the CuO_2 planes[47]. The line defects consist of CuO double chains. The separation between the chains grows linearly as n is increased and in the limit $n \to \infty$ the infinite –layer compound $SrCuO_2$ is obtained. The CuO double chains perturb both the exchange interactions and the hopping matrix elements in the CuO_2 plane. They have the effect of dividing the planes into weakly-coupled (n+1)/2–chain ladders. A study of the electronic properties of the $Cu_{n+1}O_{2n}$ planes show[47] that these can be described as frustrated quantum antiferromagnets and spin liquids. The spin gap remains on light doping with holes and singlet superconductivity has been predicted to occur on a separate but high temperature scale. The prediction is consistent with the separate energy scales for the spin gap and superconductivity in other lightly doped cuprates.

The CC model describes an isolated double-chain rather than a lattice of such chains. The simple structure of the model makes calculations easier and the results obtained provide insights on the spin gap, pairing correlations and superconductivity. The model Hamiltonian that is generally studied for the coupled chains (ladder) consists of two t-J chains with t', J' denoting the couplings between the rungs.

$$H_{CC} = J \sum_i \vec{S}_i . \vec{S}_{i+x} + J' \sum_i \vec{S}_i . \vec{S}_{i+y}$$

$$- t \sum_{i,\sigma} (C_{i\sigma}^{\dagger} C_{i+x\,\sigma} + h.c.) - t' \sum_{i,\sigma} (C_{i\sigma}^{\dagger} C_{i+y\,\sigma} + h.c.) \qquad (13)$$

The chains extend along the x-axis and the rungs connect site i with i + y ,x and y are unit vectors in the x and y directions respectiely. The usual t-J model constraint that no site can be doubly occupied is implied. The ladder has periodic boundary condition in the x-direction. At half-filling the model reduces to the Heisenberg spin ladder. The ground state consists of singlets along the rungs[49]. The lowest excitation is a triplet obtained by exciting a singlet rung to a triplet. This costs an energy proportional to the exchange coupling. Strong-coupling perturbation expansion[49] ,mean-field (MF) approaches[50] and numerical methods[51] show the evidence of a non-zero spin gap for all interchain couplings $J' > 0$. The gap is of approximate magnitude 0.5 J in the isotropic coupling limit. Exact diagonalization studies[51,52] of the doped ladders of finite size show that holes form bound pairs on the rungs to minimize the loss in exchange energy which can then propagate along the ladder. The model exhibits superconducting pairing correlations. Sigrist et al[53] have studied the doped t-J ladder (CC model) in the mean-field approximation and shown that the spin liquid state at half filling evolves into a superconducting state upon doping. The order parameter has a modified d-wave character. Exact diagonalization studies[52] besides confirming the existence of hole pairs shows that the excitation spectrum separates into QPs which carry charge +e and spin 1/2 and a triplet mode. At half filling ,the QPs do not exist but the triplet evolves continuously into the triplet band of the spin liquid.

    Recently,Bose and Gayen[54,55] have constructed a CC model in which the chains are coupled by both vertical and diagonal links (Fig.4). In contrast to the earlier CC models ,this model is

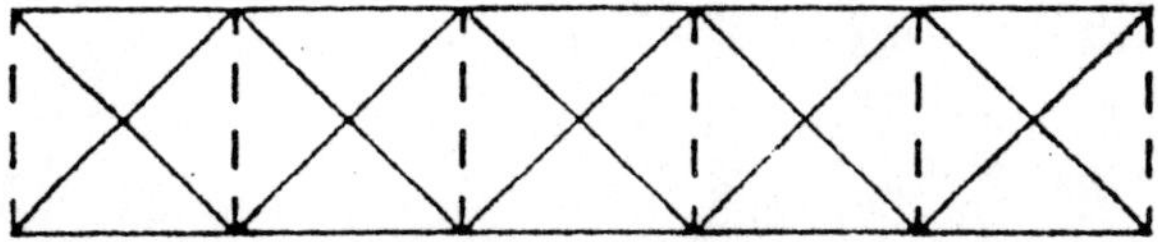

Fig.4 The coupled-chain model with both vertical and diagonal links.

exactly and analytically solvable.The model is described by the t-J Hamiltonian given in Eq.(13) with the addition of hopping and exchange interaction terms corresponding to the diagonal links.These terms have strengths t and J respectively.In the undoped state and for $J' > 2J$,the exact ground state consists of singlets along the vertical links.For $J' < 2J$,this state ,though still an exact eigenstate,may not be the ground state.If 2N is the number of sites in the system,the exact ground state energy E_g = -3J/2N.Let us now consider the case when the model is doped with a single hole.Three different types of exact eigenstates have been determined.These describe (1)the motion of a QP consisting of the hole (charge + e) and a free spin (spin 1/2) in a sea of singlets, (2) states in which the QP forms bound and anti-bound states with a localized triplet excitation and (3) scattering states of the QP against the localized triplet excitation.The anti-bound states correspond to the 'midgap'states the existence of which has been conjectured for the cuprate systems[56].The CC model described by the Hamiltonian in Eq.(13) also shows[52] the existence of a QP of charge +e and spin-1/2.The triplet excitation in this case is,however,not localized but propagating.The separation of the excitation spectrum into bound hole-spin QP and collective triplet excitation contrasts with the full spin-charge separation found in a LL.Based on the exact results obtained,the possibilty of a localization-to-delocalization transition for the hole,as J/t is varied ,has been conjectured.

We next consider the case of two holes[57].The holes are introduced in the ground state of the model in two different vertical links so that the dimers along these links are broken.The two free spins from the broken dimers combine to make the total spin of the system either S = 1,i.e.,triplet or S = 0 ,i.e.,singlet.Using the Bethe Ansatz analysis ,it can be shown that two holes form anti-bound states in a triplet spin state.The result can be extended to the case of more than two holes.The eigenvalue problem when the two holes are in a singlet state can also be solved exactly and one finds that the two holes can bind for all values of J.In this case,anti-bound states are found for sufficiently large values of J.In all these calculations,the effects of strong correlation and qantuum fluctuation have been taken into account in an exact manner.This is also one of the few calclations in which the binding of holes in a t-J type model is shown in a rigorous manner.A review of earlier work on the binding of holes in the t-J and Hubbard models is gien in Ref.6.This problem is of significance because in the superconducting state of the cuprate systems charge transport occurs through the motion of pairs of holes.

5.Concluding Remarks

In this review we have considered strongly correlated electron systems and discussed some of their interesting properties.The motion of a single hole in an antiferromagnetically interacting spin background is a non-trivial many body problem and has relevance in understanding charge transport in the cuprate systems.The hole and spin dynamics affect each other and questions like coherence of the hole motion and nature of the ground state assume significance.The cuprate systems in their metallic state exhibit anomalous properties which can not be explained by conventional Fermi liquid theory. The significance of non-Fermi liquid behaviour in 1d interacting electron systems have been discussed in Section 3. In this context, the notion of Luttinger liquids has been introduced and the generic features of such systems described. It is believed that some of the insight gained in the study of the 1d correlated systems may be valid for 2d systems also. In Section 4 ,we have discussed the possible phases of models of correlated electrons. In particular,we have reviewed recent work done on the coupled-chain models for which exact results are available. There is a striking difference between the properties of a chain and a coupled-chain AFM s = 1/2 Heisenberg model. The chain has power-law decay of the AFM correlations whereas the coupled-chain has a purely exponential decay and a finite energy gap in the spin-excitation spectrum (spin gap). As has been pointed out ,real systems like $Sr_{n-1}Cu_{n+1}O_{2n}$ exist which consists of 2d lattices of weakly-coupled (n+1)/2-chain ladders. Recent studies[58,59] have considered coupled t-J ladders and ladders with more than two chains using methods like exact diagonalization and strong-coupling expansion. The major issue that has been addressed is the survival of the spin gap. Systems with spin gap have enhanced superconducting pairing correlations. Imada[60] has suggested a mechanism of superconductivity arising from the doping of holes into the spin-dimerized insulator such as the spin-Peierls state which has a spin gap in the excitation spectrum. Further studies need to be undertaken to understand the relation between spin gap and superconductivity. In our discussion of strongly correlated electron systems we have touched upon only some of the significant issues. Even for these it has not been possible to give a detailed description of the results that have been obtained so far. For more details and

further insights one should consult the appropriate references
mentioned throughout the text.

References

1.   J.G.Bednorz and K.A.Müller, Z.Phys.B 64 , 189 (1986)

2.   P.W.Anderson ,Science 235 ,1196 (1987)

3.   B.Batlogg ,Physics Today 44 , 44 (1991) ;also in "High
     Temperature Superconductivity : The Los Alamos Symposium ",
     Eds. K.Bedell,D.Coffey ,D.Meltzer ,D.Pines and J.R.Schrieffer
     (Addison-Wesley Publishing Co. ,1990)

4.   C.Gros ,R.Joynt and T.M.Rice ,Phys.Rev.B 36 ,381 (1987)

5.   F.C.Zhang and T.M.Rice ,Phys.Rev.B 37 ,3759 (1988)

6.   E.Dagotto ,Rev.Mod.Phys. 66 ,763 (1994)

7.   S.Trugman ,Phys.Rev.B 37 ,1597 (1988)

8.   E.Dagotto and D.Poilblanc ,Phys.Rev.B 42 ,7940 (1990)

9.   C.M.Varma ,P.B.Littlewood ,S.Schmitt-Rink ,E.Abrahams and
     A.Ruckenstein ,Phys.Rev.Lett. 63,19996 (1989)

10. P.W.Anderson and J.R.Schrieffer ,Phys.Today 44 ,55 (1991)

11. P.W.Anderson ,Phys.Rev.Lett. 64,1839 (1990);Phys.Rev.B 42 ,
     2624 (1990)

12. C.Kane ,P.Lee and N.Read ,Phys.Rev.Lett. 39 ,6880 (1989)

13. I.Bose,Phys.Rev.B 43 ,13602 (1991);46,607 (E) (1992);47 ,
     11537 (1993)

14. I.Bose and S.Gayen ,Phys.Rev.B 48 ,10653 (1993) ;J.Phys. :
     Condens.Matter 6 ,L405 (1994)

15. U.Brandt and A.Gieskuus ,Phys.Rev.Lett. 68 ,2648 (1992)

16. H.Tasaki ,Phys.Rev.Lett. 70,3303 (1993)

17. D.C.Mattis and E.H.Lieb ,J.Math.Phys. 6 ,304 (1965)

18. F.D.M.Haldane ,J.Phys.C 14 ,2585 (1981)

19.H.Bethe ,Z.Phys. 71 ,205 (1931)

20.F.D.M.Haldane ,Phys.Lett. 81A ,153 (1981)

21.H.Shiba and M.Ogata ,Prog.Theor.Phys.Suppl. 108 ,265 (1992)

22.H.J.Schultz,Phys.Rev.Lett. 64 ,2831 (1990)

23.M.Ogata and H.Shiba ,Phys.Rev.B 41 ,2326 (1990)

24.P.W.Anderson,Phys.Rev.Lett. 65,2306 (1990);67,2092 (1991);67,
   3844 (1991),71,1220(1993)

25.P.W.Anderson ,Prog.Theor.Phys.Suppl. 107 ,41 (1992)

26.P.W.Anderson ,"The Princeton RVB Book " ,Chapter 6
   (unpublished)

27.M.Ogata and P.W.Anderson ,Phys.Rev.Lett. 70 ,3087 (1993)

28.Y.Ren and P.W.Anderson ,Phys.Rev.B 48 ,16662 (1993)

29.P.W.Anderson ,Science 256 ,1528 (1992)

30.M.Fabrizio and A.Parola ,Phys.Rev.Lett.70 ,226 (1993)

31.R.Valenti and C.Gros ,Phys.Rev.Lett.68,2402 (1992)

32.C.S.Hellberg and E.J.Mele,Phys.Rev.Lett. 67,2080 (1991)

33.Y.C.Chen and T.K.Lee (unpublished)

34.W.O.Putikka, R.L.Genister ,R.R.P.Singh and H.Tsunetsugu ,
   Phys.Rev.Lett. 73 ,170 (1994)

35.Y.C.Chen,A.Moreo,F.Ortolani,E.Dagotto and T.K.Lee ,Phys.Rev.B
   50 ,655 (1994)

36.J.Solyom ,Adv.Phys. 28,209 (1979)

37.V.J.Emery in Highly Conducting One-dimensional Solids ed.by
   J.T.Devreese et al ,p.327 (1979) (Plenum,New York)

38.J.Riera and E.Dagotto ,Phys.Rev.B 50 ,3215 (1994)

39.S.Haas,E.Dagotto,A.Nazarenko,J.Riera ,preprint

40.E.Dagotto,J.Riera,Y.C.Chen,A.Moreo,A.Nazarenko,F.Alcaraz and
   F.Ortolani,Phys.Rev.B 49,3548 (1994)

41.H.J.Schulz,Int.J.Mod.Phys. B 5,57 (1991)

42.J.M.Tranquada ,P.M.Gehring,G.Shirane,S.Shamoto and M.Sato
   Phys.Rev.B 46,5561 (1992)

43. S.Shamoto et al ,Phys.Rev.B 47 ,5320 (1993)

44. A.J.Millis and H.Monien ,Phys.Rev.Lett. 70 ,2810 (1993); 71, 210 (E) (1993)

45. M.U.Ubbens and P.A.Lee ,Phys.Rev.B 50 ,438 (1994)

46. H.Shiba and M.Ogata,Prog.Theor.Phys.Suppl. 108,265 (1992)

47. T.M.Rice,S.Gopalan and M.Sigrist ,Europhys.Lett.23 ,445 (1993)

48. R.S.Eccleston,T.Barnes,J.Brody and J.W.Johnson,Phys.Rev.Lett. 73,2626 (1994)

49. T.Barnes,E.Dagotto,J.Riera and E.Swanson,Phys.Rev.B 47,3196 (1993)

50. S.Gopalan,T.M.Rice and M.Sigrist, Phys.Rev.B 49 ,8901 (1994)

51. E.Dagotto,J.Riera and D.J.Scalapino ,Phys.Rev.B 45,5744 (1992)

52. H.Tsunetsugu,M.Troyer and T.M.Rice ,Phys.Rev.B 49 ,16078 (1994)

53. M.Sigrist,T.M.Rice and F.C.Zhang ,Phys.Rev.B 49 ,12058 (1994)

54. I.Bose and S.Gayen ,Phys.Rev.B 48 ,10653 (1993)

55. I.Bose and S.Gayen ,J.Phys.:Condens.Matter 6 ,L405 (1990)

56. Y.Ohta,T.Tsutsui,W.Koshibae,T.Shimozato and S.Maekawa Phys.Rev.B 46,14022 (1992)

57. S.Gayen and I.Bose, to be published

58. M.Reigrotzki,H.Tsunetsugu and T.M.Rice,J.Phys.:Condens. Matter 6,9235 (1994)

59. D.Poilblanc,H.Tsunetsugu and T.M.Rice,Phys.Rev.B 50,6511 (1994)

60. M.Inada ,J.Phys.Soc.Jpn. 60 ,1877 (1991)

A Phenomenological Theory of a High T_c Superconductor with Coexisting Magnetic Order

A Phenomenological Theory of a High T_C Superconductor
with Coexisting Magnetic Order

S.P. Singh and P. Singh

Department of Physics
G.B. Pant University
Pantnagar, U.P.
India 263 145

A phenomenological theory based on superconducting order parameter $|\psi|$ and magnetic order parameter $|\vec{M}|$ and their interaction is presented for a high Tc superconductor with coexisting magnetic order. It is found that the coupling between superconducting order parameter and magnetic order parameter suppresses superconductivity by reducing the critical transition temperature for the pure superconductor. Analytical expressions for thermodynamic and electromagnetic quantities such as specific heat, penetration depth and upper critical field are obtained. A specific heat varying linearly with temperature is obtained. The penetration depth increases while the critical magnetic field H_{c2} decreases as a result of the coupling. The fluctuations of the superconducting order parameter are considered and the fluctuation induced electrical conductivity and diamagnetic susceptibility are shown to decrease over those of the pure system. The coexistence of superconductivity and magnetic ordering is shown to be a distinct possibility.

INTRODUCTION :

The recent observation of the coexistance of superconductivity and magnetic order in some high T_c superconductors[1,2] ($YBa_2Cu_3O_x$ for $6.38 < x < 6.48$, $RBa_2Cu_3O_7$ with R = Gd, Dy, Er) has attracted the attention of both experimentalists and theorists. The problem of the study of the interplay of superconductivity and magnetism has been pursued vigorously for more than two decades with the view to get a detailed insight into the interrelationship of these two cooperative phenomenon[3,4,5,6,7]. The earlier attempt to study this problem were confined to the effect of magnetic impurities introduced in the superconducting matrix on the superconductivity in the host. The impurities thus doped went to random positions and depressed the superconducting transition temperatures[3]. The suppression of superconducting transition temperature Tc was explained by Abrikosov and Gorkov[4] treating the impurities as uncorrelated. Gorkov and Rusinov extended Abrikosov and Gorkov's theory of superconductors with magnetic impurities including the effect of spin-orbit scattering and predicted the possibility of superconductivity and ferromagnetism[5].

However it was not until 1979 that the real coexistance of superconductivity and magnetism could be established in $ErRh_4B_4$[8] and $Ho_{1.2}MoS_8$[9] with the former as a reentrant superconductor and the latter as a superconducting antiferromagnet. $ErRh_4B_4$ did show the coexistance of superconductivity and ferromagnetism but in avery narrow region with the eventual return to normal state. This re-entrant behaviour wasexplained by Sinha and Singh[7]. The coexistance of superconductivity and magnetism reported prior to 1979 was of spin glass type because of the random positions occupied by magnetic impurities[10]. A generalized Ginzburg-Landau theory for a spin glass superconductor for a recently discovered system $YBa_2(Cu_{1-x}Fe_x)O_{7+y}$ has been developed by Kishore and Singh[11]. In what followswe present a pohenominological theory for a superconductor with coexisting magnetic order by generalizing Ginzburg-Landau theory for superconductivity.

2. FREE ENERGY FUNCTIONAL AND GINZBURG –LANDAU EQUATIONS :

The Ginzburg and Landau phenomenological theory[13] in fact preceeded the famous Bardeen Cooper and Schrieffer microscopic theory[14].

The high Tc superconductors such as $YBa_2Cu_3O_x$[1] and $RBa_2Cu_3O_7$[2] (with R= Gd,Dy,Er) contain a regular lattice of magnetic ions and therefore can be treated as two subsystems — a superconductor and a magnetic one. The free energy density can be expressed as

$$f = f_n + a|\psi|^2 + \frac{b}{2}|\psi|^4 + \alpha|\vec{M}|^2 + \frac{\beta}{2}|\vec{M}|^4 + \gamma|\psi|^2|\vec{M}|^2$$

with $F = \int f\, dr$

Where f_n is the free energy density for normal state, $|\psi|$ is the superconducting order parameter and $|\vec{M}|$ is the magnetic order parameter . The second and third term constitutes free energy density for the pure superconductor and the fourth and the fifth represent free energy density for the magnetic state . The last term denotes the coupling with γ as the coupling parameter here a and α are given by

$$a = a_0\left(\frac{T}{Tco} - 1\right) \qquad \dots\dots\dots(2.2)$$

and
$$\alpha = \alpha_0\left(\frac{T}{Tmo} - 1\right) \qquad \dots\dots\dots(2.3)$$

For arbitrary variation of ψ and $\vec{M}$ we get

$$(a + b|\psi|^2 + \gamma|\vec{M}|^2) \, |\psi| = 0 \qquad \ldots\ldots(2.4)$$

$$(\alpha + \beta|\vec{M}|^2 + \gamma|\psi|^2) \, |\vec{M}| = 0 \qquad \ldots\ldots(2.5)$$

These constitute generalized Ginzburg–Landau equations whose solutions can be studied under various conditions.

It is easy to see that the free energy for the pure superconductor or pure magnetic states respectively will be given by

$$f_{sc} = -\frac{a^2}{2b} \qquad , \quad |\vec{M}| = 0 \qquad \ldots\ldots\ldots\ldots(2.6)$$

$$f_m = -\frac{\alpha^2}{2\beta} \qquad , \quad |\psi| = 0 \qquad \ldots\ldots\ldots\ldots(2.7)$$

Let T_c and T_m be the superconducting and magnetic order transition temperatures and T_{c0} and T_{m0} denote the corresponding transition temperatures for the pure superconductor and magnetic subsystems respectively.

For $|\psi| \neq 0$ and $|\vec{M}| \neq 0$ Equations (2.4) and (2.5) can be solved to get

$$|\psi|^2 = \frac{a\beta - \alpha\gamma}{\gamma^2 - b\beta} \qquad \ldots\ldots\ldots\ldots(2.8)$$

and

$$|\vec{M}|^2 = \frac{a\gamma - b\alpha}{b\beta - \gamma^2} \qquad \ldots\ldots\ldots\ldots(2.9)$$

Although several interesting situations such as

$$\text{i) } T > T_c \ (T_m) \quad \text{when} \quad |\psi| = |\vec{M}| = 0,$$

$$\text{ii) } T_c > T_m \quad \text{with} \quad |\psi| \neq 0 \text{ and } |\vec{M}| = 0 \text{ and}$$

$$\text{iii) } T_m > T > T_c \quad \text{with} \quad |\vec{M}| \neq 0' \text{ and } |\psi| = 0$$

can be discussed but we shall be more interested in the case where $T < T_c \ (T_m)$ if $T_c \ (T_m) < T_m(T_c)$ i.e. the coexistence region. In this regime we have

$$|\psi|^2 = \frac{a_0}{b}\left(1 - \frac{T}{T_{c0}}\right) - \frac{\gamma\alpha_{m0}}{b}\left(1 - \frac{T}{T_{m0}}\right) \qquad ....(2.10)$$

$$|M|^2 = \frac{\alpha_{m0}}{\beta}\left(1 - \frac{T}{T_{m0}}\right) - \frac{\gamma a_0}{b\beta}\left(1 - \frac{T}{T_{c0}}\right) \qquad .....(2.11)$$

In obtaining equation (2.10)and (2.11) we have made use of equations (2.2) and (2.3) and have retained only terms linear in γ the coupling constant. In the following sections (3) and (4) we shall discuss the thermodynamics and electromagnetic properties using the above equations. It should be pointed out however that since the free energy functional (2.1) is only valid for small values of the order parameter and in the weak coupling case, the present theory is applicable near the transition temperatures and specially when both superconducting and magnetic order transition temperatures are close to each other i.e. in the coexistence regime

3.   Thermodynamic Properties :

3.1.   Superconducting   and   magnetic   order
            transition   temperatures

                          S.P. Singh and P. Singh

The superconducting temperature Tc is obtained by letting

$$|\psi|^2 \longrightarrow 0 \text{ which accounts to}$$

$$\alpha\beta - \alpha\gamma = 0 \qquad \dots\dots\dots\dots\dots(3.1)$$

using (2.2) and (2.3) yields

$$T_c = T_{c0} - \frac{\gamma\,\alpha_0 T_{c0}}{\beta\,a}\left(1 - \frac{T_{c0}}{T_{m0}}\right) \qquad \dots\dots\dots(3.2)$$

For $T_{c0} > T_{m0}$, T_c (say T_{c2}) $> T_{c0}$

While for $T_{c0} < T_{m0}$, T_c (say T_{c2}) $< T_{c0}$

Therefore T_c become larger than T_{c0} which essentially means that the magnetic coupling increases T_c and hence support superconductivity, Which is the case of some impurities like Fe, Ni and Co in RuCe compounds [15], which is explained by healing mechanism[11-16], while in the other case Tc deceases and this is due to the pair breaking effects[17].

The magnetic ordering transition temperature Tm can be obtained by equating

$$T_m = T_{m0} - \frac{\gamma a_0\,T_{m0}}{b\,\alpha_0}\left(1 - \frac{T_{m0}}{T_{c0}}\right) \qquad \dots\dots\dots\dots(3.3)$$

If $T_{c0} > T_{m0}$, $T_m < T_{m0}$

which means that superconductivity suppresses the magnetic ordering transition temperature. In the other case if $T_{c0} < T_{m0}$ $T_m > T_{m0}$, which accounts to the fact that superconductivity is assisting in the magnetic ordering process. Fig. (1) shows the

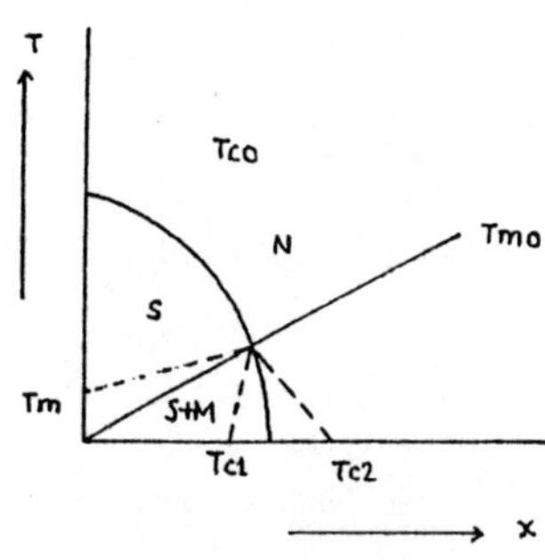

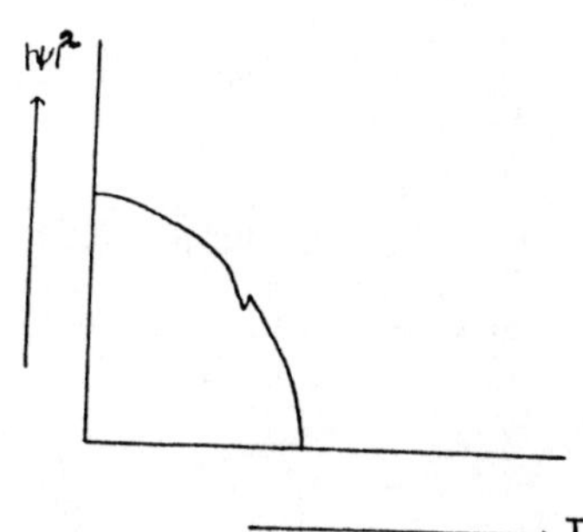

Fig.1: Variation of superconducting
and magnetic ordering transition
temp.   Tc   and   Tm   respectively with x; the impurity
concentration.

Fig.2: Superconducting   order   paramet
as a function of T.

variation of $T_c (T_m)$ Vs. x.

3.2 Superconducting and magnetic order parameters:

3.2.1 Superconducting order parameters :

From the generalised Gingburg-Landau equations the
super conducting order parameter can be expressed as

$$|\psi|^2 = \frac{a_0}{b}(1 - \frac{T}{T_{co}}) \theta(T_c - T) - \frac{\gamma \alpha_0}{b\beta}(1 - \frac{T}{T_{mo}}) \theta(T_c - T) \theta(T_m - T)$$

$$\dots\dots(3.4)$$

Here $\theta(x)$ is the Heaviside step function. The variation $|\psi^2|$ Vs.T

is shown in fig.(2) which exhibits a small depression at T_m.

3.2.2. Magnetic order parameter :

The magnetic order parameter can be obtained as

$$|M|^2 = \frac{\alpha_{mO}}{\beta}(1 - \frac{T}{T_{mO}}) \theta(T_m - T) - \frac{\gamma a_0}{b\beta}(1 - \frac{T}{T_{cO}}) \theta(T_c - T) \theta(T_m - T)$$

$$\dots\dots\dots\dots(3.5)$$

3.3. Specific Heat :

The specific heat per unit volume is given by

S.P. Singh and P. Singh

$$C = -T \frac{\partial^2 F}{\partial T^2} \qquad\qquad \ldots\ldots\ldots(3.6)$$

Using equations (2.1),(2.8) and (2.9) the specific heat is expressed as

$$C = C_n + \left[\left(\frac{a_0^2}{bT_{cO}^2} + \frac{\alpha_{mO}^2}{\beta T_{mO}^2}\right) - \frac{2\gamma\, a_0\, \alpha_{mO}}{b\beta\, T_{cO}\, T_{mO}}\right] T \qquad \ldots(3.7)$$

Where C_n is the specific heat per unit volume of the normal state corresponding to the free energy functional fn. It is clear from the above expressions that there is a sharp jump in the value of the specific heat at T_m which is suppressed somewhat by the presence of the coupling term. As shown in fig.(3) a small discontinuity appears at T_c and T_m.

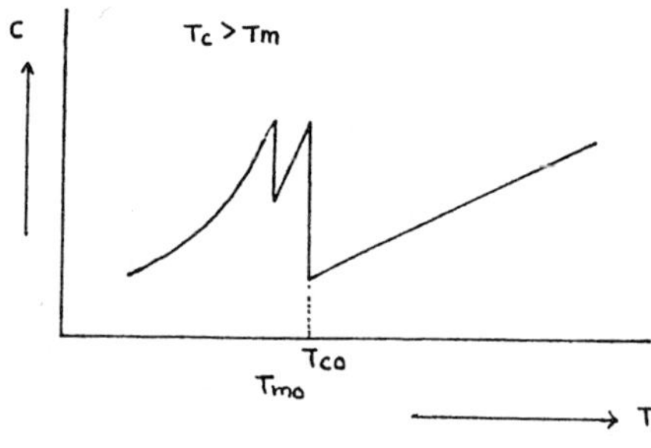

Fig.3 *Specific heat as a function of Temperature*

4. Electromagnetic properties :

To study electromagnetic properties we write the free energy functional in the presence of weak magnetic field H as

$$f(\psi,H) = a|\psi|^2 + \frac{b}{2}|\psi|^4 + \frac{1}{2m^*}\left|\left(-i\hbar\vec{\nabla} - \frac{e^*A}{c}\right)\psi\right|^2$$
$$+ \frac{H^2}{8\pi} + \alpha|\vec{M}|^2 + \frac{\beta}{2}|\vec{M}|^4 + \gamma|\psi|^2|\vec{M}|^2$$

where A is the magnetic vector potential $\overrightarrow{H} = \nabla * \overrightarrow{A}$ e^* and m^* are the effective charge and effective mass of the superconducting pairs.

4.1 London Penetration depth

Following   Ginzburg–Landau   theory   the   London penetration depth λ is given by

$$\frac{1}{\lambda^2} = \frac{4\,\pi\,e^*\,|\psi|^2}{m^*\,c^2} \qquad \ldots(4.2)$$

So the square of the penetration depth λ varies as the square of the inverse of the superconducting parameter from equation it is clear that the $|\psi| \neq 0$, we have

$$a + b|\psi|^2 + \gamma|\overrightarrow{M}|^2 = 0 \qquad \ldots\ldots\ldots\ldots(4.3)$$

which gives

$$|\psi|^2 = -\frac{a}{b} - \gamma\,|\overrightarrow{M}|^2$$
$$|\psi|^2 = |\psi_0|^2 - \frac{\gamma}{b}|\overrightarrow{M}|^2 \qquad \ldots\ldots\ldots(4.4)$$

where $|\psi_0|^2 = -\frac{a}{b}$, corresponds to the equilibrium $|\psi_0|$ for the pure superconducting state. It is quiteclear from equations (4.2) and (4.3) that the London penetration depth λ will increase as a result of the coupling of the magnetic and superconducting order parameters. λ^2 Vs. T shows a small jump at T_m as evident from fig.(4)

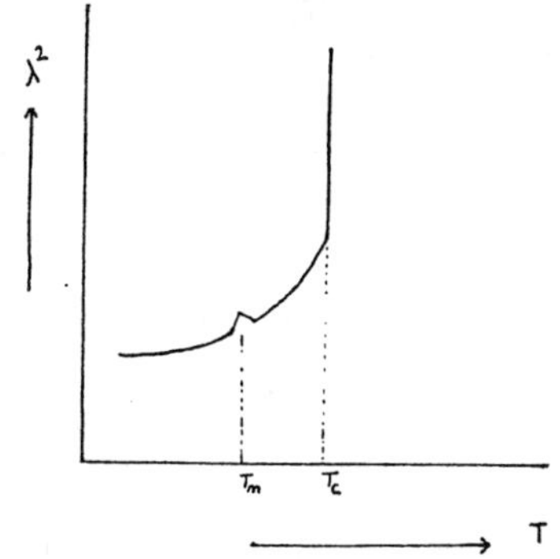

Fig.4   *Variation of square of penetration depth with temperature T.*

4.2.   Coherence length:

The superconducting coherence length ξ can be expressed as

$$\frac{1}{\xi^2} = \frac{2m^* \, b \, |\psi|^2}{\hbar^2} \qquad \dots\dots\dots\dots\dots(4.5)$$

A small inhomogeneity in the superconducting parameter is assumed in the derivation of the above expression. It is clearly seem from the expression that the inverse of the square of coherence length ξ varies as $|\psi|^2 \sim \Delta$ i.e.as superconducting order parameter

It is important to note that the Gingburg-Landau parameter,

$$K = \frac{\lambda}{\xi} = \sqrt{\frac{b}{2\pi}} \; \frac{m^* c}{\hbar \, e^*} \qquad \dots\dots\dots\dots\dots(4.6)$$

remains unaffected by the magnetic coupling.

4.3. UPPER CRITICAL FIELD :

For a type II super conductor the upper critical magnetic field H_{c2} defined as the magnetic field just below which bulk superconductivity begins to appear is given by

$$H_{c2} = \frac{\phi_0}{2\pi\xi^2} \qquad \dots\dots\dots\dots(4.7)$$

where ϕ_0 is flux quantum.

Thus the inverse of the square of the coherence length ξ behaves as the superconducting order parameter. In the present case with the magnetic coupling H_{c2} will decrease which means that coupling will try to reduce the type II behaviour.

5.　　FLUCTUATION INDUCED DIAMAGNETIC SUSCEPTIBILITY :

As the system prepares to make the transition to superconducting state diamagnetic susceptibility χ is enhanced as a result of fluctuations. The fluctuations above T_c can be observed by susceptibility measurements on tiny particles since χ depends λ which in turnis related to $|\psi^2|$. for spherical particles of radius $R \ll \lambda$ one gets

$$\chi \sim -R^2/\lambda^2 \sim \langle \psi^2 \rangle R^2 \quad \ldots\ldots\ldots\ldots\ldots\ldots(5.1)$$

So the diamagnetic susceptibility will behave as the superconducting gap parameter Δ since $1/\lambda^2$ goes as Δ.

5.1. FLUCTUATION INDUCED ELECTRICAL CONDUCTIVITY :

Following [18] and [19] the fluctuation induced electrical conductivity can be obtained as

$$\sigma_{yy}\big|_{3D} \approx \frac{e^2}{32\hbar} \left(1 - \frac{2\gamma}{b} \frac{|\vec{M}|^2}{|\psi|^2} \right) \frac{1}{\xi(0)} \left(\frac{T}{T-T_c} \right)^{1/2} \text{-----}(5.2)$$

$$\sigma_{yy}\big|_{2D} \cong \frac{e^2}{16\hbar} \left(1 - \frac{2\gamma}{b} \frac{|\vec{M}|^2}{|\psi|^2} \right) \frac{A}{d} \left(\frac{T}{T-T_c} \right) \text{-----}(5.3)$$

These expressions clearly indicate that the magnetic coupling suppresses fluctuation induced electrical conductivity. The magnetic field is known to decrease fluctuation induced electrical conductivity related to its zero field value. This[20] seems consistent with the generalunderstanding that the magnetic order generally suppresses (lowers) **superconduct**ing order parameter and also Cooper pair density and hence will reduce electrical conductivity.

CONCLUSION :

A phenomenological theory for a superconductor with coexisting magnetic order based on a free energy functional depending on superconducting order parameter ψ, magnetic order parameter $\vec{M}$ and their coupling is presented. The magnetic order decreases (increases) superconducting order parameter and superconducting transition temperatures. The results seem to be in broad agreement with experimental observations. The coherence length ξ and the penetration depth λ are also affected but the Ginzburg Landau parameter remains unaffected. The specific heat varies linearly with temperature. The upper critical magnetic field decreases as a result of magnetic order and has similar effect as the magnetic field. The fluctuation induced diamagnetism is also shown to decrease and so does the fluctuation induced electrical conductivity. It is suggested that experiments should be conducted to look for such features.

ACKNOWLEDGEMENTS

One of the authors (P.S) thanks Prof. Dr.P.Entel and Prof.J.P.Carbotte for fruitful discussion during his visit to McMaster University. S.P. acknowledge Rajeev Pandey for his assistance in numerical work.They both also thank to Prof.A.V. Narlikar for interesting discussion on the subject.

REFERENCES

1. H. Brewer,J.F. Carolan, W.N. Hardy, B.X. Yang, P. Schlager, R. Kadono, J.R.Kempton, R.F.Kiefe, S.R.Kreitzman, J.M.Luke, T.M.Riseman, D.Li, Williams, K.Chow, P.Dosanjh, B.Gowe, R.Krahn and M.Norman, Physica C 162-164, 33 (1989).

2. S.E. Brown, J.D. Thompson, J.O. Willis, R.M. Aikin, E.Zirngieblog J.L. Smith, Z. Firk and R.B. Schwore. Phys. Rev. B 36 2298 (1987).

3. B.T.Matthias, H.Suhl, E.Corenzwit, Phys.Rev.Lett. 1, 92 (1958).

4. A.A.Abrikosov and L.P.Gorkov, Sov.Phys.JETP 19,922 (1964).

5. L.P.Gorkove and A.I.Rusinov, Sov.Phys.JETP 12,1243 (1961).

6. M.V.Jaric, Phys.Rev.B 20 ,4486 (1979).

7. K.P.Sinha and P.Singh, Low Temp.Phys. 37, 389 (1979).

8. W.A.Fertig, D.C.Johnson, L.E.Dilong, R.W.McCallum, M.B.Maple and B.T.Matthias, Phys.Rev.Lett. 38, 377 (1977).

9. M.Ishikawa and ϕ.Fisher, Sol.Stat.Commun. 23, 37 (1977).

10. M.B.Maple in 'Magnetism' (G.T.Rado and H.Suhl, Eds.) Vol.V, p 289, Academic press, New York, 1966.

11. R.Kishore and P.Singh, Physica C 215, 59 (1993)

12. V.L.Ginzberg and L.D.Landau, Zh.Exsp.Ther.Fiz.20,1964 (1957).

13. J.Bardeem, L.Cooper, J.R.Schriffer, Phys.Rev. 108 1175 (1957).

14. H.Suhl, B.T.Matthiasand E.Corenzwit, J.Phys.Chem.Solids, 19, 346(1959).

15. B.N. Ganguly, U.N. Upadhyaya and K.P.Sinha, Phys.Rev. 146,317 (1966).

16.   P.Singh and R.Kishore, J.Superconductivity, 8, 9 (1995).

17.   L.G.Aslamasov and A.I.Larkin, Phys.Lett. 26A, 238 (1968).

18.   P.Singh, Preprint INPE, Brazil (1992).

Detection of X–rays by Pair Breaking in Superconductors

N. Murali Krishna and Keshav N. Shrivastava

School of Physics
University of Hyderabad
P.O. Central University
Hyderabad – 500 134, India

Two new processes occur in a superconducting film when it is used as a detector of x-rays. One of these processes is the scattering of the x-ray by a single electron which gives rise to the broadening of the x-ray line. Another process describes the breaking of a Cooper pair by the x-ray which also contributes to the width of the x-ray. The line width arising from the single electron process depends on T^4 whereas that arising from the pair breaking process varies almost as T^6 at low temperatures. Lines occur at $\hbar\omega_q \pm 2\Delta$, and at $\hbar\omega_q$ where $\hbar\omega_q$ is the energy of the x-ray and 2Δ is the gap of the superconductor

PACS numbers.07.62. +s , 74.90. +n , 78.70. Ck

Recently, it has been pointed out by Jochum et al[1] that new detectors have initiated important discoveries. Kraus et al[2] have pointed out that the rate of relaxation of particles down to the gap edge is,

$$\gamma \propto E^3/\tau_o$$

where E is the energy of excitations and τ_o is a time constant characteristic of the material. Several properties of the detectors of x-rays using superconducting junctions have been reported[3-6] but the quantum theory of widths and shifts has not yet been solved. The phonon scattering has been considered to be important[7-11]

     N. Murali Krishna and Keshav N. Shrivastava

for the processes to take place but Burstein et al[12] have considered the use of superconductors as detectors of microwaves in which the photons play an important role. In the B.C.S. theory[13], the phonon operators of the Hamiltonian are eliminated to derive the attractive interaction between the electrons so that the phonon effects are found in the coupling constant. Thus the pair formation and breaking by phonons is stronger than by photons but the frequency of the x-ray is very large so that the two, phonon and photon resonances are far apart and hence do not interfere. Therefore it is of interest to determine the spectrum of the scattered x and γ-rays . The solar x-rays can be detected using the superconductors. However, a monochromatic source of x-rays and a detector, both with an energy resolution of the order of the superconducting gap is not yet available. Whenever such a technology is developed it will be necessary to have a calculation of shifts and widths, particularly because the widths determine the resolution of the instrument. Hence our calculation is of interest in the area of astrophysics and astronomy also.

In this paper, we construct the Hamiltonian for the interaction of x-rays with Cooper pairs in a superconductor. As the x-ray hits the Cooper pair, it breaks the pair into two electrons and the x-ray is scattered. The spectrum of the scattered x-ray consists of several frequencies. One of the frequencies is the same as that of the incident x-ray, $\hbar\omega_q$, as in the Rayleigh scattering and there are sidebands at $\hbar\omega_q \pm 2\Delta$. We have performed the Bogoliubov transformation on the pair breaking interaction which we have used to calculate the self energy of the x-ray.

This self energy corresponds to a broadening of the x-ray. We find that in a three dimensional superconductor, this broadening depends on temperature as T^4 for the electron scattering process and as T^6 for the pair breaking process at low temperatures.

We consider the x-rays interacting with the conduction electrons so that the unperturbed Hamiltonian of the system is described as,

$$H_o = \sum_q \hbar\omega_q \beta_q^\dagger \beta_q + \sum_{k,\sigma} \epsilon_{k,\sigma}(c_{k\sigma}^\dagger c_{k\sigma} + c_{-k,-\sigma}^\dagger c_{-k,-\sigma}) \qquad ($$

where $\hbar\omega_q$ is the single particle energy of one x-ray. The $\beta_q^\dagger$ and β_q are the boson creation and annihilation operators for the x-ray. The single particle energy of the electrons of wave vector $\vec{k}$ and spin σ in the conduction band is $\epsilon_{k\sigma}$. The wave vector of the x-ray is $q = \omega/c$ where c is the velocity of light. The attractive interaction between the electrons is given by

$$H_a = -V \sum_{k,k'}{}' c_{k',\sigma}^\dagger c_{-k',-\sigma}^\dagger c_{-k,-\sigma} c_{k,\sigma} \tag{2}$$

This interaction is useful for the formation of spin singlet zero momentum pairs as in the B.C.S. theory[13,14] and is diagonalized by the

$$c_k = u_k d_k + v_k d_{-k}^\dagger \tag{3}$$

where k is associated with spin σ and $-k$ with $-\sigma$. The coefficients u_k and v_k are real, where u_k is an even function of k and v_k is an odd function of k, $u_k = u_{-k}, v_k = -v_{-k}$ with $u_k^2 + v_k^2 = 1$. The gap is written in terms of pair operators,

$$\Delta_k = V\Sigma_{k'}' c_{-k,-\sigma} c_{k,\sigma} \;,$$
$$\Delta_k^* = V\Sigma_k' c_{k,\sigma}^\dagger c_{-k,-\sigma}^\dagger \;. \tag{4}$$

The interaction of the x-ray photon with the conduction electrons is described by

$$H_{pe}' = i \sum_{k,k',q'} G_{k,k',q'} c_{k'}^\dagger c_k \beta_{q'}^\dagger \delta(k - k' - q')$$
$$+ i \sum_{k,k'',q} G_{k,k'',q} c_{k''}^\dagger c_{-k} \beta_q \delta(k'' - q + k) + h.c. \tag{5}$$

where $G = gV^{-1/2}$ with

$$g = \frac{e\hbar}{mc}\left(\frac{\hbar\omega_q}{\kappa}\right)^{1/2} \tag{6}$$

where κ is the dielectric constant of the medium and m is the mass of the electron. One of the terms in the second-order energy operator of the above interaction is given by

$$H_2' = - \sum_{k,k',k'',q,q'} (\epsilon_{k'} - \epsilon_k + \hbar\omega_q)^{-1} G_{k,k',q} G_{k,k'',q} c_{k''}^\dagger c_{-k} c_{k'}^\dagger c_k \beta_{q'}^\dagger \beta_q$$
$$\delta(q - k' - k'' - q') + h.c. \tag{7}$$

Using anticommutators for the electron operators, the product of the electronic operators in the above can be written as

$$c^\dagger_{k''}c_{-k}c^\dagger_{k'}c_k = \delta_{-k,k''}c^\dagger_{k'}c_k - c_{-k}c^\dagger_{k''}\delta_{k,k'} + c_{-k}c^\dagger_{k'}\delta_{k,k''} - c_{-k}c_k c^\dagger_{k''}c^\dagger_{k'} \tag{8}$$

from which leaving out the first three terms which describe single particle electron scattering by photons, we write the interaction corresponding to pair formation and breaking as

$$H'_2 = \sum_{k,k',k'',q,q'} (\epsilon_{k'} - \epsilon_k + \hbar\omega_{q'})^{-1} G_{k,k',q'} G_{k,k'',q} c_k c_{-k} c^\dagger_{k'} c^\dagger_{k''} \beta^\dagger_{q'} \beta_q \delta(q - q' - k' - k''). \tag{9}$$

Including the sum over intermediate states, one more contribution to the above interaction occurs which we calculate and include in the above so that

$$H''_2 = \sum_{k,k',k'',q,q'} \left[(\epsilon_{k'} - \epsilon_k + \hbar\omega_{q'})^{-1} + (\epsilon_{k''} - \epsilon_{-k} - \hbar\omega_q)^{-1} \right] G_{k,k',q'} G_{k,k'',q}$$
$$\times c_k c_{-k} c^\dagger_{k'} c^\dagger_{k''} \beta^\dagger_{q'} \beta_q \delta(q - q' - k' - k'') + h.c. \tag{10}$$

This means that the x-ray photons can break pairs of electrons into single electrons and pairs can be made by shining x-rays on electrons but the electromagnetic energy, $H^2/8\pi$, is small compared with $k_B T_c$. Hence this mechanism is too small to predict the transition temperatures but it can be effective for detecting x-rays, due to different frequency regime. Since the wave vector dependence in the coupling constant is small, we write the above interaction as,

$$H' = \sum_{\sigma} \sum_{k,k',k'',q,q'} D_{k,k',k'',q,q'} \beta_q \beta^\dagger_{q'} c_{k,\sigma} c_{-k,-\sigma} c^\dagger_{k',\sigma} c^\dagger_{k'',-\sigma} \delta(q - q' - k' - k'') + h.c. \tag{11}$$

where $h.c.$ stands for the hermitian conjugate of the previous terms. Using the commutators for bosons and anticommutators for fermions, the above interaction can be written as,

$$H = H'_o + H_a + H'_a + H'_s + H'_p \tag{12}$$

$$H'_o = \sum_{\sigma} \sum_{k,q} D_{k,q} - \sum_{\sigma} \sum_{k,q} D_{k,q} \beta^\dagger_q \beta_q - \sum_{\sigma} \sum_{k,q} D_{k,q}(c^\dagger_{k\sigma} c_{k\sigma} + c^\dagger_{-k,-\sigma} c_{-k,-\sigma}) \tag{13a}$$

$$H'_a = \sum_{\sigma} \sum_{k,k',q} D_{k,k',q} c^\dagger_{k',\sigma} c^\dagger_{-k',-\sigma} c_{-k,-\sigma} c_{k,\sigma} \tag{13b}$$

$$H'_s = \sum_{\sigma} \sum_{k,k'',q,q'} D_{k,k'',q,q'} \beta^\dagger_{q'} \beta_q c^\dagger_{k'',-\sigma} c_{-k,-\sigma} \delta(q - q' - k - k'')$$
$$+ \sum_{\sigma} \sum_{k,k',q,q'} D_{k,k',q,q'} \beta^\dagger_{q'} \beta_q c^\dagger_{k',\sigma} c_{k,\sigma} \delta(q - q' - k' + k) + h.c. \tag{13c}$$

$$H'_p = \sum_{\sigma} \sum_{k,k',k'',q,q'} D_{k,k',k'',q,q'} \beta^\dagger_{q'} \beta_q c^\dagger_{k',\sigma} c^\dagger_{k'',-\sigma} c_{k,\sigma} c_{-k,-\sigma} \delta(q - q' - k' - k'') + h.c. \tag{}$$

where the first term in (13a) is a constant. The second term describes a small negative shift in the x-ray energy. The third term gives the reduction in the single particle energy of electrons in the conduction band. The term given by (13b) is a positive pairing interaction and hence its effect is to reduce the strength of attractive interaction given by (2). The expression (13c) shows that the x-rays are scattered by single electrons and (13d) shows that pairs are broken by the x-rays. Upon rearranging the various terms, the Hamiltonian of the system becomes,

$$H = H_{oo} + H'_a + H'_s + H'_p \tag{14}$$

where

$$H_{oo} = \sum_{\sigma}\sum_{k,q} D_{k,q} + \sum_{q} \hbar\omega'_q \beta^\dagger_q \beta_q + \sum_{k,\sigma} \epsilon'_{k,\sigma}(c^\dagger_{k,\sigma}c_{k,\sigma} + c^\dagger_{-k,-\sigma}c_{-k,-\sigma}) \tag{15}$$

$$H'_a = -V' \sum_{k,k'} c^\dagger_{k',\sigma}c^\dagger_{-k',-\sigma}c_{-k,-\sigma}c_{k,\sigma} \tag{16}$$

with

$$\hbar\omega'_q = \hbar\omega_q - \sum_{\sigma}\sum_{k} D_{k,q} \tag{17a}$$

$$\epsilon'_{k,\sigma} = \epsilon_{k,\sigma} - \sum_{q} D_{k,q} \tag{17b}$$

and

$$V' = V - \sum_{q} D_{k,k',q} \tag{17c}$$

in which the first term is a constant and the single particle energy of the x-ray has been slightly reduced. Similarly, the single particle energy of the electrons is also reduced by a small amount. The transformation (3) substituted in (16) gives the ground state energy with single particle energies shifted by the pair breaking energy[14]. The transformed form of the scattering interaction of (13c) is found to become,

$$H'_s = \sum_{\sigma}\sum_{k,q,q'} D_{k,q,q'} v^2_k \beta^\dagger_{q'}\beta_q + \sum_{\sigma}\sum_{k,k'',q,q'} D_{k,k'',q,q'}(u_k u_{k''}\beta^\dagger_{q'}\beta_q d^\dagger_{k''}d_{-k}$$

$$+ v_k v_{k''}\beta^\dagger_{q'}\beta_q d^\dagger_k d_{-k''})\delta(q - q' - k - k'') + h.c. \tag{18}$$

There are terms with $\vec{k}$ replaced by $-\vec{k}$ and also charge non-conserving terms containing factors of the type $d_{-k''}d_{-k}$ or $d^\dagger_k d^\dagger_{k''}$ but all such terms cancel each other so that there is no charge non-conserving interaction. The pair breaking Hamiltonian may be expressed as

$$H'_p = H'_p(o) + H'_p(1) + H'_p(2) + H'_p(3) + H'_p(4) \tag{19}$$

where

$$H'_p(o) = \sum_\sigma \sum_{k,k',q} D_{k,k',q} u_k u_{k'} v_k v_{k'} (d^\dagger_k d_k - d^\dagger_{-k} d_{-k}) \beta^\dagger_q \beta_q$$
$$+ 2 \sum_\sigma \sum_{k',k'',q,q'} D_{k',k'',q,q'} v_{k''} v^3_{k'} d^\dagger_{k'} d_{-k''} \beta^\dagger_{q'} \beta_q \delta(q - q' - k' - k'') + h.c. \tag{20}$$

Upon decoupling, it is seen that this interaction gives a shift in the single particle frequency of the x-ray which depends on the number density of electrons. The second term in (20) gives the electron scattering within a band. Next we write the terms which contain a pairing operator of the form $d_k d_{-k}$ which annihilates a fermion pair as,

$$H'_p(1) = \sum_\sigma \sum_{k,k'',q} D_{k,k'',q} u^2_k u_{k''} v_{-k''} d_k d_{-k} \beta^\dagger_q \beta_q$$

$$+ \sum_\sigma \sum_{k,k',k''q,q'} D_{k,k',k''q,q'} u_k v_k v_{k'} v_{k''} d_{-k'} d_{-k''} \beta^\dagger_{q'} \beta_q \delta(q - q' - k' - k'')$$
$$- \sum_\sigma \sum_{k,k',q} D_{k,k',q} u_{k'} v_{k'} v^3_{-k} d^\dagger_{-k} d^\dagger_k \beta^\dagger_q \beta_q$$
$$- \sum_\sigma \sum_{k,k',k'',q,q'} D_{k,k',k'',q,q'} u_k u_{k'} u_{k''} v_k d^\dagger_{k'} d^\dagger_{k''} \beta^\dagger_{q'} \beta_q \delta(q - q' - k' - k'') + h.c. \tag{21}$$

We now write the interaction terms in which pairs are broken into two particles by the x-ray as,

$$H'_p(2) = \sum_\sigma \sum_{k,k',k'',q,q'} D_{k,k',k'',q,q'} u^2_k u_{k'} u_{k''} d^\dagger_{k'} d^\dagger_{k''} d_k d_{-k} \beta^\dagger_{q'} \beta_q \delta(q - q' - k' - k'')$$
$$- \sum_\sigma \sum_{k,k',k'',q,q'} D_{k,k',k'',q,q'} v^2_k v_{k'} v_{k''} d^\dagger_{-k} d^\dagger_k d_{-k'} d_{-k''} \beta^\dagger_{q'} \beta_q \delta(q - q' - k' - k'') + h.c. \tag{22}$$

The terms which do not conserve the quasiparticle number unless pairs are formed are given below,

$$H'_p(3) = \sum_\sigma \sum_{k,k',k'',q,q'} D_{k,k',k'',q,q'} [u_k v_k v_{k'} v_{k''} d^\dagger_{-k} d_{-k'} d_{-k''} d_{-k} \beta^\dagger_{q'} \beta_q$$
$$+ u_k v_k v_{k'} v_{k''} d^\dagger_k d_{-k'} d_{-k''} d_k \beta^\dagger_{q'} \beta_q + u_k u_{k'} u_{k''} v_k d^\dagger_{k'} d^\dagger_{k''} d^\dagger_{-k} d_{-k} \beta^\dagger_{q'} \beta_q$$
$$+ u_k u_{k'} u_{k''} v_k d^\dagger_{k'} d^\dagger_{k''} d^\dagger_k d_k \beta^\dagger_{q'} \beta_q] \delta(q - q' - k' - k'') + h.c. \tag{23}$$

Now we write those terms which are quadratic in pair breaking upon interaction with the x-ray as,

$$H'_p(4) = \sum_\sigma \sum_{k,k',k'',q,q'} D_{k,k',k'',q,q'} [u_k^2 v_{k'} v_{k''} d_{-k'} d_{-k''} d_k d_{-k}$$
$$- u_{k'} u_{k''} v_k^2 d_{k'}^\dagger d_{k''}^\dagger d_{-k}^\dagger d_k^\dagger] \beta_{q'}^\dagger \beta_q \delta(q - q' - k - k'') + h.c. \tag{24}$$

This completes the definition of our Hamiltonian. Considering the single particle energies and the attractive interaction as given by (16), we find that the coherence factors are given by,

$$u_k^2 = \frac{1}{2} \left[1 + \frac{\epsilon'_k}{(\epsilon'^2_k + \Delta^2)^{1/2}} \right]$$

and

$$v_k^2 = \frac{1}{2} \left[1 - \frac{\epsilon'_k}{(\epsilon'_k + \Delta^2)^{1/2}} \right] \tag{25}$$

where the single particle energy is given by (17) which is slightly shifted from that in the conduction band due to shining with x-rays.

The width of the x-ray as measured by the superconducting detector is given by $\hbar/\tau$ where τ is the life time of the x-ray due to the interaction with the superconductor as calculated from the imaginary part of the self energy,

$$\frac{1}{\tau_q} = \frac{2}{\hbar} Im\Sigma_q \tag{26}$$

using Dirac's identity

$$\lim_{\epsilon \to 0} \frac{1}{x \pm i\epsilon} = \frac{P}{x} \mp i\pi\delta(x) \tag{27}$$

where the first term on the right hand side indicates that the Cauchy's principal value has to be evaluated for determining the real part. The width of the x-ray is caused by two important processes. One of these is the electron scattering process given by (18) and the other is the pair breaking process given by (22). The total width is given by $\hbar\delta\omega_q \simeq \hbar\left(1/\tau_s + 1/\tau_p\right)$. The method of calculation of the two mechanisms occuring here is similar to that used for the radiation damping from the magnon-photon interaction[15] and recombination in electron-hole droplets in semiconductors[16].

The first term of (18) gives rise to a shift of the energy of the x-ray as $\sum_\sigma \sum_{k,q} D_{k,q} v_k^2 n_q$ to which we add also the term $\sum_\sigma \sum_{k,q} D_{k,q} u_k^2 n_q$ where $n_q = [\exp(\hbar\omega_q/k_B T) - 1]^{-1}$ is the number density of the x-ray. This term does not contribute to the life time of the x-ray. Therefore, we calcualte the self energy of the x-ray due to the second term of (18) as

$$\Sigma_q^x(1) = \sum_{k,k'',q'} \frac{2G_1^2 \zeta_1 \delta(q - q' - k - k'')}{E - \hbar\omega_{q'} - \epsilon_{k''} + \epsilon_{-k}} \tag{28}$$

where

$$G_1 = D_q u_k u_{k''} , \tag{29a}$$

and

$$\zeta_1 = (1 + n_{q'})(1 - f_{k''})f_{-k} + n_{q'} f_{k''}(1 - f_{-k}) \tag{29b}$$

and f_k is a Fermi distribution

$$f_k = [\exp(\epsilon_k - \epsilon_F)/k_B T + 1]^{-1} .$$

A factor of 2 arises due to two spin configurations. Using the Dirac's identity (27), the imaginary part of the above is found to be,

$$Im\, \Sigma_q^x(1) = 2\pi \Sigma_{k,k'',q'}\, G_1^2 \zeta_1 \delta(q - q' - k - k'')\delta(\hbar\omega_q - \hbar\omega_{q'} - \epsilon_{k''} + \epsilon_{-k}) \tag{30}$$

where we took $\hbar\omega_q$ for E. We eliminate k'' using the first δ function and then near small k, $\hbar\omega_q \simeq \hbar\omega_{q'}$, so that the second δ function gives $\frac{1}{\hbar c}\delta(q - q')$. Then using (26) we find the life time of the x-ray as,

$$\frac{1}{\tau_1} \simeq \frac{2V\omega^2}{\pi\hbar^2 c^3} \sum_k G_1''^{\,2} \zeta_1''^{\,2} \tag{31}$$

where

$$G_1'' = D_q u_k^2 , \tag{32}$$

$$\zeta_1'' = f_{-k}(1 - f_{-k})(2n_q + 1) \tag{32b}$$

and V is the volume of the superconductor. In the case of a three dimensional superconductor the summation over k leads to T^3 dependence in the inverse life time. Another factor of T ariese from n_q so that the rate given by (31) varies as T^4 at low temperatures. There is one more term in the interaction due to the symmetry in the Bogoliubov transformation given by the third term of (18). This term is of the form $D_q v_k v_{q-q'-k}$ in place of (29a) so that when we add this contribution to (31) we obtain

$$\frac{1}{\tau_s} \simeq \frac{2V\omega^2}{\pi\hbar^2 c^3} \sum_k D_q^2 \zeta_1'' \ . \tag{33}$$

as the life time of the x-ray due to electron scattering from the processes of the form given by the second and third terms of (18). The first term of (20) does not contribute to the life time of the x-ray. The contribution of the second term is small compared with that calculated above. The contribution of all of the terms of (21) is real because of pair formation so that the imaginary part of the self energy is zero and only a shift occurs. This interaction becomes completely diagonalizable when pairs are introduced using the prescription given by (4). Sidebands are not produced by this interaction. Similarly (23) and (24) do not contribute to the width of the x-ray.

The broadening of the x-ray line due to pair breaking is given by the expression (22). The self energy of the x-ray due to the first term of (22) is found to be,

$$\Sigma_q^x(2) = \sum_\sigma \sum_{k,k',k'',q'} \frac{F_1^2 \eta_1 \delta(q - q' - k' - k'')}{E + 2\epsilon_k - \epsilon_{k'} - \epsilon_{k''} - \hbar\omega_{q'}}) \tag{34}$$

where

$$F_1 = D_{k,k',k'',q'} u_k^2 u_{k'} u_{k''} \tag{35a}$$

$$\eta_1 = (n_{q'} + 1)(1 - f_{k'})(1 - f_{k''})f_k f_{-k} + n_{q'} f_{k'} f_{k''}(1 - f_k)(1 - f_{-k}) \tag{35b}$$

in which $2\epsilon_k$ appears because we assumed $\epsilon_k = \epsilon_{-k}$ in the denominator. We evaluate the sum over k'' by using the δ function and then by using (26) and (27) we obtain the life time of the x-ray due to pair breaking from the first term of (12) as,

$$\frac{1}{\tau_2} \simeq \frac{2V(\omega + \frac{2\Delta}{\hbar})^2}{\pi\hbar^2 c^3} \sum_{k,k'} F_1''^2 \eta_1''^2 \tag{36}$$

where

$$F_1'' = D_{k,k'} u_k^2 u_{k'} u_{(2\Delta/\hbar c)+k'} \tag{37a}$$

and

$$\eta_1'' = [n(2\Delta + \hbar\omega) + 1](1 - f_{k'})(1 - f_{-k'-2\Delta/\hbar c})f_k f_{-k}$$
$$+ n(2\Delta + \hbar\omega)f_{k'} f_{-k'-2\Delta/\hbar c}(1 - f_k)(1 - f_{-k}) \tag{37b}$$

Similarly, the second term of (22) gives,

$$\frac{1}{\tau_3} \simeq \frac{2V(\omega - \frac{2\Delta}{\hbar})^2}{\pi\hbar^2 c^3} \sum_{k,k'} F_1''^2 \eta_2'' \ . \tag{38}$$

where

$$F_2'' = D_{k,k'} v_k^2 v_{k'} v_{(2\Delta/\hbar c)+k'} \tag{39a}$$

$$\eta_2'' = [n(\hbar\omega - 2\Delta) + 1](1 - f_k)(1 - f_{-k})f_{-k'}f_{-k'+2\Delta/\hbar c}$$
$$+ n(\hbar\omega - 2\Delta)f_k f_{-k}(1 - f_{k'})(1 - f_{-k'+2\Delta/\hbar c}) \tag{39b}$$

The total life time of the x-ray due to pair breaking is

$$\frac{1}{\tau_p} = \frac{1}{\tau_2} + \frac{1}{\tau_3} \tag{40}$$

and the width of the x-ray is $\hbar\delta\omega = \frac{\hbar}{2}(\frac{1}{\tau_s} + \frac{1}{\tau_p})$ which is built up from the single quasiparticle scattering as well as pair breaking processes. The relaxation rates τ_2 and τ_3 suggest the occurance of three lines one each at ω and $\omega \pm 2\Delta/\hbar$. The temperature dependence in the pair breaking process is a bit stronger than T^t at low temperatures. The calculations of τ_p^{-1} involves $\Sigma_{k,k'}$ which for a three dimensional solid is equivalent to $\int\int 2(m/\hbar^2)^3 \epsilon_k d\epsilon_k d\epsilon_{k'} \propto \epsilon^3$ in agreement with the E^3 dependence suggested by Kraus et al[2]. The resolution of the detector is limited by the line width determined from the life time of the x-ray (40).

The thermal and non-equilibrium response of superconductors as radiation detectors has been studued by Zhang and Frenkel[17] and Epifani[18]. The x-ray operations of a thin film Nb superconducting strip particle detector is given by Parlato et.al[19] and by Gonsev et.al[20]. In the present work we have found the width of the x-ray as detected by a superconducting detector. Our calculation is thus of interest for the developement of superconducting films as detectors. Since lines occur at $\omega, \omega \pm 2\Delta$, we predict that there are side bands which affec the resolution of the detector. Since the gap is proportional to the transition temperature, the smaller the gap the better is the resolution of the detector. Hence the smaller the transition temperature, the better is the resolution. In the case of large T_c the lines ω and $\omega \pm 2\Delta$ will be well resolved.

We have discovered that the width of the x-ray line is determined by two processes. One of these is the process of electron scattering by x-rays and the other is the pair breaking. The width due to the former process depends on T and that due to later is slightly stronger than T^6 at low temperatures. Both the mechanisms give zero line width at $T = 0$.

We are grateful to the authorities of the University Grants Commission, New Delhi, for the award of fellowship to one of the authors (NMK).A preliminary version of this paper was shown to **Prof.Dr.R.L.Mößbauer.** We would like to express our gratitude to him for his comments which led us to think in terms of the tuning problem. Since the x-ray frequency is much larger than the phonon frequency, creation of pairs by interaction of photons is possible at very low temperatures.

References

1. J. Jochum, H. Kraus, M. Gutsche, B. Kemmather, F.v. Feilitzsch and R.L. Mössbauer, Ann. Physik **7**, 611 (1993).

2. H. Kraus, J. Jochum, B. Kemmather, M. Gutsche, F.v. Feilitzsch and R.L. Mössbauer, Nucl. Instru. Methods Phys. Res. **A315**, 213 (1992).

3. H. Kraus, Th. Peterreins, F. Pröbst, F.v. Feilitzsch, R.L. Mössbauer and V. Zacek and E. Umlauf, Europhys. Lett. **1**, 161 (1986).

4. H. Kraus, J. Jochum, B. Kemmather, M. Gutsche, F.v. Feiltzsch and R.L. Mössbauer, Nucl. Instru. Methods. Phys. Res. **A326**, 172 (1993).

5. H. Kraus, F. von Feilitzsch, J. Jochum, R.L. Mössbauer, Th. Peterreins, and F. Pröbst, Phys. Lett. **B231**, 195 (1989).

6. J. Jochum, H. Kraus, M. Gutsche, B. Kemmather, F.v. Feilzsch and R.L. Mössbauer, Nucl. Instru. Methods, Phys. Res. **A338**, 458 (1994).

7. J.R. Schrieffer and D.M. Ginsberg, Phys. Rev. Lett. **8**, 207 (1962).

8. D.M. Ginsberg, Phys. Rev. Lett. **8**, 204 (1962); See also discussion No. 33 in Rev. Mod. Phys. **36**, 215 (1964).

9. A. Rothwarf and M. Cohen, Phys. Rev. **130**, 1401 (1963).

10. B.I. Miller and A.H. Dayem, Phys. Rev. Lett. **18**, 1000 (1967).

11. A. Rothwarf and B.N. Taylor, Phys. Rev. Lett. **19**, 27 (1967).

12. E. Burstein, D.N. Langenberg and B.N. Taylor, Phys. Rev. Lett. **6**, 92 (1961).

13. J. Bardeen, L. Cooper and J.R. Schrieffer, Phys. Rev. **108**, 1175 (1957).

14. C. Kittel, Quantum theory of solids, Wiley, N.Y., 1963.

15. K.N. Shrivastava, Phys. Rev. **B19**, 1598 (1979).

16. A. Suguna and K.N. Shrivastava, Phys. Rev. **B22**, 2343 (1980).

17. Z.M. Zhang and A.Frenkel, J.Supercond. **7**,871(1994).

18. M. Epifani, J. Appl. Phys.**76**,1256(1994).

19. L. Parlato, G.Pelso, G.Pepe, R.Vaglio, C.Attanasio, A.Ruosi, S.Barbanera. M.Cirillo, R.Leoni, Nucl. Instru. Methods A**348**, 127(1994).

20. Y.P. Gousev, G.N. Goltsman, T.O. Klaassen, W.T. Wenckebacah, C.T. Foxon, J.Appl.Phys **75**, 3695(1994).

The Effect of Electric Field in Superconductors

Lydia S. Lingam

School of Physics
University of Hyderabad
P.O. Central University
Hyderabad 500 046, India

Abstract The effect of electric fields on the superconductors has been reviewed. We have studied the quantised resistivity and have made an effort to see if quantum effects can be observed. It is found that conductance as a function of voltage shows quantised steps. Scattering is found near the gap of the superconductor and the scattering rate has been calculated self-heating and hot spots are found and the effect of non-equilibrium quasiparticles has been reviewed and the charge-imbalance caused by the magnetic impurities has been studied. It is noted that when the voltage becomes equal to the gap energy kinks occur in the response of the superconductor. This is known as the Andreev-scattering. The effect of the magnetic field on the non-linear current voltage characteristics has been studied.

We have calculated the effect of the electrid dipole moment on the current voltage characteristics upto second order in the perturbation theory. The resistivity is predicted to be an isotropic function of the electric field. Several quantum effects have been found. The effect of flux quantization on the current is well demonstrated. Our results compare well with the experiments performed by Mannhart et al.

Thermoelectric effects in superconductors have been pointed out and tunneling characteristics due to the thermoelectric effect have been reviewed. The effect of the electric field on one of the coefficients of the Gunzburg-Landau model has been described.

1.Introduction

The superconductivity is caused by pairs of electrons of equal and opposite momenta and spin. The spin singlet pairs of electrons give perfect diamagnetism and are consistent with zero resistivity. The singlet zero-momentum pairs of electrons are called as the Cooper pairs [1]. The paired state explains the properties of superconductors [2]. The pairs can tunnel through thin layers of insulators [3]. This tunneling current can be used for making electronic devices.

Recently, Mannhart et al [4,5] have found that resistivity of a device can be changed by the application of a gate voltage. The current and voltage characteristics of the device can not be understood on the basis of two fluid model having normal and supercurrents. Therefore.

Shrivastava [6] has calculated the contribution of the dipole moment to the current upto second-order in the perturbation theory. This calculation requires the concept of quantized resistivity. It shows that the resistivity is asymmetric with respect to the change in sign of the gate voltage, consistent with the experimental observations. It appears that heat loss and thermal effects have to be taken into account properly for a proper understanding of the experimental data.

Artemenko and Volkov [7] have studied the effect of longitudinal electric fields on Josephson devices. The penetration of the electric field through the superconducting films has been studied by Ryazanov et al [8]. The current induced resistive state has been discussed by Huebener et al [9] and by Tinkham [10]. The current-voltage characteristics of microbridges have been reviewed by Lindelof [11] and the nonequilibrium effects are described by Pals et al [12]. The original studies of the penetration of the electric field into superconductors have been given by Landau [13] and by Pippard et al [14,15].

In this thesis, we study the quantized conductance including current-voltage relationships in superconducting film devices. We find the nonlinearity in the current-voltage characteristics in the presence of a magnetic field. We discuss the effect of relaxation and life time of quasiparticles. The effect of non-equilibrium conditions on the characteristics is described. The contribution of the electric dipole moment to the currents has been described. The heat loss has been taken into account. This calculation leads to the understanding of quantized resistivity. We have discussed the thermoelectric effects which occur because of the difference in temperatures of different layers of the devices. This work is of importance for the use of superconducting films as components of the devices for the development of electronics.

2.Quantized Conductance

We wish to study the quantized conductance from the current-voltage characteristics [16-28]. We consider the application of the electric field in the superconducting film device [4,8,29-46]. The effect of the finite life time of the quasiparticles is considered [47-49]. the effect of the disorder is also discussed [50]. The current voltage characteristics are found to exhibit steps [51] due to quantization of resistivity. The nonlinearity in these characteristics in the presence of a magnetic field has been investigated [52]. There are several studies of the current-voltage characteristics [53-60]. The effect of self-heating and hot spots have been reported [61], and a detailed theory has been developed by Artemenko and Volkov [62].

2.1 Conductance quantized at $4e^2/h$

The matrices for the scattering of electrons and holes through the NS junction are constructed from the reflection and trasmission matrices, r and t as,

$$S_N(\epsilon) = \begin{pmatrix} S_o(\epsilon) & 0 \\ 0 & S_o(-\epsilon)^* \end{pmatrix} \quad , \quad S_o \equiv \begin{pmatrix} r_{11} & t_{12} \\ t_{21} & r_{22} \end{pmatrix} \tag{1}$$

where S_o is the unitary single-electron S matrix [16,63]. The normal-superconductor (NS) interface is located at $x = 0$. The pair potential $\Delta(r)$ in the bulk of the superconductor ($x \gg \xi$) has amplitude Δ_o and phase ϕ. For $x < 0$, $\Delta(r) = 0$ in the normal part of the junction. The gap function is then of the form $\Delta(r) = \Delta_o e^{i\phi}\theta(x)$. The step function $\theta(x) = (x > 0)$ in the superconducting region and $\theta(x) = 0$ ($x < 0$) in the normal part of the NS junction. For energy $\epsilon < \Delta_o$, there are no propagating modes in the superconductor. We can define a $2N \times 2N$ S matrix for the reflection at the NS interface called [64] the Andreev scattering matrix,

$$S_A(\epsilon) = \exp\left[-i \arccos(\epsilon/\Delta_o)\right] \begin{pmatrix} 0 & e^{i\phi} \\ e^{-i\phi} & 0 \end{pmatrix} \tag{2}$$

An electron incident in lead N_1 is reflected either as an electron with scattering amplitude S_{ee} or as a hole with scattering amplitude S_{he}. Similarly, the matrices S_{hh} and S_{eh} have the elements for reflection of a hole as a hole or as an electron. For the conductance G_{NS} of the NS junction at zero temperature we need only the S matrix at the Fermi level, i.e., at $\epsilon = 0$. The general expression for the conductivity is then

$$G_{NS} = \frac{2e^2}{h} Tr(1 - S_{ee}S_{ee}^+ + S_{he}S_{he}^+) = \frac{4e^2}{h} Tr S_{he} S_{he}^+ \tag{3}$$

We write S in terms of normal S_N and Andreev S_A terms as

$$G_{NS} = \frac{4e^2}{h} Tr t_{12}^+ t_{12}(1 + r_{22}^* r_{22})^{-1} t_{21}^* t_{21}^T (1 + r_{22}^+ r_{22}^T)^{-1} . \tag{4}$$

In zero magnetic field, $S_o = S_o^T$ so that the above can be written as

$$G_{NS} = \frac{4e^2}{h} Tr \left(\frac{t_{12} t_{12}^+}{2 - t_{12} t_{12}^+} \right)^2 \equiv \frac{4e^2}{h} \sum_{n=1}^{N} \frac{T_n^2}{(2 - T_n)^2} \tag{5}$$

where T_n $(n = 1, 2, \cdots N)$ are the eigenvalues of the Hermitian matrix $t_{12} t_{12}^+$. This is for an arbitrary transmission matrix. It may be compared with the Landauer formula for the normal-metal conductance,

$$G_N = \frac{2e^2}{h} Tr t_{12} t_{12}^+ = \frac{2e^2}{h} \sum_{n=1}^{N} T_n . \tag{6}$$

The conductance formula (5) is the multichannel generalization of a formula first obtained by Blonder et al [65] and later by others [50,66] for the single channel case.

If the junction consists of a ballistic constriction with a normal state conductance quantized at $G_N = 2N_o e^2/h$. The quantization occurs because the transmission eigenvalues are either zero or one. Therefore (5) shows that the conductance of the NS junction is quantized in units of $4e^2/h$,

$$G_{NS} = 4N_o e^2/h \tag{7}$$

where N_o is an integer. In the classical limit $N_o \to \infty$, so that we recover the well known result, $G_{NS} = 2G_N$ for a classical ballistic point contact. In the quantum regime, however, the simple factor of 2 enhancement only holds for the conductance plateaus, and not to the transition region between the plateaus.

We assume ϵ_{res} as the energy of the resonant level, relative to the Fermi level in the reservoirs. Let $\Gamma_1/\hbar$ and $\Gamma_2/\hbar$ be the tunnel rates through the two barriers. The normal state conductance, G_N has the Breit-Wigner form,

$$G_N = \frac{2e^2}{h} \frac{\Gamma_1 \Gamma_2}{\epsilon_{res}^2 + \Gamma^2/4} \equiv \frac{2e^2}{\hbar} T_{BW} \tag{8}$$

Lydia S. Lingam

with $\Gamma = \Gamma_1 + \Gamma_2$. The transmission matrix which yields this conductance has the elements [67],

$$t_{12}(\epsilon) = U_1 \tau(\epsilon) U_2 \tag{9}$$

$$\tau(\epsilon)_{nm} \equiv \frac{\sqrt{\Gamma_{1n}\Gamma_{2m}}}{\epsilon - \epsilon_{res} + i\Gamma/2} \tag{10}$$

where $\sum_n \Gamma_{1n} \equiv \Gamma_1$, $\sum_n \Gamma_{2n} = \Gamma_2$ and U_1 and U_2 are unitary matrices. The matrix $t_{12}t_{12}^+$ (at $\epsilon = 0$) has eigenvalues $T_n = T_{BW}\delta_{n1}$ so that (5) gives,

$$G_{NS} = \frac{4e^2}{h}\left(\frac{2\Gamma_1\Gamma_2}{4\epsilon_{res}^2 + \Gamma_1^2 + \Gamma_2^2}\right)^2 \tag{11}$$

The conductance on resonance ($\epsilon_{res} = 0$) is maximal if $\Gamma_1 = \Gamma_2$ and is then equal to $4e^2/h$ which is twice the normal value.

2.2 Finite life time effect

The low-energy quasiparticle scattering, recombination, and branch-mixing life times and phonon pair breaking and scattering life times are calculated for superconductors by Kaplan et al [47]. These life times are related to the low frequency behaviour of the spectral weight $\alpha^2(\Omega)F(\Omega)$. The poles of the single-particle Green's function in the superconducting state are determined by

$$z^2(\omega)\omega^2 - \epsilon_p^2 - \phi^2(\omega) = 0 \tag{12}$$

where $z(\omega)$ is the renormalization parameter so that ωz is the energy variable and $\phi(\omega)$ is the gap parameter. For the complex frequency $\omega = E(\omega) - i\Gamma(\omega)$ both z and ϕ are complex, $z(\omega) = z_1(\omega) + iz_2(\omega)$ and $\phi(\omega) = \phi_1(\omega) + i\phi_2(\omega)$. For frequenices small compared with the typical phonon frequencies, we neglect the frequency dependence of z_1 and ϕ_1, setting $z_1(\omega) \simeq z_1(\Delta_o) \simeq z_1(o)$ and $\phi_1(\omega)/z_1(\omega) \cong \phi_1(\Delta_o)/z_1(\Delta_o) \equiv \Delta_o$. Here Δ_o is the usual temperature-dependent energy gap. We neglect the temperature dependence of $z_n(o)$, with these approximations, the imaginary part of the complex frequency becomes,

$$\Gamma(\omega) = \omega z_2(\omega)/z_1 - (\Delta/\omega)\phi_2(\omega)/z_1 \tag{13}$$

We replace $\Delta(\omega)$ by Δ in determining $z_2(\omega)$ and $\phi_2(\omega)$. The inverse life time $\tau^{-1}(\omega)$ of a quasiparticle of energy ω can be written as

$$
\begin{aligned}
\tau^{-1}(\omega) = {}& 2\Gamma(\omega) = \frac{2\pi}{\hbar z_1(o)}\int_o^{\omega-\Delta} d\Omega\, \alpha^2(\Omega)F(\Omega) \\
& \times\ Re\frac{\omega-\Omega}{[(\omega-\Omega)^2-\Delta^2]^{1/2}}\left(1-\frac{\Delta^2}{\omega(\omega-\Omega)}\right)[f(\Omega-\omega)+n(\Omega)] \\
& +\ \frac{2\pi}{\hbar z_1(o)}\int_{\omega+\Delta}^\infty d\Omega\, \alpha^2(\Omega)F(\Omega)R_e\left(\frac{\omega-\Omega}{[(\Omega-\omega)^2-\Delta^2]^{1/2}}\right) \\
& \times\ \left(1+\frac{\Delta^2}{\omega(\Omega-\omega)}\right)[f(-\omega+\Omega)+n(\Omega)] \\
& +\ \frac{2\pi}{\hbar z_1(o)}\int_o^\infty d\Omega\, \alpha^2(\Omega)F(\Omega)Re\left(\frac{\omega+\Omega}{[(\omega+\Omega)^2-\Delta^2]^{1/2}}\right) \\
& \times\ \left(1-\frac{\Delta^2}{\omega(\Omega+\omega)}\right)[f(\omega+\Omega)+n(\Omega)]
\end{aligned}
\tag{14}
$$

$$\tau^{-1}(\omega) = 2\Gamma(\omega) = \frac{2\pi}{\hbar z_1(o)} \int_o^{\omega - \Delta} d\Omega \, \alpha^2(\Omega) F(\Omega)$$

$$\times \quad Re \frac{\omega - \Omega}{[(\omega - \Omega)^2 - \Delta^2]^{1/2}} \left(1 - \frac{\Delta^2}{\omega(\omega - \Omega)}\right) [f(\Omega - \omega) + n(\Omega)]$$

$$+ \quad \frac{2\pi}{\hbar z_1(o)} \int_{\omega + \Delta}^{\infty} d\Omega \, \alpha^2(\Omega) F(\Omega) R_e \left(\frac{\omega - \Omega}{[(\Omega - \omega)^2 - \Delta^2]^{1/2}}\right)$$

$$\times \quad \left(1 + \frac{\Delta^2}{\omega(\Omega - \omega)}\right) [f(-\omega + \Omega) + n(\Omega)]$$

$$+ \quad \frac{2\pi}{\hbar z_1(o)} \int_o^{\infty} d\Omega \, \alpha^2(\Omega) F(\Omega) Re \left(\frac{\omega + \Omega}{[(\omega + \Omega)^2 - \Delta^2]^{1/2}}\right)$$

$$\times \quad \left(1 - \frac{\Delta^2}{\omega(\Omega + \omega)}\right) [f(\omega + \Omega) + n(\Omega)] \tag{14}$$

Here $1 \pm \Delta^2/\{\omega(\omega \pm \Omega)\}$ are called as the coherence factors, $f(\omega)$ is a Fermi distribution and $n(\Omega)$ is a Bose distribution. The first and third terms here correspond to absorption and emission of phonons to be denoted by $\tau_s^{-1}(\omega)$. The second term corresponds to a process in which the quasiparticle recombines with another quasiparticle to form a pair with excess energy emitted as a phonon to be denoted by τ_r^{-1}. The factor z_1^{-1} renormalizes the electron-phonon interaction. The fraction of the initial quasiparticle state in the Bloch state is given by the wave function renormalization factor $z_1^{-1/2}$. The remaining part of the state consists of a superposition of virtually scattered electrons and phonons. Since all parts of the spectrum of the final state are summed over, only $z_1^{-1/2}$ enters in reducing the matrix element.

We approximate $\alpha^2(\Omega) F(\Omega)$ by its low frequency form $\alpha^2(\Omega) F(\Omega) = b\Omega^2$, where b is a constant characteristic of a given material. At low temperatures the dominant behaviour is given by

$$\tau_o/\tau_s^{-1}(\Delta, t) \cong \Gamma(\frac{7}{2})\zeta(\frac{7}{2}) \left(\frac{k_B T_c}{2\Delta(o)}\right)^{1/2} \left(\frac{T}{T_c}\right)^{7/2} \tag{15}$$

$$\tau_o/\tau_r^{-1}(\Delta, t) \cong (\pi)^{1/2} \left(\frac{2\Delta(o)}{k_B T_c}\right)^{5/2} \left(\frac{T}{T_c}\right)^{1/2} \exp(-\Delta(o)/k_B T) \tag{16}$$

where

$$\tau_o = z_1(o)\hbar/2\pi b(k_B T_c)^3 \tag{17}$$

If the $b\Omega^2$ form of $\alpha^2(\Omega) F(\Omega)$ is not valid for $\Omega \simeq 2\Delta(o)$, one may use more general low temperature approximation for τ_r,

$$\tau_r^{-1}(\Delta, T) \simeq [4\pi\Delta(o)\alpha^2(2\Delta(o))F(2\Delta(o))/\hbar z_1(o)]$$

$$\times \quad \left[\frac{\pi k_B T}{2\Delta(o)}\right]^{1/2} \exp\{-\Delta(o)/k_B T\} \tag{18}$$

The scattering life time $\tau_s(\omega, T)$ increases as the temperature is lowered owing to a decrease in the thermal phonon population. A quasiparticle at the gap edge, $\omega = \Delta(T)$, cannot emit a phonon and scatter because it is in the lowest-energy quasiparticle state. For this reason

$\tau_s(\Delta(T), T)$ increases when T decrease. However, for quasiparticles with energy $\omega > \Delta(T)$, spontaneous phonon emission sets a limit to the scattering life time τ_s. This limiting value for the scattering life time is

$$\tau_s^{-1}(\omega, 0) = \frac{2\pi}{\hbar z_1(o)} \int_o^{\omega-\Delta} d\Omega \; \alpha^2(\Omega) F(\Omega) Re \left(\frac{\omega - \Omega}{[(\omega - \Omega)^2 - \Delta^2]^{1/2}} \right)$$
$$\times \; \left(1 - \frac{\Delta^2}{\omega(\omega - \Omega)} \right) \tag{19}$$

for $\Delta = \Delta(o)$ everywhere. The integral is solvable for $\alpha^2(\Omega) F(\Omega) = b\Omega^2$ for which,

$$\tau_o/\tau_s^{-1}(\omega, 0) = \left(\frac{\Delta}{k_B T_c} \right)^3 \left[\frac{1}{3}\{(\frac{\omega}{\Delta})^2 - 1\}^{3/2} + \frac{5}{2}\{(\frac{\omega}{\Delta})^2 - 1\}^{1/2} \right.$$
$$\left. - \; \frac{\Delta}{2\omega} \left\{ 1 + 4(\frac{\omega}{\Delta})^2 \right\} \ln \left\{ \frac{\omega}{\Delta} + \left[(\frac{\omega}{\Delta})^2 - 1 \right]^{1/2} \right\} \right] \tag{20}$$

For ω larger than several times Δ, the scattering rate varies as $(\omega/\Delta)^3$.

Schuller and Gray [48] suggest that the recombination time is approximately $\tau_E k_B T/\Delta$. Since the average scattering time is $\sim \tau_E$ only a fraction of about $\Delta/k_B T$ of these events can form bound pairs contributing to the condensation energy [68]. Thus, Schuller and Gray [48] have measured the inelastic electron collision time, τ_E which agrees with its theoretical [68] dependence on the mean free path.

2.3 Current-voltage steps

For small constant currents (1-10 μA) the voltage-temperature characteristics of tin whiskers with about 1 atomic % indium impurities do not show any step-like structure. At larger currents (20-100 μA) voltage steps build up which become more distinct and more numerous with increasing current. The voltage steps are mainly observed at the lower part of the characteristics. The temperature at which the residual resistance is reached becomes lower with increasing current and thus is not any longer independent of any applied current as it is the case at smaller currents.

The voltage steps are also observed in the voltage-current characteristics which show a larger transition width from the first onset of voltage to the complete normal state. With increasing current the voltage steps become more distinct. At a temperature of 3.6 K, and the very large current of 434 μA a nearly complete jump from the superconducting to the normal state has been observed. The voltage-current characteristics are found to be hysteretic and the width of the hysteresis increases with decreasing fixed temperature,

The first onset of voltage across the sample appears at the critical current, $I_c \sim (1 - T_c/T_{co})^{3/2}$. Here T_{co} is the thermodynamical critical temperature while T_c is the transition temperature at current I_c. By extrapolating $I_c(T_c)$ to $I_c = 0$, we can obtain T_{co}. After the first voltage step the voltage-current characteristics are straight lines. A sketch of the first voltage step is shown in Fig.2.1. The slope after the first voltage step $(dV/dI)_1$ is nearly indepedent of temperature. The back extrapolation shows a zero-voltage intercept of $I_o > 0$. The ratio I_o/I_c is nearly independent of temperature. The hysteretic behaviour of the whisker is characterized by the width of the hysteresis $I_c - I_R$ and the jump back voltage

V_R. For not too small currents $I_c - I_R$ grows linearly with increasing critical current, whereas V_R tends to a constant value V_{RS}, for large critical currents. The growth of the width of the hysteresis with increasing critical current is given by $d(I_c - I_R)/dI_c$.

Step like structures have also been observed by Skocpol et al [69] in tin microbridges. These authors [69] assume that each voltage step is related to a localized weak link in which phase slip processes occur at the Josephson frequency but the voltage across each phase-slip centre is affected by the diffusion of the quasi-particles which are produced by the slip process. In order to explain large transition width, it was assumed that inhomogeneities with very different critical temperatures T_{co} cause localized weak links with different critical currents. For a single voltage step,

$$V(I) = \frac{2\rho_n \Lambda}{A}(I - <I_s>) \tag{21}$$

with
$$<I_s> = 0.65 \ I_c$$
$$\Lambda = (\frac{1}{3} l l_2)^{1/2}$$
$$l_2 = v_F \tau_2 \tag{22}$$

Here I is the total current, ρ_n the normal state resistivity of the sample, A its cross sectional area, $<I_s>$ the time averaged supercurrent above I_c. Furthermore, Λ is the diffusion length over which a quasiparticle excitation travels by random walk before being inelastically scattered. The mean free path of the electrons is l and l_2 is the mean free path for the excitation concerning such an inelastic scattering process. Thus τ_2 is the relaxation time due to inelastic scattering process. Thus Skocpol et al [69] predict for the first voltage step a

zero-voltage intercept, $I_o = 0.65 \ I_c$ and a differential resistance $(dV/dI)_1 \simeq 2\rho_n \Lambda/A$ leading to a length $L_{An1} = \frac{L}{R_n}(\frac{dV}{dI})_1 \simeq 2\Lambda$ which is proportional to $l^{1/2}$.

The temperature independence of the differential resistance requires a temperature independent length Λ, so that the quasiparticle healing is controlled by the inelastic scattering processes of relaxation time τ_2. This time can be determined from the experiments because.

$$L_{An1}(l) \approx 2\Lambda = 2\left(\frac{1}{3}lv_F\tau_2\right)^{1/2}$$
$$\tau_2 = 3L_{An1}^2(l)/(4v_F l) \tag{23}$$

For $v_F \simeq 10^6 \ m/s$, $\tau_2 \simeq 4.2 \times 10^{-9} \ s$ for tin whiskers which is 5 times larger than $\tau_2 = 8 \times 10^{-10} \ s$ for tin microbridges [69].

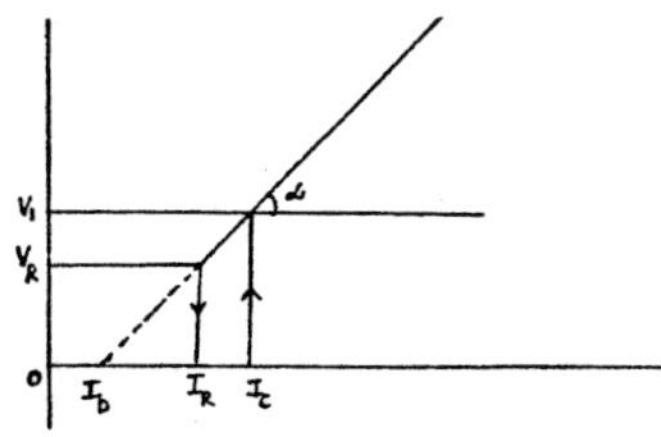

Fig.2.1 *Sketch of the first voltage step of the voltage-current characteristics.*

Lydia S. Lingam

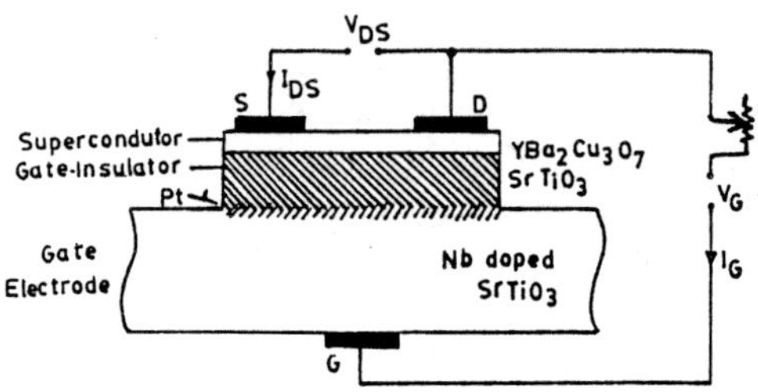

Fig.2.2 *Inverted MISFET-type sample configuration used for the study of field effects (cross section) (From Mannhart et al [4];*

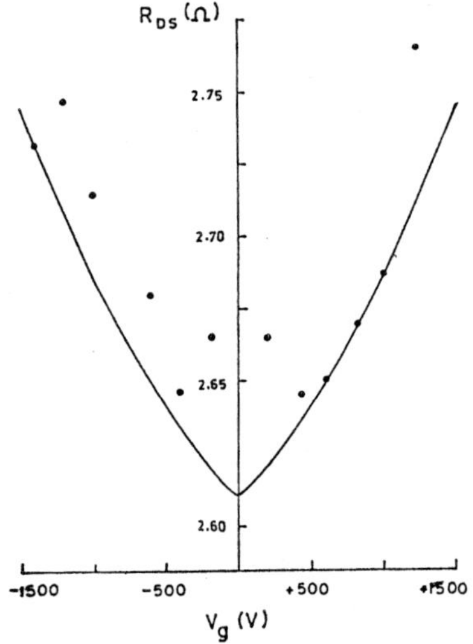

Fig.2.3 *The drain source resistivity R_{DS} as a function of gate voltage, V_g. The experimental measurements shown by dots are taken from Mannhart [4] and the curve is calculated by us for*

$$R_{DS} = aV_g + bV_g^2 + c \text{ with } a = 5.1155 \times 10^{-5} \ \Omega/v,$$
$$b = 2.5845 \times 10^{-8} \ \Omega/v^2 \text{ and } c = 2.61 \ \Omega.$$

Mannhart et al [4] have measured the drain-source resistivity for a device shown in Fig.2.2, using gold contacts [70] as a function of temperature with various gate voltages, such as 0, ± 10 V, with $I_{DS} \simeq 0.25$ A. This resistivity is found to be a continuous function of temperature and there is no indication of any steps or discontinuities. The $V_{DS}(I_{DS})$ characteristics below T_c are found to be parabolic and linear for $T > T_c$. Again there is no step like structure whatsoever for the range of values for the voltages chosen.

2.4 Self heating and hot spots

In microbridges of dimensions of about 1 μm with and $\sim$0.1 μm thickness of tin, heat is generated in localized dissipative regions and is transferred from such regions in two ways. Firstly by thermal conduction within the film and secondly by surface heat transfer across the temperature discontinuity which develops at the boundary with the substrate [61]. The thermal conductivity K of the thin film may be estimated from the electrical resistivity using the Wiedemann-Franz law which gives the ratio of the thermal and electrical conductivities and is typically of the order of 0.05 W/cmK. The heat transfer coefficient per unit area, α, to glass or sapphire substrates for temperatures near the transition temperature of tin is of the order of 2 W/cm^{2}K. The substrate may be assumed to be essentially at the ambient bath temperature T_b. The combination of conduction within the film and surface heat transfer leads to a characteristic thermal healing length given by

$$\eta = (Kd/\alpha)^{1/2} \tag{24}$$

where d is the thickness of the film. We consider a one-dimensional model appropriate to bridges which are much longer than the thermal healing length, η. We assume a bridge of length L, width W, and thickness d, with normal region of length $2x_o$ and resistivity ρ, symmetrically centered in the bridge. The temperature distribution $T(x)$ along the bridge must satisfy the heat flow equation

$$- \ K_N \frac{d^2 T}{dx^2} + \frac{\alpha}{d}(T - T_b) = \left(\frac{I}{Wd}\right)^2 \rho \quad (x < x_o)$$

$$- \ K_S \frac{d^2 T}{dx^2} + \frac{\alpha}{d}(T - T_b) = 0 \qquad (x < x_o) \tag{25}$$

where I is the current flow through the bridge. In the long-bridge approximation, we use the boundary condition $T(\pm \frac{1}{2}L) = T_b$ at the ends of the bridge, since the cooling area rapidly increases there, $T \simeq T_b$. At N/S interaface x_o, we match T and K dT/dx. For self consistency $T(\pm x_o) = T_c$. The resulting temperature distribution is given by

$$T_n(x) \ = \ T_c + (T_c - T_b)\left(\frac{K_S}{K_N}\right)^{1/2} \cot\mathrm{h}\left(\frac{x_o}{\eta_N}\right) \cot\mathrm{h}\left(\frac{\frac{1}{2}L}{\eta_S} - \frac{x_o}{\eta_S}\right)$$

$$\times \ \left\{1 - \cosh\left(\frac{x}{\eta_N}\right)\left[\cosh\left(\frac{x_o}{\eta_N}\right)\right]^{-1}\right\} \quad (x < x_o) \tag{26}$$

$$T_s(x) \;=\; T_b + (T_c - T_b)\left[\sinh\left(\frac{\frac{1}{2}L}{\eta_s} - \frac{x}{\eta_s}\right)\right]\left[\sinh\left(\frac{\frac{1}{2}L}{\eta_s} - \frac{x_o}{\eta_s}\right)\right]^{-1} \quad (x > x_o)$$

(27)

The current required for this self-consistent solution is,

$$I(x_o) \;=\; \left(\frac{\alpha W^2 d(T_c - T_b)}{\rho}\right)^{1/2}\left[1 + \left(\frac{K_s}{K_N}\right)^{1/2}\coth\left(\frac{x_o}{\eta_N}\right)\right.$$
$$\times \; \left.\coth\left(\frac{\frac{1}{2}L}{\eta_s} - \frac{x_o}{\eta_s}\right)\right]^{1/2}$$

(28)

and by Ohm's law the corresponding voltage is

$$V(x_o) = I(x_o)2x_o\rho/Wd$$

(29)

This means that centre of the film is much hotter than the edges and $V(I)$ curves are very different from the linear behaviour. The resistance of the bridge neglecting the spreading resistance at the ends is given by $R_B = \rho L/Wd$. The current which generates sufficient heat to balance surface heat transfer from a normal film at $T = T_c$ is given by

$$I_1 = \left[\alpha W^2 d(T_c - T_b)/\rho\right]^{1/2}$$

(30)

For $L \gg x_o \gg \eta$, the $I - V$ characteristics approach a constant minimum current of $(1 + K_s/K_N)^{1/2}I_1 \simeq \sqrt{2}I_1$. This current generates sufficient heat to balance surface heat transfer at the temperature which is required well inside the normal region to maintain a stable N/S interface.

In order to study short microbridges, we consider one-dimensional flow within a bridge of length $L + \frac{1}{2}W$ and radial flow beyond a radius $\frac{1}{2}W$ into the film at each end of the bridge. In the radial regions the temperature $T(r)$ satisfies,

$$-\; K\left(r^2\frac{d^2T}{dr^2} + r\frac{dT}{dr}\right) + \frac{\alpha}{d}r^2(T - T_b) = \left(\frac{I}{\pi d}\right)^2\rho \quad \text{(normal)}$$

(31)

$$-\; K\left(r^2\frac{d^2T}{dr^2} + r\frac{dT}{dr}\right) + \frac{\alpha}{d}r^2(T - T_b) = 0 \quad \text{(superconducting)}$$

(32)

The solutions of these equations involve modified Bessel functions of argument r/η, where η is given by (24). The one-dimensional and radial solutions are matched by comparing temperature and heat flow at the boundaries and $T(\infty) = T_b$. The N/S interface may occur either in the one-dimensional bridge at x_o or in the radial film at r_o. The resistance of the normal spot may be written approximately as,

$$R(r_o) = \frac{\rho}{d}\left[\frac{L}{W} + \frac{1}{2} + \frac{2}{\pi}\ln(2r_o/W)\right]$$

(33)

For the case of a small bridge with $L < \eta$ and $W < \eta$ the current reduces to,

$$I(r_o) = \left(\frac{K d^2(T_c - T_b)}{\rho}\right)^{1/2} \left[K_o\left(\frac{r_o}{\eta}\right)\right]^{-1/2} \left\{\left(\frac{\eta}{W}\right)^2 \left[K_o\left(\frac{W}{2\eta}\right)\right.\right.$$
$$+ \frac{\pi}{2}K_1\left(\frac{W}{2\eta}\right)\cot h\left(\frac{\frac{1}{2}L}{\eta} + \frac{\frac{1}{4}W}{\eta}\right)\right]^{-1} + \pi^{-2}\int_{W/2\eta}^{r_o/\eta}\frac{I_o(x)}{x}dx\right\}^{-1/2}$$

$$(34)$$

where $I_o(x)$, $K_o(x)$ and $K_1(x)$ are modified Bessel functions. The voltage across the spot then becomes,

$$V(r_o) = I(r_o)R(r_o) \tag{35}$$

where $R(r_o)$ is given by (33). The current-voltage relations are thus far from being linear at small currents. The critical currents generally follow the mean-field behaviour predicted by the Ginzburg-Landau theory,

$$I_c = Wd\left[cH_c(T)/3\sqrt{6}\pi\lambda(T)\right] \tag{36}$$
$$= I_c(o)(1 - t^2)^{3/2}(1 + t^2)^{1/2} \tag{37}$$

where $H_c(T)$ is the thermodynamic critical field and $\lambda(T)$ is the London penetration depth with $t = T_b/T_c$.

The McCumber [71] curve for $I_{min}/I_c < 0.8$ is indistinguishable from the simple dependence

$$I_{min}/I_c = I_c^{-1/2} \tag{38}$$

This is also what is expected for a junction like bridge with $I_c \propto 1 - t$ with a minimum current $I_{min} = I_h\alpha(1 - t)^{1/2}$. Thus McCumber's result describes the mean field theory as pointed out by Fulton [72].

3. Nonequilibrium Effects

The injection of a quasiparticle current causes an imbalance between the populations of the electron-like and hole-like branches of the excitation spectrum of a superconductor so that the pairs and quasiparticles acquire a potential difference. Such a non-equilibrium superconductor has been of interest to a large number of workers [73-80]. We discuss below some of the results. The effect of magnetic impurities and that of the Andreev scattering on the currents has been examined.

3.1 Applied voltage larger than gap

Rieger et al [81] have considered a current I flowing through a superconductor S of volume Ω in which the chemical potentials of the quasiparticles, μ_{qp}, and that of the pairs, μ differ by,

$$(\mu_{qp} - \mu_p)/e = I\tau_{GL}/24e^2\Omega N(o) \tag{39}$$

Lydia S. Lingam

where τ_{GL} is the Ginzburg-Landau relaxation time, and $N(o)$ the density of states at the Fermi-level for electrons. The experimental work of Clarke [73] shows that the difference btween the quasiparticle potential and μ_p/e is given by

$$V = I\tau_Q/2e^2\Omega N(o)g_{NS} \tag{40}$$

where τ_Q is the relaxation time for the electron-like and hole like imbalance and g_{NS} is the normalized conductance of an NS tunnel junction in the low voltage limit. The current through the junction used by Tinkham and Clarke [73] is

$$\delta I = \frac{G_{NN}}{e}\left[\int_\Delta^\infty \frac{E_{k>}}{(E_{k>}^2 - \Delta)^{1/2}}(v_{k>k_F}^2 - u_{k>k_F}^2)dE_{k>k_F}\right.$$
$$\left. + \int_\Delta^\infty \frac{E_{k>k_F}}{(E_{k<k_F}^2 - \Delta^2)^{1/2}}(v_{k<k_F}^2 - u_{k<k_F}^2)\delta f_{k<k_F}dE_{k<k_F}\right] \tag{41}$$

when N_p is maintained at a potential μ_p/e. Since $E(E^2 - \Delta^2)^{1/2} = E/\epsilon_k|$ and $v_{k>}^2 - u_{k>}^2 = -(v_{k<}^2 - u_{k<}^2) = |\epsilon|/E$ the expression (41) may be written as

$$\delta I = \frac{G_{NN}}{e}\int_\Delta^\infty (\delta f_{k>} - \delta f_{k<})dE_k \equiv \frac{G_{NN}Q^*}{2N(o)e} \tag{42}$$

where

$$Q^* = 2N(o)\int_\Delta^\infty (\delta f_{k>} - \delta f_{k<})dE_k \tag{43}$$

$N(o)$ is the density of states at the Fermi level. The voltage required between the two probes to make the current null is then $V = \delta I/G_{NS}$ where G_{NS} is the tunneling conductance of the junction SN_p and the measured value is therefore

$$V = Q^*/2N(o)eg_{NS} \tag{44}$$

where g_{NS}/G_{NN}. The Eq.(44) does not require the two branches to be separately in thermal equilibrium.

We consider the case of electron injection at high bias voltages ($eV > \Delta$), when majority of the excitations will be electron like. The high-energy excitations decay rapidly by phonon emission into lower-energy states. The phonons emitted with energies $> 2\Delta$ have a high probability of exciting a pair into two quasiparticles. It is assumed that Q is unaffected by this process as this process populates the two branches equally. The scattering processes tend to bring each branch separately into thermal equilibrium and also to equalize the populations of the two branches. Near T_c, Δ approaches zero and the branch mixing process becomes very slow. Hence, near T_c the definition of a chemical potential for each branch, becomes meaningful. For each branch $\delta f_k = -(\partial f_k/\partial E_k)\delta\mu$, where $\delta\mu$ is the displacement of the corresponding chemical potential from μ_p, and f_k is the Fermi function. For such δf_k (43) and (44) lead to,

$$V = (Q^*/Q)(\mu_> - \mu_<)/2e . \tag{45}$$

When chemical potentials are defined, Q and Q^* are related by

$$\frac{Q^*}{Q} = \int_\Delta^\infty -\left(\frac{\partial f}{\partial E}\right) dE \left[\int_\Delta^\infty \frac{E}{(E^2 - \Delta^2)^{1/2}} \left(-\frac{\partial f}{\partial E}\right) dE\right]^{-1}$$

$$= \frac{2f(\Delta)}{g_{NS}} \tag{46}$$

which approaches unity as T approaches T_c and (45) becomes

$$V = (\mu_> - \mu_<)/2e \tag{47}$$

For $\mu_> = \mu_<$, $V = 0$ independent of μ_p. Near T_c due to injection $\dot{Q} = I/e\Omega$, for all bias voltages, $Q^* = Q = I\tau_Q/e\Omega$ and from (44)

$$V = I\tau_Q/2e^2\Omega N(o)g_{NS} \tag{48}$$

We assume $eV_{inj} > k_B T_c$ so high energy electron like quasiparticles dominate the injected population and $\dot{Q} = I/e\Omega$.

When the bath temperature is zero, the electrons cool by spontaneous phonon emission and the probability per unit time of energy loss between ϵ and $\epsilon + d\epsilon$ is $2\epsilon^2 d\epsilon/\tau_\theta(k_B\theta)^3$. The quadratic dependence on ϵ results from combining the appropriate density of states with the square of the matrix element of the electron-phonon interaction both proportional to ϵ. The maximum energy loss is equal to the Debye energy, $k_B\theta$, τ_θ is the scatteing time at $T = \theta$. The mean energy of the injected quasiparticles is $k_B T^*$. The rate of decrease of T^* due to phonon emission is determined from,

$$T^* \approx \theta(3\tau_\theta/16 \ t)^{1/3} \tag{49}$$

For tin $\tau_\theta = 2 \times 10^{-14}$ s, $\theta = 200$ K and $T_c = 3.8$ K, the time required to cool down to T_c is 5×10^{-10} s and the approach of T^* to T is exponential.

To estimate the rate of Q relaxation, we take the coherence factor for branch mixing to be zero except for transitions involving a state within Δ of the bottom of the distribution. Near T_c where $\Delta \ll k_B T$ mixing is slow so that T^* reaches T_c before Q relaxaes. In that case $\Delta(T)/k_B T_c$ for all transitions involves branch crossing and we find,

$$\tau_Q = \frac{0.068 \ \tau_\theta(\theta/T_c)^3}{\Delta(T)/\Delta(o)} \simeq \frac{2 \times 10^{-10}}{\Delta(T)/\Delta(o)} \ s \tag{50}$$

For $T < T_c$, $\Delta \simeq \Delta(o) = 1.76 \ k_B T_c$, even before T^* has reached T_c, all phonon emission processes have $\sim 50\%$ probability of branch crossing. Hence, Q relaxes while the injected electrons are cooling through the vicinity of T_c independent of temperature. The relaxation of nonequilibrium electrons thus plays an important role in determining the current-voltage characteristics.

3.2. Magnetic impurity charge-imbalance relaxation

We consider a pair-breaking mangetic impurity that induces charge rleaxation through elastic exchange scattering causing a charge-imbalance relaxation rate. The rate of injection of quasiparticle charge is

$$\dot{Q}_i^* = \frac{2}{\Omega} \sum_k q_k \dot{f}_k|_i \tag{51}$$

where $\dot{f}_k|_i$ is the rate at which quasiparticles are injected into the state $\vec{k}$, and the sum over spin has been performed. The Eq.(51) can be written as

$$\dot{Q}_i^* = F^* I_i / e\Omega \tag{52}$$

where $F^*(\Delta, \Delta', V_i)$ is a calculable function that has the limiting form

$$F^* \approx 1 - \pi\Delta/2 \; eV_i|(eV \gg k_B T, \Delta, \Delta') \tag{53}$$

Here I_i and V_i are the current through the voltage across the injector junction and Δ' is the energy gap in the injector film. Since Q^* is uniformly distributed over the volume Ω, a relaxation rate may be defined as

$$\frac{1}{\tau_{Q^*}} = \dot{Q}_i^* / Q^* \tag{54}$$

The measured values are represented as

$$\frac{1}{F^* \tau_{Q^*}} = \frac{I_i}{e\Omega Q^*} \tag{55}$$

The value of F^* varies between 0.95 and 1.0. When a charge imbalance exists in the volume of the superconductor adjacent to the tunnel barrier, a voltage V_d must be applied across the junction to keep the current through the junction zero. This voltage is related to Q^* by

$$Q^* = 2N(o)eV_d g_{NS} \tag{56}$$

where $g_{NS}(0, T)$ is the zero-voltage conductance $G_{NS}(0, T)$ of the detector junction normalized to its value at the transition temperature, T_c. Combining (55) and (56) we obtain

$$\frac{1}{F^* \tau_{Q^*}} = \frac{I_i}{2N(o)e^2 \Omega g_{NS} V_d} \tag{57}$$

Schmid and Schön [68] considered the effect of pair breaking on $\tau_{Q^*}^{-1}$ and obtained an analytic result for $1/F^* \tau_{Q^*}$ when $\Delta/k_B T \ll (\tau_E \Gamma)^{-1/2}$ where $\Gamma = \tau_E^{-1} + 2\tau_s^{-1}$ in which τ_E is the scattering time due to magnetic exchange interaction. When $\tau_{Q^*}^{-1} \ll \tau_E^{-1}$,

$$\frac{1}{F^* \tau_{Q^*}} = \frac{\pi\Delta}{4k_B T_c \tau_E} \left[1 + \frac{2\tau_E}{\tau_s}\right]^{1/2} \quad (\tau_E \ll \tau_Q^*) \tag{58}$$

Consider the limit $\tau_s^{-1} \ll \tau_E^{-1}$, in which the exchange scattering is a weak perturbation on the inelastic scattering so that (58) can be expanded to give,

$$\frac{1}{F^*\tau_{Q^*}} = \frac{\pi\Delta}{4k_BT_c}\left[\frac{1}{\tau_E} + \frac{1}{\tau_s}\right] \quad (\tau_s^{-1} \ll \tau_E^{-1}, \tau_{Q^*}^{-1} \ll \tau_E^{-1}) \tag{59}$$

The coherence factors are different from zero only when one of the states is in the range $\Delta < E < 2\Delta$. If we assume the quasiparticles to be uniformly distributed in the energy range Δ to k_BT_c then only a fraction $\sim \Delta/k_BT_c$ of the inelastic events contribute to charge relaxation, producing a charge-relaxation rate $\Delta/k_BT_c\tau_E$. Pethick and Smith [80] have obtained the relaxation rate using exchange scattering as,

$$\frac{1}{F^*\tau_Q^*} = \frac{\Delta}{k_BT_c}\left[\frac{\pi}{4\tau_E} + \frac{1}{\tau_s}\right] \quad (\tau_s^{-1} \ll \tau_E^{-1}, \tau_{Q^*}^{-1} \ll \tau_E^{-1}) \tag{60}$$

In the limit $\tau_s^{-1} \gg \tau_E^{-1}$ (58) reduces to

$$\frac{1}{F^*\tau_Q^*} = \frac{\pi\Delta}{4k_BT_c}\left(\frac{2}{\tau_E\tau_s}\right)^{1/2} \quad (\tau_s^{-1} \gg \tau_E^{-1}, \tau_{Q^*}^{-1} \ll \tau_E^{-1}) \tag{61}$$

Thus the exchange scattering modifies the quasiparticle distribution substantially because the lower energy excess quasiparticles undergo exchange scattering to the other branch more rapidly than higher-energy quasiparticles can cool to replace them. As a result, the energy below which exchange charge relaxation is important is, increased from $\sim 2\Delta$ to an energy E^*. This energy is estimated by equating the cooling rate, $\sim \tau_E^{-1}$ with the exchange branch crossing rate, $\Delta^2/E^{*2}\tau_s$ to find $E^* \sim \Delta(\tau_E/\tau_s)^{1/2} \gg \Delta$. Thus at temperatures near T_c of the quasiparticles scattered downwards by cooling, a fraction $\sim E^*/k_BT_c$ contributes significantly to charge relaxation so that the rate is $E^*/k_BT_c\tau_E \simeq (\Delta/k_BT_c)(\tau_E\tau_s)^{-1/2}$ is in agreement with (61). The exchange relaxation rate, τ_E^{-1} enters the result not because it contributes to the charge relaxation but because it determines the rate at which quasiparticles scatter downwards into the region from which they exchange scatter to the other branch.

3.3 Andreev scattering contribution to charge imbalance

It was found by Andreev [64] that electron excitations over the barrier at the boundary of the normal and superconducting phases, a temperature drop occurs when there is a flow of heat. The additional thermal resistance of a superconductor in the intermediate state increases exponentially as the temperature is lowered and does not depend on the electron mean free path. At the voltage,

$$neV = 2\Delta \tag{62}$$

where Δ is the gap energy and n is an integer sudden changes in the resistance dV/DI are found to occur. Such a subharmonic gap structure is often observed in the current-voltage characteristics of superconducting microbridges and nonideal tunnel junctions. We can get the value of e/Δ from these structures the same way as k_BT_c/Δ from the infrared measurement. Klapwijk et al [82] suggested that subgap structure is the result of multiple Andreev reflections between the two normal metal-superconductor (NS) interfaces of a SNS microbridge. When an electron is incident on a superconductor from a normal metal, it can

                      Lydia S. Lingam

be trasmitted only if its energy E (measured relative to pair chemical potential, μ_s), lies outside the gap region of the superconductor. Otherise it may be Andreev reflected as a hole at an energy $-E$. A pair is formed in this process and propagates into the superconductor. For an abrupt, clean NS interface the probability of Andreev reflection is

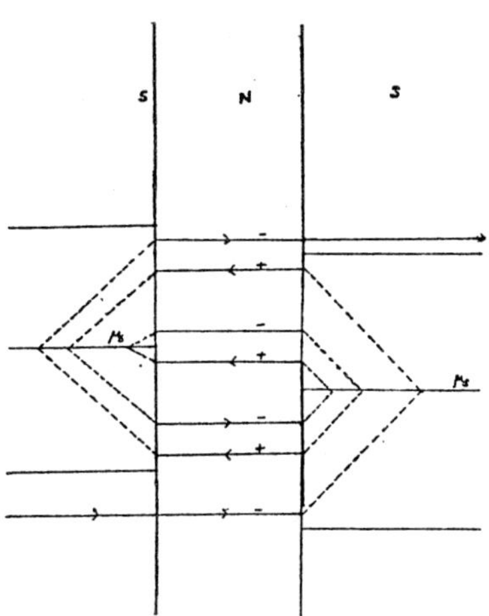

Fig.3.1 *Multiple Andreev reflections in an SNS junction. A voltage is applied between the superconductors. Since an electron Andreev reflects to become a hole, the particle gains energy with each pass across the normal metal. After seven passes it has gained enough energy to escape.*

$$A_{NS}(E) = \frac{1 - \{1 - (\Delta/E)^2\}^{1/2}}{1 + \{1 - (\Delta/E)^2\}^{1/2}} \qquad (63)$$

outside the gap and unity inside the gap. A voltage V is applied raising the Fermi level on the left to eV with respect to Fermi level on the right. Four terms then contribute to the electron current,

$$f_o(E - eV)A_{NS}(E)dE, \qquad \int f_o(E - eV)[1 - A_{NS}(E)]dE \ ;$$

$$-\int (1 - f_o(E - eV)]A_{NS}(E)dE, \qquad -\int f_o(E)N_s(E)T_{SN}(E)dE \ . \qquad (64)$$

where f_o is the equilibrium Fermi distribution, A_{NS} is the Andreev reflection coefficient and T_{SN} is the SN transmission coefficient. N_s is the superconductor density of state divided by the density of states in a normal metal [77]. Detailed balancing requires that

$$N_s(E)T_{SN}(E) = 1 - A_{NS}(E) \qquad (65)$$

which shows that the last of (64) is of the form $-\int f_o(E)[1 - A_{NS}(E)]dE$. Calculating all the terms it is found that $I(V)$ slightly deviates from a linear behaviour. Peshkin and Buhrman [77] have pointed out that first and third of (64) should be corrected to become,

$$\int f_o(E - eV)A_{NS}(E)f_o(-E - eV)dE \ ,$$

and

$$-\int [1 - f_o(E - eV)]A_{NS}(E)[1 - f_o(-E - eV)]dE \tag{66}$$

because the occupation of states was not taken into account in (64). The total summed current is not seriously affected by this correction. Consider a one-dimensional planar sandwich SNS junction with an applied voltage 2 eV$> \Delta$. If the thickness of the normal metal, d is small compared to the elastic scattering length, l, the current will be given by $I_f - I_r$, the difference of forward and reverse currents.

For the density of states as N/mv_F^2, the forward current is

$$\begin{aligned}
I_f &= e(N/mv_F)A(eV - \Delta) \\
&= (Ne\tau_{el}/m)(A/l)(eV - \Delta)
\end{aligned} \tag{67}$$

We define,

$$R_{min} = mv_F/Ne^2A \ . \tag{68}$$

If the elastic scattering in the normal metal is not entirely neglegible ($l \sim d$), the actual resistance of the normal metal R_N to first order is given by

$$R_N = R_{min}(1 + d/l) \tag{69}$$

All of I_f as defined above contributes to a nonequilibrium accumulation of charge imbalance density Q^* in the superconductor. If the relaxation length for charge imbalance is λ_Q, in a one-dimensional system, we have a total charge imbalance of

$$\int Q^*(v)dv = Q^*(o)\lambda_Q A \tag{70}$$

in the superconductor, where $Q^*(o)$ is the charge imbalance density at the interface ($z = 0$). If the relaxation time of the charge imabalance is $\tau_Q \gg l/v_F$, equilibrium between the injected current and the decay current will occur when the net current flow into the supercurrent satisfies,

$$\tau_Q I_t = (I_f - I_r)\tau_Q = Q^*(o)\lambda_Q A \ . \tag{71}$$

We have left out the energy dependent effective charge from the left hand side of (71). If $Q^*(o)$ is nonzero it will create a reverse current,

$$I_r = Q^*(o)Av_F/2 \tag{72}$$

which reduces the total current, I_t, for which

$$I_t(1 + \tau_Q v_F/2\lambda_Q) = (NeA/mv_F)(eV - \Delta) \tag{73}$$

 Lydia S. Lingam

which means that the actual resistance of the NS interface is given by $R_{min}(1 + \rho_Q)$ with

$$\rho_Q = I_r/I_t = \tau_Q v_F/2\lambda_Q \tag{74}$$

so that ρ_Q describes the charge-imbalance contribution to the resistance. We consider the R_{min} to be the resistance due to the normal region, since that is the total resistance when the superconductor is unperturbed. Therefore, the dynamical resistance of the superconductor is given by

$$R_s = \rho_Q R_{min} \tag{75}$$

for $(eV > \Delta)$. This is the branch imbalance resistance first described by Pippard et al [14]. If the reverse currents due to both NS interfaces in the SNS junction are considered. we find,

$$R_{SNS} = (1 + 2\rho_Q)R_{min} \tag{76}$$

If $d < L$, we have to first order

$$R_{SNS} = (1 + d/l + 2\rho_Q)R_{min} \tag{77}$$

In the dirty limit, $v_F \tau_Q \gg l$, the motion of the quasiparticles away from the interface is diffusive. so that

$$\lambda_Q = (\tau_Q v_F l/3)^{1/2} \tag{78}$$
$$\rho_Q = (3v_F \tau_Q/4l)^{1/2} \gg 1 \tag{79}$$

showing that nonequilibrium effects are important in this region. For $eV > 2\Delta$. the dynamic resistance of an SNS junction has an additional term $2\rho_Q R_{min}$ which represents quasiparticle dissipation in the superconducting electrodes. This extra resistance is related to that seen in NS interfaces as $T \to T_c$ where thermal excitations are responsible for injecting quasiparticles into the superconductor. In this case the applied voltage, rather than thermal excitations is responsible for the injection.

The flux of electrons transmitted into N from $S1$ at energy E is [82],

$$f_o(E - eV)[1 - A_{NS}(E - eV)] \tag{80}$$

and that of holes is

$$[1 - f_o(E - eV)][1 - A_{NS}(E - eV)] \tag{81}$$

The electrons of (79) will always traverse the normal region at least once. If they are Andreev reflected against $S2$, and then again Andreev-reflected against $S1$ they will traverse it again. The double reflection process may then be repeated untill the electrons are completely depleted by transmission into the superconductors as shown in Fig. 3.3.1. Similar considerations apply to the holes.

When these multiple contributions to current are considered the total currents from incident electrons and from incident holes are found to be,

$$\int f_o(E - eV)[1 - A_{NS}(E - eV)](1 + A_o A_1 + A_o A_1 A_2 A_3 + \cdots)dE \tag{82}$$

and

$$\int -[1 - f_o(E - eV)][1 - A_{NS}(E - eV)](A_o + A_oA_1A_2 + \cdots)dE \qquad (83)$$

respectively. Here $A_m = A_{NS}(E + meV)$ and we have used the property that $A_{NS}(E) = A_{NS}(-E)$. These reverse currents from $S2$ into N must also be considered. They are

$$\int -f_o(E - eV)[1 - A_{NS}(E - eV)](1 + A_oA_{-1} + A_oA_{-1}A_{-2}A_{-3} + \cdots)dE \qquad (84)$$

and

$$\int [1 - f_o(E - eV)][1 - A_{NS}(E - eV)](A_o + A_oA_{-1}A_{-2} + \cdots)dE \qquad (85)$$

The terms (81) till (84) have to be added and integrated. The results of dV/dI versus voltage, eV/Δ, show peaks in dynamic resistance at the correct voltages. The peaks at $n = 1, 2, \cdots 8$ are easily seen. At low temperature, $T \ll T_c$, the subgap structure is seen to disappear. The calculated structure is neither so sharply defined in voltage, nor so large in amplitude as the structure experimentally observed.

The Andreev condition is comparable with several other phenomena as follows:

$2eV$	$=$	$\hbar d\phi/dt$	Josephson [3]	(86a)
$2eV$	$=$	$n(2\Delta)$	Andreev [64]	(86b)
$2eV$	$=$	$\hbar\omega_\omega$	Shapiro[83]	(86c)
$\hbar\omega_m$	$=$	$nj_c\phi_o$	Shrivastava [83]	(86d)

where ω_n is the microwave frequency and j_c is the critical current. It is clear that interesting new phenomena should occur when $nj_c\phi_o$ becomes comparable to 2Δ. Due to Josephson weak links the current j_c becomes a variable so that there will be currents when the two energies become comparable.

3.4 Nonlinear $I(V)$ with magnetic field

We consider the d.c. resistance of the interface between the normal and superconducting phases at current densities of the order of the Ginzburg-Landau critical value in the presence of a constant magnetic field [52]. At small magnetic fields, the resistivity reduces with increasing magnetic field,

$$R = R_o \left(1 - \frac{\beta}{\pi} \frac{H^2}{H_o^2}\right) \qquad (87)$$

but for large magnetic fields, it increases with increasing field,

$$R = R_o \left(1 - \frac{H^2}{H_o^2}\right)^{-1/2} \left(1 + \frac{\beta}{\pi} \frac{H^2}{H_o^2}\right) \qquad (88)$$

where $\beta = 8(T_c - T)/\pi\gamma$ and $\gamma \approx T^3/\theta_{Df}$ with θ_{Df} as the diffusion coefficient the critical field for a cylinder with a small radius $r_o << \lambda$, is given by $H_o = 4\lambda H_c/r_o$. For $\eta \gg 1$, large magnetic fields,

$$R \approx R_o \left(1 - \frac{H^2}{H_o^2}\right)^{-1/2} [1 + I(\beta)]^{-1/2} \qquad (89)$$

in the entire range of currents where $I(\beta) = 2\beta/\pi$ for $\beta \ll 1$ and $\beta^{1/2}$ for $1 \leq \beta \leq k_B^2 T^2/\Delta^2$. In the temperature range, $\gamma/T \ll (T_c - T)/T_c \ll (\gamma/T)^{1/2}$, where $\beta \gg 1$, $p = \beta(1 - \Delta^2) \gg 1$ and $I = p^{1/2}$, the voltage on the SNS interface is given by

$$V = \frac{\Delta_o^2}{4eT} \left(\frac{j}{j_c}\right)^{1/2} \left(\frac{28\zeta(3)}{27\pi}\right)^{1/2} \left(\frac{T}{\gamma}\right)^{1/2} \ln\left[\frac{j^2}{j_c^2}\frac{(T_c - T)}{\gamma}\right] \tag{90}$$

Thus in this temperature range the current voltage characteristics are nonlinear. The dimensionless gap is given by

$$\Delta = \overline{\Delta}\Delta' \tag{91}$$

where

$$\overline{\Delta}^2 = \frac{8\pi^2 T^2}{7\zeta(e)}\frac{(T_c - T)}{T_c} \tag{92}$$

Thus the gap vanishes at T_c. The variation in the resistivity indicates that the voltage-current characteristics are nonlinear.

3.5 Levels in the gap

Temperature and field dependence of critical currents have been measured by Dupart et al [18]. The former is well described by a model of quantized quasiparticle levels in the normal regions. Supercurrents show interference patterns in the presence of a magnetic field which become less marked as the superconducting critical temperature of S_n is approached. An unexpected change of a few orders of magnitude in differential resistivity above the critical current is observed at a certain temperature in exceptionally clean samples.

For finite Δ in the superconductor and zero in the normal state, and T not too small, the expression for j_c following Bardeen [85], Kulik et al [86] is found to be

$$j_c = \frac{6n_e he}{md_N^*} \exp\left[-\frac{2d_N^* k_B T}{\xi_o \Delta(o)}\right] \tag{93}$$

with

$$d_N^* = d_N + \pi\xi_o \tag{94}$$

where d_N is the width of the normal region and n_e is the electronic density. The excitation spectrum of a system of alternating regions of lengths d_N ($\Delta = 0$) and $d_S(\Delta \neq 0)$ has been studied. It consists of a series of allowed and forbidden energy bands. If the energy E is measured from the Fermi level, $\phi = \sin^{-1}(E/\Delta)$, the allowed energies are found to be,

$$E - E_n \leq \Delta e^{-A_s}/(A_N + 1) \tag{95}$$

with

$$E_n = (n + \frac{1}{2})\pi\Delta(T)/(A_N + 1) \tag{96}$$

and

$$A_{SN} = (d_{S,N}/\pi\xi_o|\cos\theta|)\Delta(T)/\Delta(o) \tag{97}$$

where θ is the angle between the electron trajectory and the direction of the current.

In actual samples, d_s and d_N are larger than ξ_o so that at not too high temperature $A_s \gg 1$. Considering $E \ll \Delta(T)$, $\phi \simeq \sin\phi = E/\Delta$, $\cos\phi \simeq 1$, the expression,

$$\cos\left[(k_z - k_{zF})a_E\right] = e^{A_s \cos\phi}(\cos\phi + A_N \sin\phi)/\cos\phi$$
$$+ \; e^{-A_s \cos\phi}\cos(\phi - A_N \sin\phi)/\cos\phi \qquad (98)$$

can be simplified to give,

$$\cos\left[(A_N + 1)E/\Delta\right] = (-1)^n \sin\left[(A_N + 1)E/\Delta - (n + \tfrac{1}{2})\pi\right]$$
$$\simeq \; e^{-A_s}\cos ka \ll 1 \qquad (99)$$

where $a_E = d_s + d_N$. Thus quantized energies appear. In an infinite potential well with an effective width,

$$d_N^* = d_N + \pi\xi_o|\cos\theta|\Delta(o)/\Delta(T) \qquad (100)$$

which gives

$$j_c = \frac{6n_e\hbar e}{md_N}\exp\{-2[d_N + \pi\xi_o\Delta(o)/\Delta(t)]k_BT/\xi_o\Delta(o)\} \qquad (101)$$

for the maximum supercurrent. This is equivalent to the previous result (92) for the effective normal depth,

$$d_N^* = d_N + \pi\xi_o\Delta(o)/\Delta(T) \qquad (102)$$

except near T_c. Thouless [87] has found that the gap parameter of a strong coupling superconductor varies as

$$\delta = \Delta(T)/\Delta(o) = \tanh(\delta/t) \qquad (103)$$

where $t = T/T_c$. For $\Delta(o) \simeq 2k_BT_c$, the current becomes,

$$j_c = \frac{6n_e\hbar e}{md_N}\exp\left[-\{4a_E\tanh(\delta/t)/3\xi_o + \pi\}t/\delta\right] \qquad (104)$$

so that the j_c can be plotted as a function of t/δ and ξ_o can be determined. Dupart et al [18] have obtained the value of ξ_o for lead by these types of graphs. They find $\xi_{ocal} \simeq 0.4 \; \mu\,m$ for $a_E \simeq 4.2 \; \mu\,m$ with $j_c\,(T = 4.2\;K) \simeq 1.25 \times 10^7 \; Am^{-2}$.

4.Electric Dipole Moment

Recently, Mannhart et al [4] have found that the density of mobile carriers in a superconductor can be changed by the application of an electric field. Thus the critical current density of $YBa_2Cu_3O_{7-\delta}$ films can be controlled by an electric field. Changes of relative conductivity of the order of one in 10^4 have been produced earlier by Glover, III and Sherrill [88] but the recent results [41] are much larger in magnitude. The carrier densities in the high temperature superconductors are about 100 times less than in the low transition temperature superconductors, even then field effects have been measured in films [31,33,89,90].

4.1 Anisotropic Resistivity

The effect of the electric field on the oxygen ion is to change their position in the basal plane so that the coordination of Cu^{2+} ions changes. The relaxation effects owing to the motion of the oxygen ions have been observed [91]. An activation energy of 0.97 ± 0.03 eV has been measured [92] for oxygen ion by tracer diffusion. Several measurements [93-100] show that permanent electrical dipole moments are present in $YBa_2Cu_3O_{7-\delta}$. The dipole moments calculated [101] for the planar oxygen atoms is about 0.003 e Å thereby showing that the lattice is polarizable by the application of an electric field. Since many oxygen ion sites are vacant, the negatively charged oxygen ions jump from the occupied sites to the empty sites upon application of a small electric field. The experimental work of Mannhart et al shows that the drain source resistance increases when a gate voltage of $+10$ V is applied and decreases when a gate voltage of -10 V is applied.

In this section, we discus the mechanism of change of resistivity upon the application of an electric field in a polarizable superconducting device near the onset temperature. As the system is cooled, this resistivity approaches zero. We find that in the case of a finite electric dipole moment, the resistivity is quantized.

We assume that there are vacancies of oxygen ions so that when the electric field $|\vec{E}|$ is applied, the system develops an electric dipole moment,

$$p = -e \sum_i r_i \tag{105}$$

where r_i is the radius vector of the i-th ion. The potential energy of the ion due to the dipole moment is given by

$$V^{(1)} = -E. <p> \tag{106}$$

The quadratic polarization effects give rise to a change in energy expressed by the usual second-order perturbation theory,

$$V^{(2)} = -\sum_n \frac{<0|E.p|n><n|E.p|0>}{E_n - E_o} \tag{107}$$

the change in energy due to the electric field is then given by the sum of the previous two terms,

$$V = -|\vec{E}| \left[<p> + \sum_n \frac{<0|E.p|n><n|p|0>}{E_n - E_o} \right] \tag{108}$$

The current, j, is related to the electric field $\vec{E}$ through the conductivity, σ per unit length as,

$$j = \sigma E \tag{109}$$

The flux is quantized in units of $\phi_o = hc/2e$ so that $nj\phi_o/c$ with $n = 1, 2, 3, \ldots$, an integer represents the current in terms of energy. We have taken current in units of e/s, ϕ_o in terms of Gcm^2 and c is the velocity of light in cm/s. Here e is the electronic charge, so that $j\phi_o/$ is measured in erg. Accordingly $n\sigma|E|\phi_o/c$ represents the energy of a flux quantized system with electric field $\vec{E}$ so that,

$$-\frac{n\sigma E\phi_o}{c} = -|\vec{E}| \left[<p> + \sum_n \frac{<0|E.p|n><n|p|0>}{E_n - E_o} \right] \tag{110}$$

Therefore we can write the conductivity as

$$\sigma = \frac{c<p>}{n\phi_o}\left[1 + \sum_n \frac{<0|E.p|n><n|p|0>/<p>}{E_n - E_o}\right]$$ (111)

When the sign of $\vec{E}$ is changed the second term in the above changes sign so that for one direction of E we have,

$$\sigma_+ = \sigma_o(1+\gamma)$$ (112)

whereas for the direction of E reversed, we get

$$\sigma_- = \sigma_o(1-\gamma)$$ (113)

Thus the ratio of the conductivity for the two directions of $\vec{E}$ is

$$\frac{\sigma_+}{\sigma_-} = \frac{1+\gamma}{1-\gamma}$$ (114)

Similarly, the ratio of resistivity for the two directions of the electric field is $(1-\gamma)/(1+\gamma)$ where

$$\sigma_o = \frac{c<p>}{n\phi_o}$$ (115)

$$\gamma = \sum_n \frac{<0|E.p|n><n|p|0>/<p>}{E_n - E_o}$$ (116)

The conductivity is the inverse of resistivity so that

$$\rho_+ = 1/\sigma_+ \quad \text{and} \quad \rho_- = 1/\sigma_- .$$ (117)

We now compare our theory with the experimental results [4]. At ~ 90 K, the increase in the drain source resistance upon the application of gate voltage of $+10$ V is 0.11 kΩ while the decrease in the same resistance upon reversing the gate voltage to -10 V is 0.18 kΩ so that the ratio of ρ_+/ρ_-=0.61, when compared with the calculated conductivity expression (114) gives $(1-\gamma)/(1+\gamma)$=0.61 so that γ=0.24 indicating that there is a 24% correction to the energy of the system due to the second-order dipole moment. When the sign of $\vec{E}$ changes, the sign of γ also changes so that the resistivity is either increased or reduced depending upon the direction of the electric field. This feature of experimental measurements is also fully in accord with the expressions (112) and (113). The quantized resistivity,

$$\rho_+ = \rho_o(1+\gamma)^{-1}$$ (118a)

$$\rho_- = \rho_o(1-\gamma)^{-1}$$ (118b)

has a prefactor of $\rho_o = n\phi_o/c<p>$ so that when $<p>=0$, the system has infinite resistivity or it becomes an insulator. Thus for our theory a non-zero $<p>$ is needed. This requirement is met in a polarizable material where $<\sum_i r_i>\neq 0$. Once the system is polarized, the polarization decays unless there are pinning centres. In this way pinning becomes necessary to hold the dipole moments.

Lydia S. Lingam

The energy in (106) is repulsive but that in (107) may be attractive. Since the second term is only about 24% of the first term, we assume that for a polar material (108) is repulsive. This means that the transition temperature of the BCS superconductor is reduced upon application of an electric field.

Unlike the microwave absorption [102, 103] the present effect of electric field is not caused by weak links but weak links and vacant oxygen sites increase the dipole moments and hence increase the conductivity or the carrier concentration. It will be of interest to study the problem of flux flow in which case the linear electric field shifts the vortices. Similarly, the problem of flux-lattice melting [104] in an electric field may be of interest for the development of superconducting devices. Since ϕ_o in (115) is quantized, the conductivity is also quantized. The conductance per square quantized in units of e^2/h has also been considered by Levy et al [105] who found the variation of conductance with gate voltage. Thus there is a change in the drain source resistivity of a superconducting film upon the application of an electric field.

This change in resistivity is caused by an electric dipole moment, which we have calculated upto second-order in the perturbation theory. The flux quantized resistivity requires the existence of vacant oxygen sites and pinning centres in the system. The calculated results are in resonable agreement with the experimental measurements performed by Mannhart et al [4] in a thin film of $YBa_2Cu_3O_{7-\delta}$.

4.2 Correction to McCumber relations

The electric field applied to a superconducting film produces a normal current, $\sigma V(t)$, and a supercurrent $J_o \sin \phi(t)$, where σ is the conductivity of the junction, $V(t)$ is the time dependent potential and J_o is the maximum value of the Josephson current. In the case of a Josephson junction, the phase difference, $\phi(t)$, between the wave functions on the two sides of the junction is subject to the condition,

$$\hbar d\phi(t)/dt = 2eV(t) \tag{119}$$

for the flow of the supercurrent. Therefore, the current through the junction is given by

$$I(t) = \frac{\sigma\hbar}{2e}\frac{d\phi(t)}{dt} + J_o \sin \phi(t) \tag{120}$$

McCumber [106] considered a capacitive contribution $CdV(t)/dt$ to the current so that (120) is modified to

$$I(t) = \frac{\hbar C}{2e}\frac{d^2\phi(t)}{dt^2} + \frac{\hbar\sigma}{2e}\frac{d\phi(t)}{dt} + J_o \sin \phi(t) \tag{121}$$

Similarly, the consideration of the inductive contribution to the voltage modifies (119) to,

$$2eV(t) = 2eL\frac{dI(t)}{dt} + \hbar\frac{d\phi}{dt} . \tag{122}$$

These expressions describe the voltage versus current relationships very well. For large values of $\beta_c = \omega C/\sigma$ with $\omega = 2eV(t)/\hbar$, the voltage becomes a linear function of the current but for smaller values of β_c there is a threshold current below which the voltage is zero. There is only one value of the voltage for one value of the current.

Mannhart [5] has found that one value of the resistivity of the drain-source channel R_{DS} as a function of the gate voltage is associated with two values of the gate voltage, V_g, so that $R_{DS}(V_g)$ is parabolic which can not be understood on the basis of the McCumber's current expression. We have found [107] that the atoms move from the occupied to the empty sites upon the application of an electric field so that there is a dipole moment. This dipole moment causes a contribution to the current so that the expression (120) for the current in terms of a normal and a supercurrent is not adequate in materials with a large number of oxygen sites vacant.

In this section, we calculate the current produced by the motion of the oxygen ions from the occupied to the empty sites which corrects the expression (119) for the current.

The calculated result is in accord with the observation of resistivity as a function of gate voltage. In this way we are able to correct the expression for the current and understand the measurements.

4.3. The normal state

We assume that there are vacant as well as occupied sites of O^{2-} ions. When an electric field is applied through the gate voltage, oxygen ions move from the occupied to the empty sites. The coordinates of the i-th electron are r_i which makes an angle of θ from the direction of the electric field $\vec{E}$ which is normal to the surface of the superconducting film. The first order energy of the system due to the dipole moment is $-\vec{E}.\vec{p} = -E < p > \cos \theta$. A non-zero value of $< p >$ is possible owing to oxygen ion defects in the film. The potential energy of the ion due to the dipole moment is given by (108). In the normal state, this potential energy gives rise to a current J_n which corresponds to the energy of $e\rho J_n$. Thus the application of the electric field E gives rise to the current versus voltage relation of

$$J_n = \frac{1}{e\rho} \left[E < p \cos \theta > + \sum_n \frac{< 0 | Ep \cos \theta | n > < n | Ep \cos \theta | 0 >}{E_n - E_o} \right] \tag{123}$$

The resistivity ρ as a function of E from this result is predcited to be parabolic.

We approximate the second term of (123) as, $\gamma_2 E^2$ so that the total normal current becomes

$$I_n(t) = \sigma V(t) + \frac{\sigma}{e} < p \cos \theta > V(t) + \frac{\sigma}{e} \gamma_2 V^2(t) \tag{124}$$

where

$$\gamma_2 \simeq \sum_n \frac{< 0 | p \cos \theta | n > < n | p \cos \theta | 0 >}{E_n - E_o} \tag{125}$$

which shows that due to the motion of the ions from the occupied to the empty sites, there is a current, proportional to the square of the voltage. It was estimated earlier [6] that such a correction coming from the last term of (124) can be as large as 24% of the previous term. The current-gate-voltage relationship is then predicted to be of the form of (124) so that the resistivity as a function of gate voltage becomes,

$$\rho = aV_g + bV_g^2 \tag{126}$$

112 Lydia S. Lingam

with

$$a = \frac{(1+ <p\cos\theta>/e)}{I(t)} \tag{127}$$

and

$$b = \gamma_2/[eI(t)] \tag{128}$$

which predicts that the resistivity as a function of gate voltage is parabolic. The result is consistant with the measurements of impedence [107].

Mannhart [5] has measured the normal state resistance of the drain source channel R_{DS} of a film of $YBa_2Cu_3O_{7-\delta}$ as a function of gate voltage which is parabolic in accord with that predicted by the expression (126).

4.4. The superconducting state

In the superconducting junction, the quantized energy is $nj\phi_o/c$. Since there is a finite conductivity, $j = \sigma E$, the first-order energy due to the dipole moment becomes, $n\phi_o E/(cp)$ so that the current due to the application of the electric field becomes,

$$j_1 = \frac{Ec <p\cos\theta>}{n\phi_o} \tag{129}$$

Because of the Josephson relation (119) the above becomes,

$$j_1 = \frac{<p\cos\theta>}{2\pi n}\frac{d\phi(t)}{dt} \tag{130}$$

so that the expression for the current (119) given by McCumber is modified to,

$$I(t) = \left(\frac{\sigma\hbar}{2e} + \frac{<p\cos\theta>}{2\pi n}\right)\frac{d\phi(t)}{dt} + J_o\sin\phi(t) \tag{131}$$

The second-order correction from (123) in the superconducting state is given by

$$j_2 = \frac{c|E|^2\gamma_2}{n\phi_o}. \tag{132}$$

Using the Josephson expression this contribution to the current becomes,

$$j_2 = \frac{\hbar\gamma_2}{4\pi ne}\left(\frac{d\phi(t)}{dt}\right)^2 \tag{133}$$

so that the total current becomes,

$$I(t) = \left\{\frac{\sigma\hbar}{2e} + \frac{<p\cos\theta>}{2\pi n}\right\}\frac{d\phi(t)}{dt} + \frac{\hbar\gamma_2}{4\pi ne}\left(\frac{d\phi(t)}{dt}\right)^2 + J_o\sin\phi(t) \tag{134}$$

Thus two types of corrections to the McCumber relation are generated. The first correction is caused by the dipole moment and is linear in the voltage, $d\phi(t)/dt$. The second correction is quadratic and has square of the voltage, $(d\phi(t)/dt)^2$. The current versus voltage thus becomes a parabola. Similarly, the resistivity versus voltage also becomes a parabola. Thus the work of Mannhart et al [4] has led us to find new correction to the current relations.

4.5. Heat Loss

We suppose that the current is reduced by the resistivity of the substrate so that the quantum energy becomes $-\frac{n\phi_o}{c}\frac{|E|}{\rho}$. The energy is utilized partly in moving the atoms and partly in heating the film. The first-order energy of the system due to the dipole moment is $-\vec{E}.\vec{p} = -|E| <p> \cos\theta$ and the heating loss is $j^2\rho = |E|^2/\rho$. Therefore

$$- E <p> \cos\theta = -\frac{n\phi_o}{c}\frac{|E|}{\rho} + \frac{|E|^2}{\rho} \tag{135}$$

which has two solutions. One solution is $E = 0$ which means that the current is zero, and the other solution is,

$$E = \frac{n\phi_o}{c} - \rho <p><\cos\theta> \tag{136}$$

which shows that the resistivity is

$$\rho = \left(\frac{n\phi_o}{c} - |E|\right)\left(-e < \sum_i r_i ><\cos\theta>\right)^2 . \tag{137}$$

Since $\phi_o = hc/2e$, the above corresponds to quantization of resistivity per unit length in a unit radius. When $E = 0$, $<\cos\theta> = 1/4$, the minimum value of the resistivity is $4nh/2e^2 < \sum_i r_i >$. This resistivity should be larger than the smallest quantized resistivity. Since the charge of the Cooper pair is $2e$, the minimum quantized resistivity is $h/4e^2$. Therefore the minimum resistivity is consistent with the minimum quantized resistivity. It may be noted that the system responds with the velocity of light and the role of the dipole moment is limited to producing the resistivity.

We now calculate the current from the above relation without the use of $j = \sigma|E|$. The energy balance equation is

$$j^2\rho - \frac{n\phi_o}{c}j + \frac{1}{4}E <p> = 0 \tag{138}$$

which has two solutions for j,

$$j_\pm = \frac{n\phi_o}{2c\rho} \pm \frac{1}{2\rho}\left\{\left(\frac{n\phi_o}{c}\right)^2 - \rho E <p>\right\}^{1/2} \tag{139}$$

For $E = 0$, one of the solutions of, the above gives

$$\rho j_+(o) = n\phi_o/c \tag{140}$$

which shows that there is a minimum quantized voltage of $n\phi_o/c$ and the other solution is $j_-(o) = 0$, which shows that there is no current when applied voltage is zero. The quantity within the square root sign is real and positive only as long as the electric field term is less than $h/4e^2 \simeq 6\ k\Omega$. The solutions are of the form,

        Lydia S. Lingam

$$\left(\frac{\rho j_{\pm}}{E_o} - \frac{1}{2}\right)^2 = \frac{1}{4}\left\{1 - \frac{\rho <p> E}{E_o^2}\right\} \tag{141}$$

where $E_o = n\phi_o/c$ is a constant. This describes the current, $j_{\pm}$, voltage, E, relationship as a parabola. Our result is consistent with the measurements of V_{DS} and I_{DS} which describe a parabolic behaviour for $T < T_c$ as measured by Mannhart [5].

We have calculated the second-order energy due to the electric dipole moment which explains the change in resistivity upon the application of a gate voltage. The energy correct upto second order is

$$V = -|E| <p><\cos\theta> -|E|^2\gamma_2 \tag{142}$$

where

$$\gamma_2 = \sum_n \frac{<0|p|n><n|p|0><\cos\theta>^2}{E_n - E_o}. \tag{143}$$

The expression (138) is then correlated to become,

$$j^2\rho - \frac{n\phi_o}{c}j + \frac{1}{4}E <p> +\gamma_2 E^2 = 0. \tag{144}$$

This is the modified relation between j and E which we write as,

$$\frac{j\rho}{E}\left(j - \frac{n\phi_o}{c\rho}\right) = -\frac{1}{4} <p> -\gamma_2 E \tag{145}$$

It is clear that the current has been reduced by a quantum voltage $n\phi_o/c$ by the amount $n\phi_o/c\rho$ and the voltage has been increased by the dipole moment.

In order to apply electric field to superconducting thin films of YBa$_2$Cu$_3$O$_{7-\delta}$ inverted metal-insulator-semiconductor field effect transistor (MISFET) type heterostructures were made by Mannhart [5]. If one of the electrodes is replaced by a superconducting film interesting new electronics would be developed. In the case of one such structure, a gate electrode is made of Nb-doped SrTiO$_3$. This electrode is coated with platinum. SrTiO$_3$ is then deposited on Pt and then a film of superconducting YBa$_2$Cu$_3$O$_7$ is also quoted on it Such a stack of superconductor - SrTiO$_3$-Pt-Nb:SrTiO$_3$ forms a MISFET type structure Four gold contact pads were made on the superconductor and fifth contact pad is put on Nb SrTiO$_3$. The voltage applied between the superconductor and Nb: SrTiO$_3$ is called the gate voltage. the voltage across the gold points on the superconductor is called the drain-source (DS) voltage and the current is also measured within the same circuit. At a temeprature of 101 K $(T > T_c)$ the drain source voltage, V_{DS} and the drain-source current, I_{DS} are linear functions of each other. However at low temperature $(T < T_c)$ the currents become very large and tend to reach a saturation value as $T \to 0$. The normal state linearity between V_{DS} and I_{DS} is then completely destroyed and $I_{DS}(V_g)$ also becomes parabolic.

In Fig.2.3 we show the experimental measurements of Mannhart [5] along with our calculated curve $R_{DS} = 2.61 + 5.115 \times 10^{-5} V_g + 2.5845 \times 10^{-8} V_g^2$. The calculated curve is in reasonable agreement with the measurements.

Study of epitaxial YBa$_2$Cu$_3$O$_{7-\delta}$ films by the scanning-tunneling microscope revealed that there are nanometer-sized holes on the surface [108]. The holes are arranged at random on a self similar surface. Similar holes were found on silica gel which when studied by the Raman scattering of light showed fractal behaviour [109]. Therefore, it is expected that the resistivity of the film in the normal state will describe a fractal behaviour quite unlike a metal in as far as the temperature dependence is concerned. Near the transition temperature in the superconducting state the Josephson current is proportional to the gap Δ, the dispersion relation, $j_c = \pi\Delta/2e\rho$. This current gives rise to a voltage of $j_c\rho$ which is small compared with the gate voltage and hence the gap effect is not observable in the characteristics. The pinning of flux lines causes a thermally activated resistivity, $\rho = \rho_o \exp(-U/k_B T)$. Therefore, part of the temperature dependence in the current is caused by pinning. The effect of the flux-lattice melting [110] in the superconducting state also reduces the currents. In the case of a.c. gate voltage, the device shows a small current due to a capacitive contribution as in (121). Such a contribution has been measured by Jenks and Testardi [29] according to which the capacity varies with temperature showing a maximum near the transition temperature. However, in the case of a d.c. gate voltage [4] such a term plays no role at all. We have found that there are currents due to the dipole moments generated in the superconductor by the application of the electric field. We have calculated this effect upto second-order in the perturbation theory. The current-gate-voltage characteristics are found to become parabolic rather than linear. Similarly, the resistivity as a function of gate voltage also becomes parabolic.

5.Thermoelectric effects

5.1   Seebeck effect

Thermoelectric effects do not occur in superconductors. However, if we make a junction, then such effects are easily observed [111-117]. In a normal metal the current is given by

$$j = \sigma(E - \nabla\mu) + b\nabla T \tag{146}$$

where $\vec{E}$ is the electric field strenght, μ is the chemical potential appropriately normalized and ∇T is the temperature gradient. If the conducting circuit is not closed, $j = 0$, so that

$$E = \nabla\mu - \frac{b}{\sigma}\nabla T \tag{147}$$

We may neglect the change in chemical potential so that

$$E = -S\nabla T \tag{148}$$

where $S = b(T)/\sigma(T)$ is called the Seebeck coefficient. It defines the thermo-emf in the circuit consisting of conductors I and II which are having the temperature difference ∇T. From the above it is clear that the thermo-emf in such a circuit is.

$$\epsilon = \int_{T_1}^{T_2} \left\{ \left(\frac{b}{\sigma}\right)_{II} - \left(\frac{b}{\sigma}\right)_I \right\} dT \tag{149}$$

When both the metals I and II become superconductors, ϵ becomes zero. In the normal state the electric field is given by (147). In the superconducting state the field $E - \nabla\mu$ is equal to zero in the thermal equilibrium. If $E - \nabla\mu \neq 0$, the superconducting electrons would be accelerated due to $\partial(\Lambda j_s)/\partial t = E - \nabla\mu$ where j_s is the density of the superconducting current. At the same time, the temperature gradient must give rise to a normal current with density

$$j_n = b_n \nabla T + \sigma_n \left(E - \nabla\mu_n \right) \tag{150}$$

In the superconductor, the electrons form pairs so that the gap in the single-particle dispersion is 2Δ. The density of heat flow is $q = H\nabla T$ where $H = H_{ph} + H_{el} + H_c$ in which H_{ph} is the thermal conductivity due to the lattice phonons, H_{el} is the thermal conductivity due to normal electrons and H_c is the thermal conductivity connected with the circulation. According to Wiedemann-Franz law $H_{el} = (\pi^2 k_B^2/3e^2)T\sigma_n$ where $\sigma_n = j_n/E$ is the electric conductivity of the normal electrons. The circulatory heat flow, $q_c \sim H_c\nabla T \sim j_n(\Delta/e)$ and $H_c = b_n(T)\Delta(T)/e$. Using Wiedemann-Franz law we obtain,

$$\frac{H_c}{H_{el}} \simeq \frac{3eS_n\Delta}{\pi^2 k_B^2 T} \sim \frac{\Delta(T)}{E_F} \sim \frac{k_B T_c}{E_F} \tag{151}$$

where

$$S_n = (b_n/\sigma_n) = \pi^2 k_B^2 T(3eE_F)^{-1} \tag{152}$$

has been used for the Seebeck coefficient. In conventional superconductors $T_c \sim 1 - 10 \ K$ and $E_F \sim 3 - 10 \ eV$ so that the circulatory thermal conductivity is small. However, for high-temperature superconductors $T_c \sim 100 \ K$ and $E \sim 0.1 \ eV$, we obtain $H_c/H_{el} \simeq 0.1$. Since the thermoelectric coefficient b (or S) depends on the derivative of the electronic distribution function with respect to momentum or energy, $H_c/H_{el} \simeq 1$, i.e., the effect can be large. For p- or d-wave pairing, the coefficient b_n increases considerably in the superconducting state because it is multiplied by a factor of $E_F/k_B T_c$. For a completely superconducting circuit, there is a certain current, $j \neq 0$, but only near the surface of the metal. The total magnetic flux ϕ through contour C is given by,

$$\Phi = n\Phi_o + \frac{4\pi}{c} \int_{T_1}^{T} \left\{ (b_n\lambda_L^2)_{II} - (b_n\lambda_L^2)_I \right\} dT \tag{153}$$

where $\phi_o = hc/2e$. In the bulk of the specimen $(d \gg \lambda_L)$, on the contour C, we have $j = j_s + j_n$, $j_n = b_n\nabla T$ and

$$j_s = (e\hbar n_s/2m) \left[\nabla\phi - (2e/\hbar C)A \right] ,$$

where n_s is the concentration of the superconducting electrons $(\lambda_L^2 = mc^2/4\pi c^2 n_s)$, φ is some scalar phase and curl $A = H$. Integrating the equality $j_n = -j_s$ along the contour C we obtain (153) because $\oint A dS = \int H dS$ and $\oint \nabla\varphi ds = 2\pi n$. Let us assume that $(b_n\lambda_L^2)_{I}$ $\gg (b_n\lambda_L^2)_I$ and $\lambda_{II}^2 = \lambda_{L_{II}}^2(o) \left(1 - T/T_c \right)^{-1}$. Then near T_c the temperature dependent part of this flux is

$$\phi_T = \phi - n\phi_o \approx \frac{4\pi}{c} b_{n_{II}}(T_c)\lambda_{L_{II}}^2(o)T_c \ln\frac{T_c - T_1}{T_c - T_2} \tag{154}$$

For tin $\lambda_{L_{II}}(o) \simeq 2.5 \times 10^{-6}$ cm, $b_n(T_c) \simeq 10^{11}$ cgse/K and $\phi_T \sim 10^{-2}\phi_o \ln(T_c - T_1/T_c - T_2)$. Such a flux when measured [116,117] was found to be larger than predicted

5.2 Temperature dependent resistivity

Battersby and Waldram [114] have calculated the resistivity of a SNS' sandwich as

$$R_{SNS} = R_{Cu} + (2\lambda_D \rho^s/A)2f(\Delta)$$
$$G_{SNS}R_{SNS} = (GR)_{Cu} + (2\lambda_D \rho^s/A)\left[G^s - G^N\{1 - 2f(\Delta)\}\right] \tag{155}$$

where R_{Cu} and $(GR)_{Cu}$ represent contributions to the resistance and thermoelectric coefficient due to the bulk copper. The remaining terms correspond to the excess voltage due to charge imbalance, where ρ^s is the resistivity and G^s is the thermoelectric coefficient of the superconductor assumed continuous at T_c. The thermoelectric coefficient of copper is G^N, the interface area is A. The Fermi function is $f(\Delta)$ and quasiparticle diffusion length is λ_D. Near T_c the term in G^N is small. For Pb/Cu interface τ_D is essentially equal to λ_{Q^*}, the diffusion length for charge imbalance. For $\Delta \ll k_B T$

$$\lambda_{Q^*}(T) = \lambda_{Q^*}^o \left[\frac{\Delta(o)2f(\Delta)}{\Delta(T)}\right]^{1/2} \tag{156}$$

which reduces to van Harlingen's [115] form for λ_{Q^*} very close to T_c. Combining (155) with (156) we predict an intertfacial contribution to the resistance with a temperature dependence similar to those of the theories of Hsiang and Clarke [19] but a contribution to the thermopower which differs from that of van Harlingen [115]. The experimental results [114] on Pb/Cu/Pb sandwiches are in reasonable agreement with the above theory.

5.3 Tunneling in thermoelectric effect

A calculation of the thermoelectric power which takes into account the tunneling current with Bogoliubov coherence factors properly has been performed by Hirsch [113]. It is claimed that an intrinsic thermoelectric effect across normal metal-insulator-superconductor (NIS) as well as superconductor-insulator-superconductor, SIS, tunnel junctions occurs within the model of hole superconductivity. The effect arises due to the energy-dependent gap in the dispersion relation.

The quasiparticle energy in the superconducting state is given by

$$E_k = [(\epsilon_k - \mu)^2 + \Delta_k^2]^{1/2} \tag{157}$$

where μ is the chemical potential. We define an energy shift

$$\nu = \frac{1}{a}\frac{\Delta_m}{D/2}\Delta_o \tag{158}$$

where D is the bandwidth, Δ_m is an energy parameter [120],

$$a = \left[1 + \left(\frac{\Delta_m}{D/2}\right)^2\right]^{1/2} \tag{159}$$

and

$$\Delta_o = \frac{\Delta(\mu)}{a} \tag{160}$$

Lydia S. Lingam

The gap function is now defined as

$$\Delta_k = \Delta_m \left(-\frac{\epsilon_k}{D/2} + c \right) \equiv \Delta(\epsilon_k) \tag{161}$$

and the factors arising from the Bogoliubov transformation, called the coherence factors, are given by

$$u_k^2 = \frac{1}{2} \left(1 + \frac{\epsilon_k - \mu}{E_k} \right) \tag{162a}$$

$$v_k^2 = \frac{1}{2} \left(1 - \frac{\epsilon_k - \mu}{E_k} \right) \tag{162b}$$

The minimum quasiparticle energy occurs at $\epsilon_k = \mu + \nu$ rather than at $\epsilon_k = \mu$ and the factors u_k and v_k are not symmetric around the minimum quasi-particle energy so that there is a branch imbalance causing a net quasiparticle charge per site of eQ^*, with

$$Q^* = \frac{2}{N} \sum_k (u_k^2 - v_k^2) f(E_k) = \frac{4\nu}{Da} \int_{\Delta_o}^{\infty} dE \frac{f(E)}{\sqrt{E^2 - \Delta_o^2}} \tag{163}$$

where N is the number of sites and f the Fermi distribution. The NS current is given by

$$I_{NS} = \frac{4\pi e}{N} |T|^2 N_k(o) N_q(o) \sum_{\text{branches}} \int_{\Delta}^{\infty} \eta(E_k) \{J_{NS}\} dE_k \tag{164}$$

where $|T|^2$ is the square of the matrix element of the tunneling interaction, $N_k(o)$ and $N_q(o)$ are the occupation numbers of the quasiparticles and $\eta(E_k) = E_k(E_k^2 - \Delta^2)^{-1/2}$. The dimensionless quantity in the integrand is found to be,

$$\begin{aligned}
J_{NS} &= \frac{1}{2}\Big\{ (1 + \frac{\nu}{E_k})[f_n(E_k - eV) - f_s(E_k)] \\
&+ (1 - \frac{\nu}{E_k})[f_s(E_k) - f_n(E_k + eV)] \Big\}
\end{aligned} \tag{165}$$

for processes involving quasiparticles of energy E_k with V as the voltage on the normal metal and f_n and f_s are the Fermi distributions in the normal and in the superconducting state. For $f_n \neq f_s$, a thermoelectric current is predicted even if $V = 0$. In this case the dimensionless factor (165) becomes,

$$J_{NS} = \frac{2\nu}{E_k} [f_n(E_k) - f_s(E_k)] \tag{166}$$

Thus positive current flows from the hotter to the colder side when there is a temperature gradient. The thermoelectric voltage is obtained from (165) by expanding the Fermi distributions and retaining only the linear terms as

$$V_t = \frac{\nu}{e} \frac{(T_s - T_n)}{T_n} \tag{167}$$

with T_n and T_s as the temperatures of the normal and superconducting sides of the junction, respectively. Thus this voltage measures the asymmetry parameter ν and hence the gap-function slope $\Delta_m/(D/2)$ which measures Δ_o independently. The energy ν depends on the band width and varies smoothly from a constant value at $T = 0$ to zero at $T = T_c$. The charge imbalance has a peak at a temperature slightly below T_c but goes to zero both at $T = 0$ and at $T = T_c$. The thermal voltage V_t approaches zero linearly at $T \to T_c$ due to linear dependence of ν on $T_c - T$. The total quasiparticle current is obtained by integrating (165) over all the states as given by (164). Thus,

$$I_{NS} = \frac{1}{eR} \int_{\Delta_o}^{\infty} \frac{E}{\sqrt{E^2 - \Delta_o^2}} \Big[f_n(E - eV) - f_n(E + eV)$$
$$+ \frac{\nu}{E} \{ f_n(E - eV) + f_n(E + eV) - 2f_s(E) \} \Big] \tag{168}$$

for constant density of states and tunneling matrix element. Here R is the resistance of the junction in the normal state. The zero voltage thermoelectric current that results if the normal metal is at zero temperature is simply the current that would be obtained in the normal state by applying a potential difference $V = D/e$ across the junction multiplied by the charge imbalance.

6. Ginzburg-Landau model

6.1 Screening Lengths

We introduce the Thomas-Fermi screening length of free electrons because of the charged nuclei. The electrostatic potential φ of a charge distribution is found from the Poisson's equation, $\nabla^2\varphi = -4\pi\rho$. In the case of a plane wave ∇^2 leads to,

$$k^2\varphi(k) = 4\pi\rho(k) \tag{169}$$

The chemical potential is written by Thomas and Fermi as,

$$\mu = \epsilon_F(x) - e\varphi(x) \simeq \frac{\hbar^2}{2m}[3\pi^2 n(x)]^{2/3} - e\varphi(x)$$
$$\simeq \frac{\hbar^2}{2m}[3\pi^2 n_o]^{2/3} \tag{170}$$

This expression is valid for electrostatic potentials that vary slowly in comparison with the wave length of an electron. Expanding the above by a Taylor series we find,

$$\frac{d\epsilon_F}{dn_o}[n(x) - n_o] = e\varphi(x) \tag{171}$$

From last of (170) $d\epsilon_F/dn_o = 2\epsilon_F/3n_o$ so that

$$n(x) - n_o = \frac{e}{2} n_o \frac{3\varphi(x)}{\epsilon_F} \tag{172}$$

The left-hand side is the induced part of the concentration, thus the Fourier components of this equation are

$$\rho_{ind}(k) = -(3n_o e^2/2\epsilon_F)\varphi(k) \tag{173}$$

Using (169), this becomes

$$\rho_{ind}(k) = -(6n_o \pi e^2/\epsilon_F k^2)\rho(k) \tag{174}$$

and by the definition of the dielectric constant

$$\epsilon(k) = 1 - \frac{\rho_{ind}(k)}{\rho(k)} = 1 + \frac{k_s^2}{k^2} \tag{175}$$

so that the square of the screening wave vector is found to be

$$k_s^2 = 6\pi n_o e^2/\epsilon_F = 4(3/\pi)^{1/3} n_o^{1/3}/a_o = 4\pi e^2 \mathcal{D}(\epsilon_F) \tag{176}$$

where a_o is the Bohr radius and $\mathcal{D}(E_F)$ is the density of states for a free electron gas. The approximation (175) for $\epsilon(0, k)$ is called the Thomas-Fermi dielectric function and $1/k_s$ is the Thomas-Fermi screening length. For Cu with $n_o = 8.5 \times 10^{22}$ cm^{-3}, the screening length is 0.55 Å.

We now define the Debye screening length which measures the distance upto which the electric field penetrates a solid. In terms of density of states $\mathcal{D}(E_F)$ and the dielectric constant, ϵ this length is given by

$$l_D = \left(\frac{4\pi \mathcal{D}(F)e^2}{\epsilon}\right)^{-1/2} \tag{177}$$

Thus the Debye screening length given above differs from the Thomas-Fermi screening length only by the square root of the dielectric constant. Thus physically speaking the two lengths should be one and the same.

6.2 Ginzburg-Landau model for the electric field effect

The electric field, directed perpendicular to the surface of the superconductor increases the carrier density n within a thin layer of thickness of the order of the Debye screening length. As a result of application of the electric field, the surface density of states $\mathcal{D}(F)$ and the coupling constant of the electron-phonon interaction also increase. Usually the coherence length, ξ, is very large so that $l_D/\xi \ll 1$. Therefore, the shift in T_c upon the application of the electric field is very small. However in the modern superconductors ξ is small and l_D is large owing to small density of states and large dielectric constant $\epsilon \sim 20 - 30$, large electric field effects are expected.

The Ginzburg-Landau free energy G, in the inhomogeneous medium can be expressed as

$$G = \int \left[\frac{(B-H)^2}{8\pi} + N(o)\left\{ \left| \xi_j \left(\nabla - \frac{2ie}{c}A\right)_j \Delta \right|^2 \right. \right.$$
$$\left. \left. - \alpha|\Delta|^2 + \frac{\beta}{2}|\Delta|^4 \right\} \right] d^3r \tag{178}$$

where Δ is the order parameter which can be the gap of the superconductor. B and H are the magnetic induction and external magnetic field, respectively and the subscript $j = x, y, z$. The electric field E is perpendiuclar to the $y - z$ surface. It can be accounted by expanding α the coefficient of the $|\Delta|^2$ term in powers of $\delta n / n_o$, where n_o is the carrier density far from the surface and δn is the deviation of the charge density at the surface due to E,

$$\alpha = \tau + \frac{1}{g} \frac{\delta n}{n_o} \tag{179}$$

where $\tau = (T_c - T)/T_c$ and T_c is the transition temperature at $E = 0$. The superconducting coupling constant of the attractive interaction is g. The form of the expansion (179) is justified because $\Delta(r)$ is small compared with the Fermi energy. The number of electrons in the conduction band change because of the overlaping band structure.

In the Thomas-Fermi approximation we get

$$\frac{\delta n}{n_o} = \frac{E}{E^*} \exp(-x/l_D) \tag{180}$$

where $E^* = 4\pi e l_D n_o/\epsilon$. If Δ_o is the order parameter in absence of the electric field we can expand it as $\Delta = \Delta_o + \delta\Delta$. Near the vortex core, Δ_o is space dependent, vanishing at the centre of the vortex. If a vortex is located at (x_o, y_o) we can use the approximation,

$$|\Delta_o(x,y)|^2 = \frac{[(x - x_o)/\xi_\perp]^2 + [(y - y_o)\xi_\parallel]^2}{\{1 + [(x - x_o)/\xi_\perp]^2[(y - y_o)/\xi_\parallel]^2\}} \frac{\tau}{2\beta} \tag{181}$$

where $\xi_\parallel$ and $\xi_\perp$ are the components of the anisotropic coherence length. The correction to the Ginzburg-Landau free energy due to the application of the electric field E is found to be,

$$\begin{aligned}
\Delta G &= -\frac{N(o)}{g} \frac{E}{E^*} \int_o^\infty dx \int_{-\infty}^\infty dy \exp(-x/l_D) \\
&\times \left[|\Delta_o(x,y)|^2 - |\Delta_o(\infty)|^2 \right]
\end{aligned} \tag{182}$$

where $\Delta_o^2(\infty) = \tau/2\beta$ is the order parameter at $x = \infty$. Solving the integral in (182), we find

$$\Delta G = \frac{N(o)\tau\pi E}{2\beta g E^*} \xi_\parallel I(\xi_\perp/l_D) \tag{183}$$

where

$$I(a) = \int_o^\infty \frac{\exp(-xa/\xi_\perp)dx}{\sqrt{1 + [(x - x_o)/\xi_\perp]^2}} \tag{184}$$

The energy correction ΔG as a function of $x_o/\xi_\perp$ peaks at a particular value $x_o^m/\xi_\perp$ and decreases as a function of distance going to zero at large distances. The energy barrier for a vortex due to the electric field is $\Delta G(x_o^m)$. Since $\tau = (T_c - T)/T_c$, $\Delta G(x_o^m)$ varies linearly with temperature. For small $\xi_\perp/l_D \sim 1$ for the modern high temperature superconductors the barrier $\Delta G(x_o^m)$ becomes important.

        Lydia S. Lingam

Conclusions

The effect of electric fields on superconductors has been reviewed in this work. It is found that the quantized resistivity plays an important role in determining the current. We have considered the effect of quasiparticle scattering on the formation of bound pairs of electrons and hence on the current. Steps are found on the current voltage characteristics which are affected by the mean free path. Formation of hot spots has been noted. Charge-imbalance and non-equilibrium effects have been reviewed. Considerable structure in the current-voltage characteristics below the gap energy is caused by the Andreev scattering and by other phenomena which scatter at the microwave frequency. The effect of the electric dipole moment on the current has been calculated and the anisotropy in the resistivity as a function of gate voltage is assigned to this effect. The current-voltage relations have been corrected. The thermoelectric effect on the tunneling currents has been discussed. A Ginzburg-Landau model which incorporates the effect of the electric field in the free energy has been discussed. From this study it is quite clear that the study of the electric field effects in superconductors is important for the development of junctions and devices.

References

[1] L. Cooper, Phys. Rev. **104**, 1189 (1956).

[2] J. Bardeen, L. Cooper and J.R. Schrieffer, Phys. Rev. **108** 1175 (1957).

[3] B.D. Josephson, Phys. Lett. **1**, 251 (1962).

[4] J. Mannhart, D.G., Schlom, J.G. Bednorz and K.A. Muller, Phys. Rev. Lett. **67**, 2099 (1991).

[5] J. Mannhart, Mod. Phys. Lett. **B6**, 555 (1992).

[6] K.N. Shrivastava, J. Phys.: Condens. Matter **5**, L597-600 (1993).

[7] S.N. Artemenko and A.F. Volkov, Usp. Fiz. Nauk **128**, 3 (1979); Soviet Phys. Uspekhi **22**, 295 (1979).

[8] V.V. Ryazanov, V.V. Schmidt and L.A. Ermolaeva, J. Low Temp. Phys. **45**, 507 (1981).

[9] R.P. Huebener and J.R. Clem, Rev. Mod. Phys. **46**, 409 (1974).

[10] M. Tinkham, Rev. Mod. Phys. **46**, 587 (1974).

[11] P.E. Lindelof, Rep. Prog. Phys. **44**, 949 (1981).

[12] J.A. Pals, K. Weiss, P.M.T.M. van Attekum, R.E. Horstman and J. Wolter, Phys. Reports **89**, 324 (1982).

[13] I.L. Landau, Pisma Zh. Eksp. Teor. Fiz. **11**, 437 (1970); Sov. Phys. JETP Lett. **11**, 295 (1970).

[14] A.B. Pippard, J.G. Shepherd and D.A. Tindall, Proc. R. Soc. (London) A**324**, 17 (1971).

[15] G.L. Harding, A.B. Pippard and J.R. Tomlinson, Proc. R. Soc. (London) A**340**, 1 (1974).

[16] C.W.J. Beenakker, Phys. Rev. B**46**, 12841 (1992).

[17] J. Clarke, Phys. Rev. B**4**, 2963 (1971).

[18] J.M. Dupart, J. Rosenblatt and J. Baixeras, Phys. Rev. B**16**, 4815 (1977).

[19] T.Y. Hsiang, Phys. Rev. B**21**, 945 (1980).

[20] T.Y. Hsiang and J. Clarke, Phys. Rev. B**21**, 956 (1980).

[21] J.M. Suter, F. Rothen and L. Rinderer, J. Low Temp. Phys. **20**, 429 (1975).

[22] J.M. Suter and L. Rinderer, J. Low Temp. Phys. **31**, 33 (1978).

[23] P. Nedellec, L. Dumoulin and E. Guyon, J. Low Temp. Phys. **24**, 663 (1976).

[24] V.P. Galaiko, J. Low Temp. Phys. **26**, 483 (1977).

[25] T.M. Klapwijk, M. Sepers and J.E. Mooij, J. Low Temp. Phys. **27**, 801 (1977).

[26] S.N. Artemenko, A.F. Volkov and A.F. Sergeev, J. Low Temp. Phys. **44**, 405 (1981).

[27] B.I. Ivlev, N.B. Kopnin and C.J. Pettick, J. Low Temp. Phys. **41**, 297 (1980).

[28] M. Tinkham, Phys. Rev. B**6**, 1747 (1972).

[29] W.G. Jenks and L.R. Testardi, Phys. Rev. B**48**, 12993 (1993); (See, also [89]).

[30] W.J. Gallagher, A.W. Kleinsasser and S.I. Raider, IBM Tech. Discl. Bull. **32**, 394 (1989).

[31] A.F. Hebard, A.F. Firoy and R.H. Eick, IEEE Trans. Mag. **23**, 1279 (1987).

[32] A.F. Hebard and A.T. Fiory, Proc. Intl. Wksp. Novel Mechanisms of Superconductivity, Berkeley, CA, 1987, Plenum Press, N.Y.

[33] A.F. Fiory, A.F. Hebard, R.H. Eick, P.M. Mankiewich, R.E. Howard, and M.L. O'Malley, Phys. Rev. Lett. **65**, 3441 (1990).

[34] A.T. Fiory and A.F. Hebard, Physica B**135**, 124 (1985).

[35] J. Mannhart, A. Kleinsasser, J. Ströbel and A. Baratoff, (preprint); Physica C**216**, 401 (1993).

[36] J. Mannhart, D.G. Schlom, J.G. Bednorz and K.A. Muller, Physica C**185-189**, 1745 (1991).

 Lydia S. Lingam

[37] J. Mannhart, J.G. Bednorz, K.A. Muller and D.G. Schlom, Z. Phys. B. condensed Matter **83**, 307 (1991).

[38] J. Mannhart, J.G. Bednorz, K.A. Muller, D.G. Schlom and J. Ströbel, J. Alloy Comp. **195**, 519 (1993).

[39] J. Mannhart and A. Kleinsasser, Mat. Res. Soc. Symp. Proc. **275**, 549 (1992).

[40] J. Mannhart, J.G. Bednorz, A. Catana, Ch. Gerber and D.G. Schlom, NATO ASI course on "Materials and Crystallographic Aspects of High T_c Superconductivity" Erice, Italy, 1993, Kluwer Press.

[41] X.X. Xi, C. Doughty, A. Walkenhorst, C. Kwon, Q. Li and T. Venkatesan, Phys. Rev. Lett. **68**, 1240 (1992).

[42] Y. Gim, C. Doughty, X.X. Xi, A. Amar, T. Venkatesan and F.C. Wellstood, Appl. Phys. Lett. **62**, 3198 (1993).

[43] B. Yurke and G.P. Kochanski, Phys. Rev. **B41**, 8184 (1990).

[44] C. Kwon, Qi Li, X.X. Xi, S. Bhattacharya, C. Doughty, T. Venkatesan, H. Zhang, J.W. Lynn, J.L. Peng and Z.Y. Li, N.D. Speneer and K. Feldman, Appl. Phys. Lett. **62**, 1289 (1993).

[45] A.T. Findikoglu, C. Doughty, S.M. Anlage, Q. Li, X.X. Xi, and T. Venkatesan, Appl. Phys. Lett. **63**, 3215 (1993).

[46] S.N. Mao, X.X. Xi, Q. Li, I. Takeuchi, S. Bhattacharya, C. Kwon, C. Doughty, A. Walkenhorst, T. Venkatesan, C.B. Whan, J.L. Peng and R.L. Greene, Apply. Phys. Lett. **62**, 2425 (1993).

[47] S.B. Kaplan, C.C. Chi, D.N. Langenberg, J.J. Chang, S. Jafarey and D.J. Scalabino. Phys. Rev. **B14**, 4854 (1976).

[48] I. Schuller and K.E. Gray, Solid State Commun. **23**, 337 (1977).

[49] J. Clarke and J.L. Paterson, J. Low Temp. Phys. **15**, 491 (1974).

[50] C.J. Lambert, J. Phys. Condens. Matter **3**, 6579 (1991).

[51] R. Tidecks and J.D. Meyers, Z. Phys. **B32**, 363 (1979).

[52] B.I. Ivlev, N.B. Konin and C.J. Pethick, Zh. Eksp Teor. Fiz. **79**, 1017 (1980); Sov. Phys. JETP **52**, 516 (1980).

[53] Y. Krahenbu, J. Low Temp. Phys. **35**, 569 (1979).

[54] U. Schulz and R. Tidecks, Solid State Commun. **57**, 829 (1986).

[55] J.G. Shepherd and H. Small, Proc. R. Soc. **A326**, 421 (1972).

[56] H.W. Lean and J.R. Waldram, J. Phys. Condens. Matter **1**, 1285 (1989).

[57] H.W. Lean and J.R. Waldram, J. Phys. Condens. Matter 1, 1299 (1989).

[58] J.M. Warlaumont and R.A. Buhrman, IEEE. Trans. Mag. **MAG-15**, 570 (1979).

[59] G.J. Dolan and J.E. Lukens, IEEE-Trans. Mag. 13, 581 (1977).

[60] I.I. Eru, S.A. Peskovatskii and A.V. Poladich, Phys. Stat. Solidi A**53**. K185 (1979).

[61] W.J. Skocpol, M.R. Beasley and M. Tinkham, J. Appl. Phys. 45, 4054 (1974).

[62] S.N. Artemenko and A.F. Volkov, Zh. Eksp. Teor. Fiz. 72, 1018 (1977); Sov. Phys. JETP 45, 533 (1977).

[63] C.W.J. Beenakker, Phys. Rev. Lett. 67, 3836 (1991); 68, 1442 (E) (1992).

[64] A.F. Andreev, Zh. Eksp. Teor. Fiz. 46, 1823 (1964); [Sov. Phys. JETP 19, 1228 (1964)]; 51, 1510 (1966) [24, 1019 (1967)].

[65] G.E. Blonder, M. Tinkham and T.M. Klapwijk. Phys. Rev. B**25**, 4515 (1982).

[66] A.L. Shelankov, Fiz. Tverd. Tela 26, 1615 (1984); Sov. Phys. Solid State 26. 981 (1984).

[67] M. Büttiker, Phys. Rev. B**41**, 7906 (1990).

[68] A. Schmid and G. Schön, J. Low Temp. Phys. 20, 207 (1975).

[69] W.J. Skocpol, M.R. Beasley and M. Tinkham, J. Low Temp. Phys. 16, 145 (1974).

[70] T.W. Jing, Z.Z.Wang and N.P. Ong, Appl. Phys. Lett. 55, 1912 (1989).

[71] D.E. McCumber, J. Appl. Phys. 39, 3113 (1968).

[72] T.A. Fulton, Phys. Rev. B**7**, 1189 (1973).

[73] M. Tinkham and J. Clarke, Phys. Rev. Lett. 28, 1366 (1972); J. Clarke, Phys. Rev. Lett. 28, 1363 (1972).

[74] M.L. Yu and J.E. Mercereau, Phys. Rev. B**12**, 4909 (1975).

[75] T.R. Lemberger and J. Clarke, Phys. Rev. B**23**, 1088 (1981).

[76] T.R. Lemberger, Phys. Rev. B**24**, 4105 (1981).

[77] M.A. Peshkin and R.A. Buhrman, Phys. Rev. B**28**, 161 (1983).

[78] B.R. Fjordboge, P.E. Lindelof and J. Clarke, J. Low Temp. Phys. 44, 535 (1981).

[79] R. Gross, H. Seifert, R.P. Huebener and K. Yoshida, J. Low Temp. Phys. 54, 277 (1984).

[80] C.J. Pethick and H. Smith, J. Phys. C**13**, 6313 (1980).

[81] T.J. Rieger, D.J. Scalapino and J.E. Mercereau, Phys. Rev. Lett. **27**, 1787 (1971).

[82] T.M. Klapwijk, G.E. Blonder and M. Tinkham, Physica B**109,110**, 1657 (1982).

[83] S. Shapiro, Phys. Rev. Lett. **11**, 80 (1963).

[84] K.N. Shrivastava, solid State Commun. **68**, 259 (1988); **68**, 1019 (1988); Supercond. Sci. Technol. **4**, S430 (1991); Phys. Stat. Solidi B**164**, K51 (1991).

[85] J. Bardeen and J.L. Johnson, Phys. Rev. B**5**, 72 (1972).

[86] I.O. Kulik, Zh. Eksp. Teor. Fiz. **57**, 941 (1969); Sov. Phys. JETP **30**, 944 (1970); C. Ishii, Prog. Theoret. Phys. **44**, 1525 (1970).

[87] D.J. Thouless, Phys. Rev. **117**, 1256 (1960).

[88] R.E. Glover, III and M.D. Sherrill, Phys. Rev. Lett. **5**, 248 (1960).

[89] U. Kabasawa, K. Asano and T. Kobayashi, Jpn. J. Appl. Phys. **29**, L86 (1990).

[90] X.X. Xi and T. Venkatesan, APS News [3] **2**, 44 (1990).

[91] L.F. Rybalchenko, V.V. Fisun, N.L. Bobrov, I.K. Yanson A.V. Bondarenko and M.A. Obolenskii, Sov. J. Low Temp. Phys. **17**, 105 (1991).

[92] S.J. Rothman, J.L. Routhbort and J.E. Baker, Phys. Rev. B**40**, 8852 (1989).

[93] L.R. Testardi, W.G. Moulton, H. Mathias, H.K. Ng and C.M. Rey, Phys. Rev. B**37**, 2374 (1988).

[94] V. Muller, C. Hucho, D. Maurer, K. deGroot and K.H. Rieder, Physica **165**, 1271 (1990).

[95] S. Mihailovic and I. Poberaj, Physica C**185**, 781 (1992).

[96] S.K. Kurtz, A. Bhalla and L.E. Cross, Ferroelectrics **117**, 261 (1991).

[97] A. Bussman-Holder, A.R. Bishop, A. Migliori and Z. Fisk, Ferroelectrics **128**, 105 (1992).

[98] V. Muller, C. Hucho, K. DeGroot, D. Winau, D. Maurer and K.H. Rieder, Solid State Commun. **72**, 997 (1989).

[99] S.D. Conradson, I.D. Raistrick and A.R. Bishop, Science **248**, 1394 (1990).

[100] K.A. Muller, Z. Phys. B**80**, 193 (1990).

[101] R. Baetzold, Phys. Rev. B**38**, 11304 (1988).

[102] K.N. Shrivastava, J. Phys. C**20**, L789 (1987).

[103] K.N. Shrivastava, Solid State Commun. **85**, 227 (1993).

[104] K.N. Shrivastava, Phys. Rev. B41, 11168 (1990).

[105] A. Levy, J.P. Falck, M.A. Kastner, W.J. Gallagher, A. Gupta and A.W. Kleinsasser. J. Appl. Phys. 69, 4439 (1991).

[106] D.E. McCumber, J. Appl. Phys. 39 3113 (1968).

[107] G.A. Wiekinson, Phys. Rev. B4, 2174 (1971).

[108] J. Mannhart, D. Anselmetti, J.G. Bednorz, Ch. Gerber, K.A. Müller and D.G. Schlom. Supercond. Sci. Technol. 5, S125 (1992).

[109] K.N. Shrivastava, Phys. Stat. Solidi B153 547 (1989).

[110] K.N. Shrivastava, Phys. Rev. B41, 11168 (1990).

[111] R.H. Dee, and A.M. Guenault, Solid State Commun. 25, 353 (1978).

[112] V.L. Ginzburg, Supercond. Sci. Technol. 4, S1 (1991).

[113] J.E. Hirrch, Phys. Rev. Lett. 72, 558 (1994).

[114] S.J. Battersby and J.R. Waldram, J. Phys. F. 14, L109 (1984).

[115] D.J. Van Harlingen, J. Low Temp. Phys. 44, 163 (1981).

[116] D.J. Van Harlingen, Physica B109, 110, 1710 (1982).

[117] V.L. Ginzburg and G.F. Zharkov, Sov. Phys.-Uspekhi, 21, 381 (1978).

[118] B. Arfi, M. Bahlouli, C.J. Pethick and D. Pines, Phys. Rev. Lett. 60, 2206 (1988); Phys. Rev. B39, 8959 (1989).

[119] P.J. Hirschfeld, Phys. Rev. B37, 9331 (1988).

[120] J.E. Hirsch and F. Marsiglio, Phys. Rev. B39, 11515 (1989); F. Marsiglio and J.E. Hirsch, Phys. B41, 6435 (1990).

[121] C. Kittel, Introduction to Solid State Physics, Fifth Edition, Wiley, New York 1976.

[122] B. Ya Shapiro, Zh. Eksp. Teor. Fiz. 88, 1676 (1985), Sov. Phys. JETP 61, 998 (1985).

[123] B. Ya Shapiro, Phys. Lett. A105, 344 (1984).

[124] L. Burlachkov, I.B. Khalfin and B. Ya Shapiro, Phys. Rev. B48, 1156 (1993).

Normal State Resistivity of Highly Correlated $Bi_2Sr_2CaCu_2O_{8+x}$

Lydia S. Lingam and Keshav N. Shrivastava

School of Physics
University of Hyderabad
P.O. Central University
Hyderabad 500 134, India

Abstract .We have found that for the electric field larger than about half of the break down field, $E > \frac{1}{2}E_{BD}$, the measured resistivity varies as the square root of the electric field, $\rho \propto E^{1/2}$. The resistivity calculated for the field-driven movement of the mobile charge carriers in the normal state of a superconductor is found to vary as the square of the electric field, $\rho \propto E^2$. Thus the mechanism of the field driven movement of charge carriers is not in agreement with the experimental measurements of the normal state resistivity of $Bi_2Sr_2CaCu_2O_{8+x}$. We predict that the resistivity of a system with conduction electrons localised along a one-dimensional chain varies as the square root of the applied electric field. The mechanism of the field induced changes in the density of states of the mobile charge carriers along a one-dimensional chain is in agreement with the experimental measurements.

In this paper , we report our calculation of the resistivity as a function of electric field due to the field driven movement of the mobile charge carriers in the normal state of a superconductor. The superconducting state was given earlier [1]. We find that the resistivity owing to this mechanism varies as the square of the electric field. In order to compare our calculation with the experimental measurements [2] , we made an effort to plot the measured resistivity as a function of the applied electric field. We then discovered that for the electric field larger than about one half of the breakdown voltage, $E > \frac{1}{2}E_{BD}$, the measured resistivity varies as $E^{1/2}$. Thus the mechanism of field-driven movement of the mobile charge carriers in the normal state is not in agreement with the experimental data. We then calculated the resistivity of a system localized in one dimension which varies with the square root of the electric field. This mechanism based on the density of states is in agreement with the experimental measurements. The displacement vector in terms of the dielectric constant and the electric field is given by

$$\vec{D} = \epsilon \vec{E}, \tag{1}$$

     Lydia S. Lingam and Keshav N. Shrivastava

and hence the polarization is identified by

$$\vec{P} = \frac{\epsilon - 1}{4\pi}\vec{E}.$$

(2

The solid consists of positive and negative charge concentrations. When the electric field is applied the positively charged background is deformed so that, as an example, the charge density along x-direction is given by,

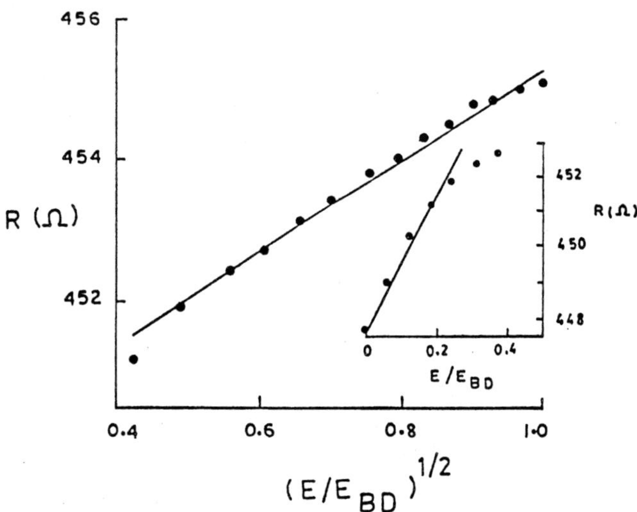

Fig.1:We have plotted the resistivity as a function of the square root of the electric field at a temperature of 100 K. The experimental measurements in $Bi_2Sr_2CaCu_2O_{8+x}$ are taken from Frey et. al [2]. We find that resistivity varies as $E^{1/2}$. For small fields, $E/E_{BD} \leq 0.2$, the resistivity shown in the inset is linear, $\rho \propto E$.

$$\rho^{+}(x) = n_0 e + \rho_{ext}\sin(k.x)$$

(3a)

Similarly, the negative charge density is given by

$$\rho^{-}(x) = -n_0 e + \rho_{ind}\sin(k.x)$$

(3b)

Here $\rho_{ext}(k)\sin(kx)$ gives an electrostatic field, that is, the external field applied to the electron gas and $\rho_{ind}(k)$ is the amplitude of the charge density variation induced in the electron gas. The Thomas-Fermi screening length is given by $1/k_s$. This length is related to the static dielectric function as

$$\epsilon(o, k) = 1 + \frac{k_s^2}{k^2}.$$

(4

Eliminating the dielectric constant from (2) and (4), we find that the polarization can be expressed as,

$$4\pi\vec{P} = \vec{E}\frac{k_s^2}{k^2}$$

(5

We define,

$$k_s = (3\pi^2 n_0)^{1/3} = \left(\frac{4\pi P k^2}{|E|}\right)^{1/2} \tag{6}$$

The current is given by

$$nev = \sigma F \tag{7}$$

where v is the velocity of the electrons and σ the conductivity, $\sigma = \frac{1}{R}$, which is the inverse of the resistivity. From (6) we can write the electron concentration as

$$n_0 = \frac{1}{3\pi^2}\left(\frac{4\pi P k^2}{\vec{E}}\right)^{3/2}, \tag{8}$$

so that from (7), the resistivity is found to become,

$$R = \frac{3\pi^2 E^{5/2}}{ev(4\pi P k^2)^{3/2}}. \tag{9}$$

In the electric field, the velocity of the electrons may be determined from the relation $\frac{1}{2}mv^2 = eE$ and substituted in (9) so that for $k^{-3} = Ad$ the resistivity becomes

$$R = \frac{3\pi^2 m^{1/2} E^2 Ad}{\sqrt{2}(4\pi e P)^{3/2}}. \tag{10}$$

which varies as E^2. In a previous paper [1] we found that in the energy of the system, one term is linear in E and another term is a small contribution proportional to E^2. In the normal state, this energy gives rise to a current determined by $e\rho J$ so that ρ may vary with one term proportional to E and another term proportional to E^2.

We consider the electrons in a Fermi sphere of volume $\frac{4}{3}\pi k_F^3$. The volume element in the k-space is $(2\pi/L)^3$. The total number of orbitals is $N = k_F^3 L^3/(3\pi^2)$. Therefore, the Fermi wave vector is $k_F = \left(\frac{3\pi^2 N}{L^3}\right)^{1/3}$ and the Fermi energy is $\epsilon_F = \frac{\hbar^2}{2m}\left(\frac{3\pi^2 N}{L^3}\right)^{2/3}$. The density of states for a three dimensional solid is given by

$$\frac{dN}{d\epsilon} = \frac{L^3}{2\pi^2}\left(\frac{2m}{\hbar^2}\right)^{3/2}\epsilon^{1/2} \tag{11}$$

Similarly for a two dimensional solid, $N = k_F^2 L^2/2\pi$ so that the Fermi wave vector becomes $k_F = (2\pi N/L^2)^{1/2}$, and hence the density of states is given by,

$$\frac{dN}{d\epsilon} = \frac{L^2}{2\pi}\frac{2m}{\hbar^2} \tag{12}$$

which is a constant independent of the carrier energy, ϵ.

In one dimension, $k_F = \pi N/L$ so that $\epsilon = \frac{\hbar^2}{2m}\left(\frac{\pi N}{L}\right)^2$. The number of carriers in a line becomes

$$N = \frac{L}{\pi}\left(\frac{2m\epsilon}{\hbar^2}\right)^{1/2} \tag{13}$$

and the density of states becomes

$$\frac{dN}{d\epsilon} = \frac{L}{2\pi}\left(\frac{2m}{\hbar^2}\right)^{1/2}\epsilon^{-1/2}. \tag{14}$$

Thus for d= 3, 2 and 1 we find that the density of states varies with energy as $\epsilon^{1/2}, \epsilon^0$ and $\epsilon^{-1/2}$ respectively. It has been reported by Klein et al[3] that the conductivity varies as the density of states, according to ,

$$\frac{\delta G}{G} = \int \frac{\delta N}{N_o} \frac{d}{d\epsilon}[-f(\epsilon + eV)]d\epsilon \tag{15}$$

where δG is the change in the conductivity, G, upon the application of the electric voltage, V. Thus the conductivity is predicted to vary as $V^{1/2}, V^0$ and $V^{-1/2}$ for the dimensionality d $= 3$, 2 and 1, respectively. For conduction along a chain the conductivity is thus expected to vary as $V^{-1/2}$ and hence the resistivity,

$$\rho \propto E^{1/2} \tag{16}$$

where E is the applied electric field. For a d-dimensional system, the resistivity is proportional to $E^{(1-d/2)}$ so that for one dimensional conduction in layered systems, where the conduction is along the c-axis, the resistivity varies as the square root of the voltage or the applied electric field.

The resistivity of $Bi_2Sr_2CaCu_2O_{8+x}$ as a function of electric field has been measured by Frey et al [2]. For small values of the electric field, E, the curve is linear as expected. For slightly larger values of E, near $\frac{E}{E_{BD}} \sim 0.2$, there may be a small term proportional to E^2. For the field larger than about half of the breakdown voltage, the variation of resistivity as a function of the electric field is shown in Fig 1. It is quite clear that we find that $R \simeq E^{1/2}$ which agrees with the resistivity of a system of electrons, the density of states of which is localized in one dimension along a chain.

In conclusion, we find that the experimental variation of resistivity as a function of increasing electric field shows three separate mechanisms. (a) For small electric fields, the resistivity is linear function of electric field. (b) For intermediate electric fields, $E \simeq \frac{1}{5}E_{BD}$, the resistivity has a small contribution varying as E^2 and (c) for $E > \frac{1}{2}E_{BD}$, the resistivity varies as $E^{1/2}$. The first two findings are in agreement with the theory and the third mechanism, $\rho \propto E^{1/2}$, is new and agrees with that of a localized system in one dimensional chain.

References

[1] K.N. Shrivastava , J.Phys.Condens.Matter 5(1993) L 597.

[2] T. Frey , J. Mannhart , J.G. Bednorz and E.J. Williams , Phys. Rev. B51(1995) 3257.

[3] T. Klein , O.G. Symko , D.N. Davydov and A.G.M. Jansen , Phys. Rev. Lett. 74(1995)365

Carbon Superconductors

N. Murali Krishna and Keshav N. Shrivastava

School of Physics
University of Hyderabad
P.O. Central University
Hyderabad – 500 046, India

Abstract Several years work on the molecule C_{60} has been reviewed and all its properties have been studied from the literature. An effort is made to understand the method of preparation, electronic energy levels, vibrations and rotations. The superconducting compounds $K_x C_{60}$, $Rb_x C_{60}$, $Cs_x C_{60}$, $Ba_6 C_{60}$ and their combinations such as $Cs_2 Rb_1 C_{60}$ are studied and their critical temperatures are correlated using ionic polarizabilities and the electron-phonon interaction.

The electron-phonon interaction has been calculated and the attractive potential is arrived at using the normal modes of oscillations. It is shown as to how the critical temperature depends on the unit cell size. This calculation automatically introduces the layer structure in the determination of the transition temperature. We are able to understand the variation of the transition temperatures of all the C_{60}-alkali metal superconductors.

The ESR and NMR results are also have been reviewed and studied using the coherence length in determining the line width.

1.Introduction

One of the forms of carbon molecule, C_{60}, has sixty atoms per molecule. The size of about 7.1Å indicates that this cannot be a linear molecule. It is not a planar molecule because the square root of 60 is not an integer. A ring structure is also not consistent with the known size of the molecule. Therefore it is thought that the sheet winds into a spherical structure. It so happens that a model of 20 hexagons and 12 pentagons forms a football with 60 apex points so that it can provide a first working model for the C_{60} molecule. For a long time, it was thought that the spherical model as shown in Fig.1.1, of C_{60} was correct and so the molecule was named as a fullerene after the name of Buckminster Fuller who had solved the problem of stability of sticks and balls forming a sphere which was used as a design for the U.S. pavilian in expo '67 in Montreal.

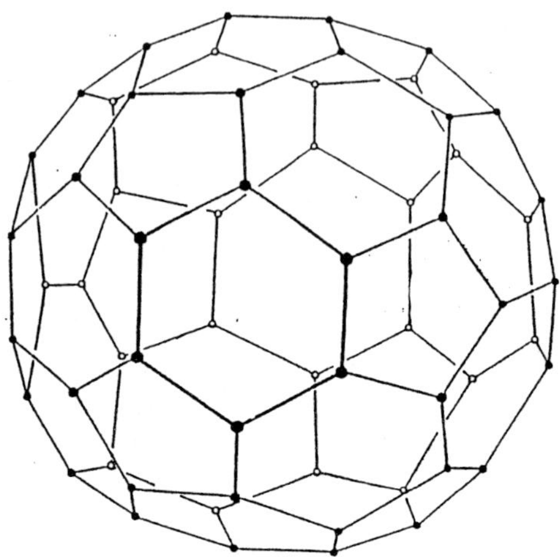

Fig.1.1. C_{60} molecule

Several authors have calculated the energy of a system of particles on the surface of a sphere to find the stability criteria [1.1-1.11]. The spherical bubbles surely satisfy the stability requirements but are not known to satisfy the crystallographic criterion of periodic symmetry.

The clusters of carbon atoms upto fifteen atoms were seen quite early [1.12-1.15] but C_{60} appears to be particularly stable. Ions of various charged states $C_{60}, C_{60}^{\pm}, C_{6}^{\pm}$ have been published [1.16-1.20]. The scanning tunneling microscope pictures [1.21-1.28] indicate the spherical nature of C_{60}.

The C_{60} forms compounds [1.29-1.33] such as $C_{60}O, C_{70}O, PtC_{60}$ and OsC_{60}. The metallic compounds [1.34,1.35] FeC_{60}^{+} and CrC_{60}^{-} are also known. $C_{60}La$ has been studied [1.36] for its oscillations and $C_{60}H_{36}^{-}$ has also been found [1.37].

Many non-carbon clusters are also known. For example, clusters of cadmium [1.3-1.39] and lead [1.38] sulphides, CdS, PbS, clusters of CH_2 molecules [1.40], (CH_2), clusters of water molecules [1.41], $(H_2O)_{57}, (H_2O)_{60}$, clusters of molecular nitrogen [1.42] $(N_2)_n$, and its oxide [1.43], $NO_2^{\pm}$, boron and hydrogen compound clusters [1.44]. $B_{32}H_{32}^{2-}$ indium phosphide [1.45] clusters, and clusters of metals such as nidoium [1.46], aluminium [1.47], antimony [1.48], bismuth [1.48] and selenium [1.49].

In this work, an effort is made to study the existing literature on C_{60}, its method of preparation, characterization and its normal state. Upon doping C_{60} becomes super conducting. The correlation of superconducting transition temperature with atomic weight of the dopant is also studied [1.50]. In the case of K, Rb and Cs doped C_{60}, the T_c increases with the size effect of the dopant. We also interpret some electron spin resonance spectra which show that at liquid nitrogen temperatures the normal state of C_{60} is insulating. The insulator to the superconducting transition is interesting. The variation of T_c with atomic weight shows that electron-phonon interaction is important and hence that pairing of electrons is caused by a mechanism not very different from BCS theory. As a function of concentration of K in $K_x C_{60}$, the metallic region is very small. In BCS theory, only a metal to the superconductings transition is allowed with the insulating region playing no role in the pairing process. We find that the polarizability of the system plays an important role in determining the super conducting critical temperature.

1.1.Preparation of C_{60}

A simple method of preparation of C_{60} compound is described by Kratschmer et al., [1.51]. Figure 1.2 shows the steps in the syntehsis and extraction of solid C_{60}. Rods of spectrographic grade graphite are butted together, and a high current, of the order of 100 amps, is passed through the rods. Both ac and dc currents have been successful. The commonly used rods are about 6 mm in diameter. Carbon vapourizes in the vicinity of the contact, producing a carbon plasma, that condenses into a graphite soot. This soot collects on the surface provided or on the walls of the chamber. A helium gas pressure of about 100 torr is maintained. Recent reports of various groups show that the yield of C_{60} and C_{70} at this pressure of helium gas is as much as 40%. The resulting black soot is gently scraped from the collecting surfaces inside the evoparation chamber and is dispersed in benzyene.

The material giving raise to the spectral features attributed to C_{60} dissolves to produce a wine-red to brown liquid, depending on the concentration. The liquid is then separated from the soot by filteration or by spinning it down in a centrifuge and dried using gentle heat, leaving a residue of dark brown to black crystalline material. An alternative concentration procedure is to heat the soot to 400°C in vacuum or in inert atmosphere, thus subliming the C_{60} out of soot. Some hydrocarbon contaminants which are present in the soot can be effectively removed by an ether wash before the concentration process.

The C_{60} and C_{70} compounds may be separated by liquid chromatography. The colour of pure C_{60} solution is magenta, while that of pure C_{70} solution is orange. In this method, fullerenes other than C_{60} are also present in large quantity. The preparation of fullerenes C_{60} to C_{266} in very high yield by a plasma discharge is reported by Parker et al., [1.52]. Hexane appears to limit the amount of higher fullerenes. Heptane extraction following Hexane extraction appears to preferentially extract C_{60}, C_{70} and C_{84}. Several important papers [1.52-1.61] have described methods which deal with fullerenes. The C_{60} is also found as a TNT product, [1.62]. The solubility of C_{60} and C_{70} with characterization has been described by Ajie et al., [1.63].

Single crystals of C_{60} of size- 100 μm were grown [1.64] by sublimation of C_{60} powder in vacuum for 6 to 24 hours. very thin film of C_{60} sublimed in ultrahigh vacuum on to a freshly cleared mica substrate has been studied by helium - atom scattering [1.65]. The samples of condensible compounds and soot from hydrocarbon combustion flames yields [1.66, 1.67] about 0.3%. C_{60}/C_{70} of the initial hydrocarbon [1.68]. The laser method [1.69-1.74] produces positively charged ions C_{60}^{+}.

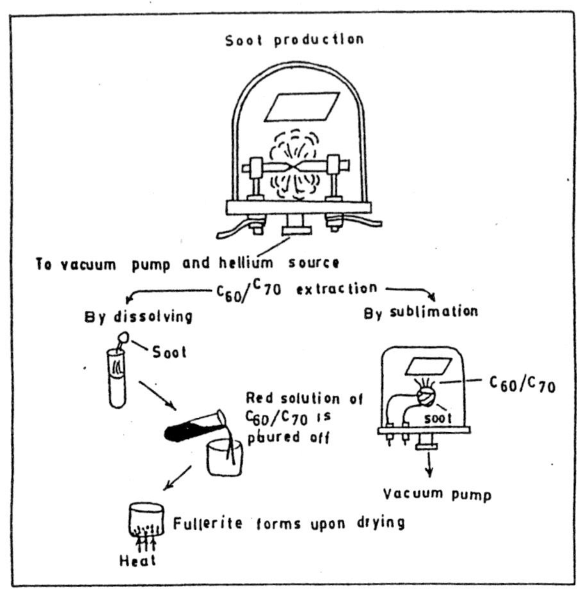

Fig.1.2: Systhesis of C_{60}

1.2. Properties of C_{60}

Density	1.70 g/cm³
Crystal Structure	F.C.C.
Nearest neighbour distance	10.04Å
Cage diameter	7.1Å
Lattice constant	14.198Å
Index of refraction	2.2 at 630 mm of wave length
Infra-red active modes	1429, 1183, 577, 528 cm⁻¹
Bulk modulus	18 giga pascals
Ionisation potential	7.6 eV
Cohesive energy	
Per C_{60} molecule	1.5eV
Per atom	7.4eV
Electrical conductivity	Non conductor
Electron band gap	1.5eV
Effective mass of conduction band electron	1.3 m_e
Superconducting T_c	
K_3C_{60}	18 K
Rb_3C_{60}	28 K
Cs_2RbC_{60}	33 K
Ba_6C_{60}	7 K

1.3. The electronic energy levels of C_{60}

The electronic energy levels of C_{60} have been calculated by many workers [1.75-1.126]. Manousakis [1.127] has calculated the electronic structure of C_{60} with in the tight binding approximation. The hybridization between the π-electronic atomic orbitals is described by the hamiltonian

$$H = - \sum_{<ij>,\sigma} t_{ij} \left(C_{i\sigma}^{\dagger} C_{j\sigma} + h.c. \right)$$

where t_{ij} is the carbon - to- carbon hopping matrix element and $C_{i\sigma}^{\dagger}$ is the operator for the creation of one electron at the i-th site with spin σ. The electronic wave function in the tight binding approximation is

$$|\psi_{m,\sigma} > = \frac{1}{\sqrt{5}} \sum_{n=1}^{5} \exp\left(\frac{2\pi imn}{5}\right) |n, \sigma >$$

where $|n, \sigma >$ describes an electron state which is the linear combination,

$$|n, \sigma > = \sum_{i=1}^{12} a_i C_{i\sigma}^{\dagger} |0 >$$

created from vaccum. Here sum is over all 12 atomic orbitals localised at site i of the n-th slice of the icosahedron. The possible values are $m = 0, \pm 1, \pm 2$.

For each value of m, there are 12 eigenstates, amongst which, some are degenerate. We consider the i-th atom within the first slice and the one located diameterically apposite on the icosahedron. We map the second atom by $\frac{4\pi}{5}$ rotation to a third atom which is its equivalent with in the first slice. If the difference between the phase φ_i of the i-th atom, $a_i = \sqrt{\rho_i} \exp[i\phi_i]$ and the phase of the 3rd atom including the rotation angle of $\frac{4\pi}{5}$ is a multiple of 2π, the state has positive (even) parity. If the difference in phase is an odd multiple of π, the state has negative (odd) parity.

It is found that the ground state of the dimerized state is singly degenerate, while the fermi level for 60 electrons is five fold degenerate. In the case of negative ion C_{60}^{-}, there is a three fold degenerate level just above zero with $m = 0, \pm 1$. The difference in energy between C_{60}^{-} and C_{60} is 0.75659 t.

1.4. Vibrations and Rotations

The vibration and rotations of C_{60} have been studied in detail [1.128-1.133]. The John-teller destortions from spherical symmetry are indicated [1.134-1.137]. The importance of electron phonon interaction in C_{60} [1.138] has been investigated. The low temperature simple cubic structure transforms to a face centered cubic structure at 249 K. The C_{60} is rotating above 249K and frozen below this temperature. The Raman bands [1.139] of C_{60} and C_{70} have also been interpreted in terms of John-Teller oscillations.

1.5.Astrophysical applications

Positively charged ions are found in the upper atmosphere. It is expected that interstellar space has positive ions. Therefore C_{60}^+ and C_{70}^+ have become interesting from the view point of astrophysical applications [1.140-1.155]. Webster [1.155] has made an effort to understand the infra-red from $C_{60}H_{60}$. There is some effort to find C_{60} in geological samples also [1.156].

1.6.Normal state of C_{60}

The energy levels of C_{60} have become available through the optical studies [1.157-1.166], electron energy loss [1.164] and the photoacoustic attenuation. There are a large number of studies relating C_{60} to $C_n(n \neq 60)$ or other clusters [1.167-1.200]. Neutron scattering [1.201] at this time is not very conclusive. There is some effort to put La atoms [1.202] inside the carbon cage. ESR results show that [1.203-1.205] the system is insulating and non-metallic.

2.Synthesis of Alkali fulleride Superconductors

Carbon C_{60} compound is found to be insulating and non-metallic. But the Bell labs group has first announced that 'upon exposing C_{60} films to potassium vapour, the electrical conductivity rose by several orders of magnitude'. They have also made detailed comments over the variation of the conductivity of potassium doped C_{60} compounds with the amount of potassium dopant. Shortly after this finding, Hebard et al [2.1], the Bell labs group announced that the potassium doped C_{60} films were superconducting with a transition temperature of 18K. They based this claim on a measured transition of the electrical resistance to zero and an observation of the Meissner effect - the expulsion of the magnetic field from the sample. A group at the Univeristy of California, Los Angeles, confirmed the results [2.2] and reported the optimum potassium concentration to be K_3C_{60}, corresponding to one potassium atom sitting in each of the tetrahedral and each of the octahedral interstitial sites that form when the pseudospheres of C_{60} arrange themselves in an f.c.c. lattice. After this, in a short span of time, a transition temperature of 28 K was reported for C_{60} doped with rubidium. These experiments have established solid C_{60} as the first three-dimentional organic supercondcutor.

2.1.Synthesis of Alkali-Fullerides

A number of groups have worked out the synthesis of alkali doped fulleride compounds in different ways [2.2-2.5].

The synthesis of K_3C_{60} as reported by Holczer et. al. [2.2] is as follows.

"In the mixing stage, a given quantity of potassium is sealed with C_{60} under vacuum in a pyrex tube of diameter of about 5-9 mm. This tube will be heated for 20 to 24 hours at a temperature of about $200^\circ C$ and as a result of which K appears to be completely absorbed by C_{60} powder. The material will be collected, put in a capillary tube of about 1.5 mm in diameter and sealed under 1 bar pressure of He gas. At this stage, the shielding diamagnetism measurements were made. The results were given elsewhere [2.2].

In the second state, which is called as the diffusion stage, the sample which is in the capillary tube is heated again for about 22 hours at a temperature of $200°C$ and again the shielding diamagnetism measurements were repeated. In the 3rd stage, called as the relaxation stage, the sample was a gain heated for about 6-8 hours, but this time at $250°C$ and it was observed after this, that an equilibrium stage has been achieved. It was also observed that no further changes in the shielding diamagnetism measurements were observed upon a subsequent heating of the smaple."

Another method of synthesis as reported by R.C. Haddon et. al. [2.5] is also reporetd to be giving good yield of K_3C_{60} compound which was found to be superconducting with $T_c = 18K$.

2.2. Superconductivity in Doped Carbon C_{60}

The samples of C_{60} and C_{70} are found to be insulating. Upon doping with alkali metals, the resulting material becomes [2.5] conducting for a certain range of the concentration of the dopant. The material K_3C_{60} is superconducting [2.1] with $T_c = 18K$. The resistivity of a film of thickness $960Å$ goes to zero upon cooling. The on set temperature is about 18K with the resistivity going to zero at about 8 K. The suseptibility becomes negative below T_c as it should for a perfect diamagnet. The author [2.1] claimed to find well defined Meissner effect. However, later studies [2.6] showed that it was a type II superconductor with Abrikosov phases. At $T = 0$, $H_{c1}(0) = 132$ Oe and $H_{c2}(0) = 49 \times 10^4$ Oe are the lower and upper critical fields. For $H_{c1} < H < H_{c2}$ the system shows a mixed state with zero resistivity. For K_3C_{60}, the critical fields are expressed in length units, with the London penetration depth, $\lambda_L = 2400Å$ and the super conducting coherence length, $\xi = 26Å$. In Rb_3C_{60} the critical temperature is $T_c \sim 29.6K$ with $H_{c1}(0) \sim 260$ Oe and $H_{c2}(0) = 78 \times 10^4$ Oe.

A scanning tunneling micro scope study [2.7] of Rb_3C_{60} gave the super conducting gap energy $\Delta = 6.6 \pm 0.4$ meV at 4.2K, corresponding to a reduced energy gap to T_c ratio of $\frac{2\Delta}{K_B T_c} \simeq 5.3$. The BCS value of this ratio is expected to be about 3.53. Since the integration of the phase space in the real material differs from that of simple three dimentional space, the value of 5.3 may be considered to be understandably in agreement with 3.53 in the ideal material.

The linear compressibility $d(\ln a)/dp$ of C_{60} is about the same as the interlayer compressibility of graphite, consistent with Van-der-Walls intermolecular bonding. The volume compressibility [2.8] is $\frac{-d(\ln V)}{dp} = 7.0 \pm 1 \times 10^{-12} \, cm^2/dyne$. This means that C_{60} is 40 times more compressible than diamond and 3 times more compressible than grpahite. The isothermal bulk modulus is [2.9] $K_o = 18.1 \pm 1.8 GPa$ and $dK_o/dp = 5.7 \pm 0.6$. The pressure dependence of the superconducting transition temperature of $K_x C_{60}$ determined using solid helium pressure technique is [2.10] found to be $dT_c/dp = -0.63 \pm 0.08 \, K/Kbar$. Part of the large values found [2.11] for K_3C_{60} and Rb_3C_{60} is caused by [2.12,2.13] the higher compressibility of the substances and part due to K_F.

The K_3C_{60} compound has a [2.14] face-centered cubic structure with $a = 14.24 \pm 0.01Å$ and $T_c \simeq 18°K$. The Rb and Cs combinations have larger transition temperatures [2.15,2.16] Rb_3C_{60} has $T_c = 29K$, $CsRb_2C_{60}$ has $T_c = 31K$ and $Cs_2Rb_1C_{60}$ has $T_c = 33K$. The compound A_6C_{60} (A=alkali metal) is a body centered cubic and A_4C_{60} is a body centered tetragonal but these structures are not superconducting [2.17], although, T_c seems to increase with increasing cell size [2.18]. Various combinations of Tl and Hg were attempted [2.19] but the largest T_c in $(Rb_xK_{1-x})_3C_{60}$ was found to be only about 29K. The micro wave absorption in these materials show that there are weak links [2.20-2.25] which may be caused by Hydrogen content [2.26,2.27].

Although $T_c \simeq 41.5K$ is found [2.28] in $Rb_{2.7}Tl_{2.2}C_{60}$ it may be noted that the micro wave absorption can be detected even if only a small part of the sample is superconducting. However, this finding was proved to be wrong in later studies.

2.3. Band Structure of Superconducting and Insulating Alkali Fullerides

To understand the phenomena of conductivity in crystalline solids, we should first know its electronic structure. The C_{60} molecule can be thought of as a "spherical porcupine" in which 60 electrons occupy π orbitals that have mostly P-like character and project radially from the surface of the molecule. Hückel calculations for isolated C_{60} molecule [2.29] show the relative positioning in energy of these molecular orbitals. The π electron pair up to fill 30 energy states. The degeneracy of the angular momentum $L = 5$ level, which contains 11 states, is removed by the icosahedral symmetry of the molecule, resulting in 3 separate levels: the h_u, or highest occupied molecular orbital (HOMO), containing 10 electrons, the triply degenerate t_{1u} lowest unoccupied molecular orbital (LUMO), which can accommodate 6 electrons; and a similar triply degenerate t_{2u} orbital (the LUMO+1) at higher energy.

The explanation, then, for the change in conductivity with doping is that the t_{1u} LUMO fills with electrons in proportion to the number of intercalated alkali atoms. Thus for a symmetric band at half filling (three electrons in the band), the density of states at fermi level N_F is maximum and the conductivity is also maximum. However, when the band is completely filled, with six electrons, there are no empty states into which these electrons can scatter and the material becomes insulator again, as it was when the band is empty.

To accommodate additional alkali atoms, the $f.c.c.$ structure with a half filled band and stoichiometry A_3C_{60} transforms to a more open structure which has been identified in crystalline powder samples by X-ray diffraction [2.30] to be a body centered cubic. As already mentioned earlier, experiments have identified a body-centered tetragonal phase with stoichiometry A_4C_{60}; $[A = K, Rb, or\ Cs]$. Thus attempts to fill the conduction band of A_xC_{60} beyond half filling $(x = 3)$ result in insulating $b.c.t.$ and $b.c.c.$ phase that can incorporate, four and six alkali atoms per C_{60}, respectively.

Also, the first principle orthogonalized linear combination of atomic orbitals in the local density approximation (LDA) has been used [2.31, 2. 32] to study the electronic structure of K_xC_{60} with $f.c.c.$ lattice for x=1 to 3 and a $b.c.c.$ lattice for $x = 6$. The pure C_{60} electronic structure has been calculated [2.33] in the tight binding approximation. First principle electronic structure calculations for K_6C_{60} which does not have a superconducting phase has been performed by Erwin and Pederson [2.34]. It is found [2.35] that K atoms are fully ionised and C_{60} is stabilized in an unusual high charge state with six additional electrons. Several $XPS, XAFS$ studies have appeared [2.36-2.40]. The is a strong coulomb interaction and the bands are quite complicated. It is not clear as to where the attraction between electrons can come about. Most of the referred publications are consistent with our discussion [2.41-2.50].

3. Electron Phonon Interaction

The phenomena of superconductivity in the alkaline (A=K, Rb, Cs) doped fullerene A_3C_{60} was explained by many a number of groups in different ways in which some are qualitative and some are semi-quantitative. We had a comparative study of some appealing models which accounted satisfactorily, to a suitable extent, to the observed high T_c of alkaline doped fullerene A_3C_{60} compounds.

The first question we have to answer, at the outset, was whether this superconductivity which was observed in A_3C_{60} compounds is solely due to the conventional phonon mechanism, and if so, which phonons are responsible for the relatively high T_c of these compounds.

It is interesting in this context to note that Holczer et. al demonstrated that K-doped C_{60} has only a single stable superconducting phase, K_3C_{60}, with $f.c.c.$ structure.

Based on the electronic structure of K_3C_{60}, we estimate various contributions to the electron-phonon interaction. If we model the system by a negative U Hubbard model and study the stability criteria of the charge density wave (CDW) and the superconducting states (SC) at the half filling, we found that SC is more stable than CDW.

To start with, if a neutral C_{60} molecule is considered, each C atom will have 4 outer electrons. Out of these 4 electrons three form strong σ-bonds are so we need not consider their relevancy at low temperatures. The remaining 60 π-electrons are well described by a tight binding model. As already described in the previous chapter, the LUMO state is a three-fold degenerate t_{1u} state, separated from the HOMO state by $\sim$ 2eV gap. If the C_{60} molecule were undoped, the inter-molecular interaction will be a weak Van-der-Waals interaction and the gap remains, to keep the molecule as an insulator. But, doping with alkali atom introduces additional electrons and for K_3C_{60}, the average valence is C_{60}^{3-}. By treating C_{60} on the whole as a unit and calculating the effective intra-molecular interaction U_{eff}, which is caused by the internal degrees of freedom, Zhang et. al., [3.1] proved that U_{eff} (dimer.)=-0.026 eV in the c-c dimerization mode and the attractive U_{eff} is $\sim$ 10 meV from John-Teller distortion. They [3.1] also claim that the K^+-ion optic phonons couple strongly to carriers on C_{60}-molecules and can lead to an effective attraction at low energies. The above conclusion can be drawn by working on the polyacetylene model of

Su-Schrieffer-Heeger [3.2] for a single C_{60} molecule.

$$H_{\text{single}\,C_{60}\,\text{molecule}} = -\sum_{(n+1,n)\sigma} \left(t_{n+1,n} a^\dagger_{n+1,\sigma} a_{n\sigma} + h.c. \right)$$

$$+ \sum_{(n+1,n)} \frac{1}{2} K_c \left(u_{n+1} - u_n \right)^2 . \tag{3.1}$$

where

$$t_{n+1,n} = t_o - \alpha(u_{n+1} - u_n) \tag{3.2}$$

The summation is over all the $c - c$ bonds within one molecule. $a_{i\sigma}$ is the annihilation operator for $C\pi$-electron. u_{n+1} is the displacement of $(n + 1)$-th carbon along the bond connecting two pentagons. t_o is the electron hopping integral without the displacement, K_c is the $c - c$ Spring constant and α is the electron-phonon interaction.

The displacement is chosen as $\pm u_o$, which implies that the 60 bonds forming the pentagons are long and the other 30 are short. The above Hamiltonian is diagonalised numerically by using the parameters of polyacetylene viz., $t_o = 2.5eV, \alpha = 4.1eV/\mathring{A}$ and $K_c = 21eV/\mathring{A}^2$, to find the optimal values of u_o for C_{60}^n. [see Table 3.1]. The results are in good agreement with the experimental data.

TABLE - 3.1

C_{60}^{n-}	C_{60}^{-}	C_{60}^{-}	C_{60}^{2-}	C_{60}^{3-}	C_{60}^{4-}	C_{60}^{5-}	C_{60}^{6-}
u_o (Å)	0.0145	0.0120	0.0095	0.0070	0.0045	0.0020	$<5\times10^{-5}$
$\Delta\Phi_n$ (eV)	-0.4405	-0.3018	-0.1895	-0.1033	-0.0432	-0.0089	--

Let us consider the Hamiltonian of K_3C_{60} [3.3] in a simplified model of a single half filled band attractive Hubbard model within mean field theory.

The Hamiltonian is

$$H = -t \sum_{(i,j,\sigma)} \left(C_{i\sigma}^+ C_{j\sigma} + h.c.\right) + U \sum n_{i\uparrow}n_{i\downarrow} \tag{3.3}$$

where the summation is carried over all the nearest-neighbors $(n.n.)$ pairs in a $f.c.c$ lattice $n_{i\sigma} = C_{i\sigma}^+ C_{i\sigma}$

Decoupling the U term in (3.3) we obtain the energy per electrons.

$$E_{CDW} = \frac{2}{N} \sum_{\bar{k},s=\pm1}^{occ} \epsilon_{CDW}(\bar{k},s) + \frac{u}{4}(1 - \delta^2) \tag{3.4}$$

where δ is the order parameter, with $< n_i >= 1 \pm \delta$, with

$$\begin{aligned}
\epsilon_{CDW}(\bar{k},s) &= \gamma_k t + S\sqrt{t^2\beta_k^2 + U^2\delta^2/4}, \\
\gamma_k &= = 4\cos(k_x a/2)\cos(k_y a/2) \quad \text{and} \\
\beta_k &= -4\{cos(k_x a/2) + \cos(k_y a/2)\}\cos(k_z a/2)
\end{aligned}$$

In (3.4), the summation is carried over $\frac{N}{2}$ Lowest energy states $(\bar{k},s)$, with N being the electron number.

The optimal value of E_{CDW} can be obtained by

$$\frac{\partial E_{CDW}}{\partial \delta} = 0$$

To study the super conducting phase, we decoupled

$$n_{i\uparrow}n_{i\downarrow} = (\Delta C_{i\downarrow}C_{i\uparrow} + h.c.) + \sum_\sigma n_{i\sigma}/2 - (\frac{1}{4} + \Delta^2)$$

where $\Delta = (C_{i\uparrow}^\dagger C_{i\downarrow}^\dagger)$ is the real SC order parameter

$$\Longrightarrow E_{SC} = -\frac{1}{N}\sum_{\bar{k}} \epsilon_{SC}(\bar{k}) + (\frac{1}{4} - \Delta^2)U.$$

with

$$\epsilon_{SC}(\bar{k}) = \sqrt{(U\Delta)^2 + (\epsilon_o(\bar{k}) - \mu)^2}$$

and

$$\epsilon_o(\bar{k}) = t(\gamma_k + \beta_k) \text{ is the band dispersion}$$

Here, the summation is carried over the entire Brillouin Zone; Δ and μ, the chemical potential, are given by

$$\frac{\partial E_{SC}}{\partial \Delta} = 0 \text{ and } \frac{\partial E_{SC}}{\partial \mu} = 0 \text{ , which are the self consistency equations.}$$

In can be concluded from the above observations that for all $U < 0$, SC is more stable than CDW and the results are plotted on the graph is shown in Fig.3.1. This is a consequence of $f.c.c.$ lattice, which is consistent with our discussion.

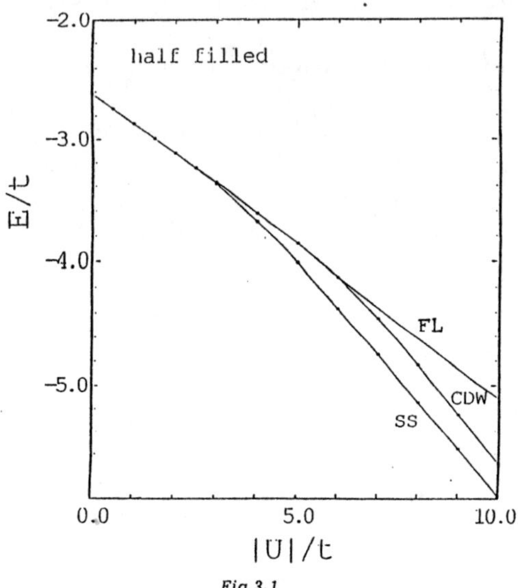

Fig.3.1

We can further note that the superconductivity is the result of a favorable combination [3.4] of high Debye temperature, due to high frequency vibrations of the C_{60} ball, and two different energy scale factors determining the value of the electron-phonon coupling constant. One of the factors is that the electron-phonon interaction is dominated by three particular on-ball modes with the energy scale of the scattering length set by the large on-ball hopping matrix element. The other factor is that kinetic energy scale of electrons is limited by the small inter-ball hopping resulting in larger density of states.

Local density calculations [3.5] of the structural and electron properties of C_{60} and K_xC_{60} show that the valence electron of K is transferred to the lowest unoccupied bands of the C_{60}. Molecular solid analysis of the electron-phonon interaction shows that the P-hybridization of the conduction electron wave function of K_xC_{60} allows new channels of scattering that are not present in graphite accounting for higher transition temperatures in K_xC_{60} than in grphites. The fermi surface [3.6] of K_3C_{60} has two sheets; the first is the free-electron-like and the second is multiply connected forming two interlocked symmetry-equivalent non-touching pieces. The calculated London penetration depth is $\lambda_L=1600$Å, Comparing the fermi velocity with the experimental coherence length leads to a superconducting pairing strength $\lambda \sim 5$ indicating very strong coupling. Partial nesting in the second fermi surface sheet may favour coupling to short-wave length optic modes. The observed transition temperatures compared with the coupled Eliashberg equations show that the Coulomb interaction is important [3.7] since the K^+ doping counteracts [3.8] the distortion in C_{60} and T_c is proportional [3.9] to T_F. The oriental disorder [3.10,3.11] present in C_{60} is not found in K-doped C_{60}. Phillips [3.12] has found that T_c is proportional to ionic polarizability. The rotations depend on the charge state. C_{60}^{3-} are static on the time scale of the line shape measurements. The ^{13}C spin-lattice relaxation rate [3.13] in the normal state of K_3C_{60} is found to be characteristic of a meta giving support to the BCS theory. The vacancies appear to play an important role [3.14] in determining the solid phases of Rb_xC_{60}. the superconducting phase occurs only near $x \simeq 3$.

3.1.Interpretation of The Transition Temperature, T_c

According to BCS theory, the transition temperature from normal to superconducting state is given by $K_BT_c = \hbar\omega \exp[-\frac{1}{N(o)V}]$, where ω is the phonon cut off frequency, $N(o)$ is the density of states near the fermi surface and V is the matrix element of attractive interaction between electrons. The larger is the V, the larger is the critical temperature. Therefore Phillips [3.12] assumed that

$$T_c = a + x(bz + c\alpha) \qquad (3.5$$

where Z is the ionic charge and α is the polarizability. In the case of Chevrel phase for which the parameters are given in Table 3.2, the Eq.(3.5) gives a reasonable fit. For A_xC_{60}

$$T_c \simeq g\alpha \qquad (3.6$$

TABLE - 3.2

The polarizability α for B_y (Mo_6S_8) compounds.

B	Y	$\alpha(A^3)$	$T_c(K)$
Sn	1.2	3.4	13
Pb	0.9	4.9	15.2

For $g = 14K/A^3$, the critical temperature depends entirely on α, with the values of α for Cs, the Eq. (3.6) gives $T_c = 47K$ for Cs_xC_{60}. This value of 47K is a little too high. Therefore, S. Koka et al. [3.15] introduced the size effect in Eq.(3.6) by including the ionic radii. Now, with this kind of approach, the value of T_c of Cs_xC_{60} was calculated and is in accord with the experimental value. This formula also indicates that the phonon exchange part of the BCS theory is correct but the polarizability is also important [3.16].

A study [3.17] of the magnetic superconductors of the type $ErRh_4B_4$ and RMo_6S_8 (R= rare earth) shows the importance of magnetic exchange in reducing the superconducting transition temperatures. However, when the $Cu - Cu$ separation is plotted as a function of T_c, it is found [3.18] that the mechanism of T_c changes at about 50 K. The alkali metal carbon superconductors show type II superconductivity. IN K_xC_{60} for $x = 3$, the transition temperature is 18K. As x increases beyond 3.4, the T_c reduces to zero. for $x > 3.4$ only a ferromagnetic phase is expected. In the BCS theory, phonon-exchange mechanism is correct. However, the transition temperatures also depend on the polarizabilities so that a suitable modification in the theory is needed.

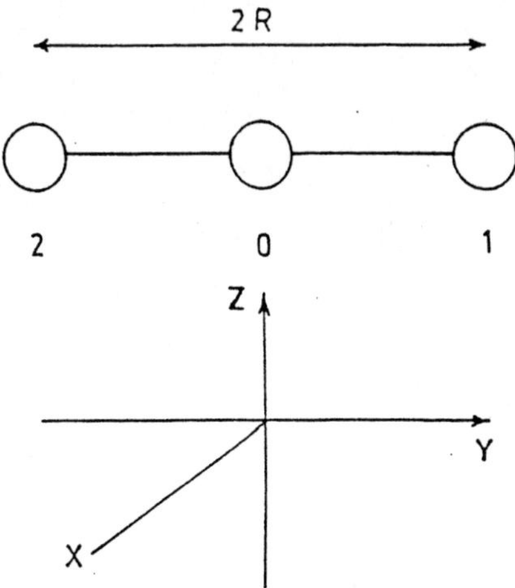

Fig. 3.2. Normal modes in a three atom system for the electron-phonon interaction.

We consider the normal mode of a three atom system. The first atom moves on the right hand side so that this vibration $+y_1$. At this time the second atom moves to the left hand side so that the amplitude of oscillation is $-y_2$. This normal mode is $y_1 - y_2$. The potential then becomes

$$V' = \left(\frac{\partial V}{\partial y}\right)_o (y_1 - y_2) \tag{3.7}$$

In the the second quantized notation

$$y_1 = \left(\frac{\hbar}{2m\omega_k}\right)^{1/2} (a_k + a^\dagger_{-k})e^{ik.R_1} \tag{3.8}$$

$$y_2 = \left(\frac{\hbar}{2m\omega_k}\right)^{1/2} (a_k + a^\dagger_{-k})e^{ik.(R_1-2R)} \tag{3.9}$$

so that

$$|y_1 - y_2| = \left(\frac{\hbar}{2m\omega_k}\right)^{1/2} (a_k + a^\dagger_{-k})e^{ik.R_1}(1 - e^{-2R}) \tag{3.10}$$

In the B.C.S. theory,

$$k_B T_c = 1.14\hbar\omega \exp[-1/N(o)V] \tag{3.11}$$

thus we see that the T_c depends on $2R$. However the V' of (3.7) appears in second order in V of (3.11) so that for small wave vectors

$$V' = \sum_{kq\sigma} v_{kq}(\hbar/2m\omega_q)^{1/2}C^\dagger_{k\sigma} C_{k-q,\sigma}(a_q + a^\dagger_{-q})(q_x a + q_y b + q_z c) + h.c. \tag{3.12}$$

and the square of the matrix element becomes

$$|M_q|^2 = |v_{kq}|^2 \left(\frac{\hbar}{2m\omega_q}\right)(q_x a + q_z b + q_z c)^2 \tag{3.13}$$

In the case of C_{60} all the compounds are cubic so that $a = b = c$, so that

$$\ln T_c \simeq -\frac{v^2}{(3.6)^2\rho_F V\omega_D^2 q^2} + \ln(1.14\hbar\omega_D/k_B) \tag{3.14}$$

This means that a plot of $\ln T_c$ with a^{-2} is predicted to be linear. We show in Fig. 3. such a plot which shows that our theoretical idea is in accord with the experiments. W have used the experimental values from Fleming et al [3.19] to plot the $T_c(a^{-2})$ graph Kortan et. al. [3.20] have found that Ba_6C_{60} is also a superconductor with $a = 11.171$ and $T_c \simeq 7K$. Its smaller unit cell indicates a smaller T_c and hence we see that this theor is in accord with the experimental data.

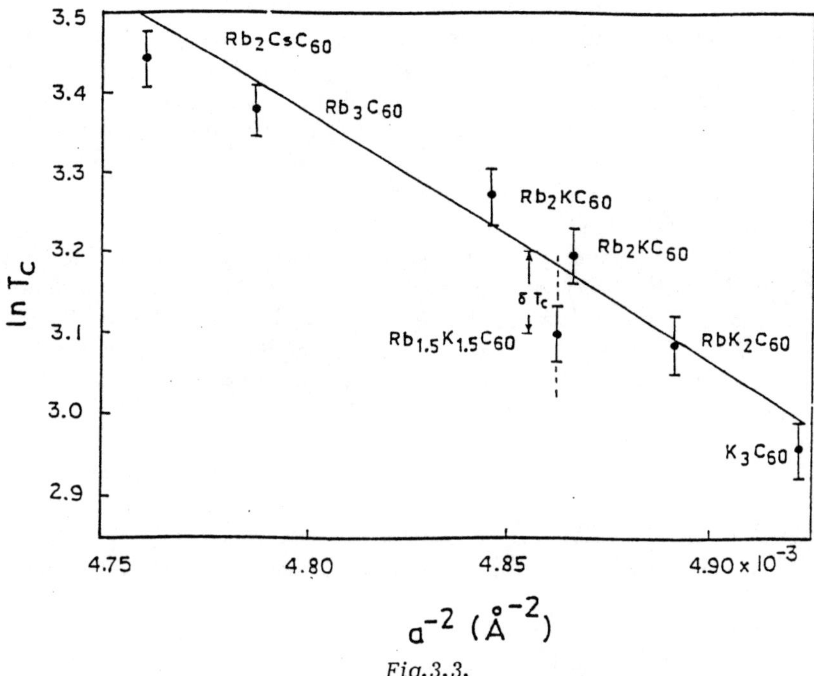

$$a^{-2} \ (\overset{\circ}{A}{}^{-2})$$

Fig.3.3.

4.Electron Spin Resonance and Nuclear Magnetic Resonance Results

4.1.Electron Spin Resonance

In the superconductors, there are pairs of electrons with the spin of one of the electrons ↑ and that of the other ↓ so that the net spin of a pair is zero when observed from a distance. Therefore, the electron spin resonance seen in the normal state vanishes upon cooling below T_c. In this way the E.S.R. intensity measurement has been used to measure the T_c. The E.S.R. intensity has also been used to demonstrate that the superconducting state is a zero spin state.

In the case of a free electron the spin is $S = 1/2$. In a magnetic field the Hamiltonian is. $\mathcal{H} = g\mu_g H_z.S_z$ so that there are two eigenvalues $\pm\frac{1}{2}g\mu_B H_z$ and the separation between the two levels is $g\mu_B H_z$. The transitions are excited by the application of a modulation field, h_x, in a direction perpendicular to the direction of the applied magnetic field. The perturbation caused by the modulation field is of the form, $H' = g\mu_B h_x S_x$. Since $S_x + iS_y = S_\pm$, the modulation field causes transitions. The transition probability is of the form $(2\pi/\hbar)| < b|H'|a > |^2 \rho_f$. The microwaves of frequency ν are absorbed when $h\nu = g\mu_B H$. The absorption vanishes when the electrons get paired to form spin zero states. However, very near the transition temperature there are short range effects so that there is a g-shift which is positive in the paramagnetic state and negative in the diamagnetic (superconducting) state. Therefore. the g-shift has been used to determine whether the state is paramagnetic or superconducting when the temperature is varied. Very often the g-shifts are anisotropic. Particularly in the case of Cu^{2+} the $g_{\parallel}$ and $g_{\perp}$ are so widely different that two separate lines are seen in the spectrum. The $g_{\parallel}$ and $g_{\perp}$ when measured near the T_c can be used to determine the anisotropy of the electrons which participate in the pairing process. Thus $g_{\parallel}$ is relatively insensitive while $g_{\perp}$ shows negative shift. Therefore it is interpreted that the superconducting electrons are in the ab-plane. Thus the planar structure of the superconducting state has been inferred from the measurement of the shift of the $g_{\perp}$ value. The superconductors have a gap of 2Δ in the dispersion relation. Since the microwave energy, $\hbar\omega$, is more than a hundred times smaller than 2Δ, the microwaves are not absorbed by the superconductors. Any absorption shows that the real material is of mesoscopic nature and Josephson weak links exist. In this way, electron spin resonance has been very informative about the superconducting materials.

The electron spin resonance signals have been observed in several systems of the $YBa_2Cu_3O_7$ type containing Cu^{2+} [4.1-4.21], Mn^{2+} [4.22], Gd^{3+} [4.23-4.27] and Tm^{2+} [4.28] ions in Bi containing supercondcutors [4.29-4.33] and in carbon superconductors [4.34-4.38]. Conduction electron spin resonance has been reported [4.39-4.40]. In some cases, a free radical is introduced in the superconductor in which case the line width $\Delta H \sim \phi_o/\lambda^2$ is used to determine the London penetration depth [4.41-4.44]. In some of the systems a local moment is not formed at all so that the lines are broadened by the coherence length [4.45-4.48].

In some materials E.S.R. signals are visible in the normal state but disappear in the supercondcuting state due to the broadening of lines caused by the coherence. Mehran and Anderson [4.45] suggest a broadening of the form,

$$\Delta H = \Delta H_N(\xi/a) \tag{4.1}$$

may be caused by coherence. Here ΔH_N is the linewidth in the normal state, ξ is the coherence length and a is the interatomic distance within the crystallographic unit cell. In the case of type-II superconductors, K.N. Shrivastava suggested [4.49] that the width may be given by flux quantization,

$$\Delta H = \Delta H_N \left(\frac{\phi_o}{B}\right)^{1/2} a^{-1} \tag{4.2}$$

where B is the magnetic induction and ϕ_o is the unit flux. Since ξ determines the upper critical field, the value of B in the above will become H_{C2} and the broadening is then minimized. Chakravarty and Orbach [4.46] have suggested that the general expression for the E.S.R. line width is,

$$\Gamma_\perp = \frac{1}{2\chi_\perp} Re \int_o^\infty dt \exp(-\omega_o t)\langle f^\dagger(t)f^\dagger(0)\rangle \tag{4.3}$$

where the correlation function is given by

$$f^\dagger(t) = \sum_{i\neq j}\sum_{\alpha\beta} g_{\alpha\beta}^{ij} S_i^\alpha(t) S_j^\beta(t) \tag{4.4}$$

Here, $\alpha, \beta = \pm$ and 0 for the spin raising and lowering operator $S^{\pm}$ and $S^o = S^z$. In terms of coherence length the above linewdith is found to be

$$\Gamma_{\perp}^{S} = \frac{0.94}{Z_c^3 Z_{\chi}^2} \frac{A_{K_o} a^4}{\hbar J} \left(\frac{\xi}{a}\right)^3 \left(\frac{T/2\pi\rho_s}{1 + T/2\pi\rho_s}\right)^4 \tag{4.5}$$

where the renormalization factors $Z_c \sim 1.18$, $Z_{\chi} = 0.44$, $K_o \sim \pi/a$, J is the exchange interaction and the stiffness constant $\rho_s = 0.18J$, A_{ko} is the wave vector dependent hyperfine interaction. Compared with the Mehran and Anderson's [4.45] expression, this result is having an extra factor of $(\xi/a)^2$. Since the correlation length diverges, $\xi = (1 - T/T_c)^{-\nu}$, the linewidth also diverges near T_c. Otherwise, the ESR signal will be so broad in the coherent state that it will not be visible (Table 4.I).

Table - 4.1

Line width coherence length relationship

S.No.	Line width	Coherence length	Year	References
1.	Line width (general)	$\Delta F \xi_o / \ell$	1988	Shrivastava [50,51]
2.	Line width	$\Delta H \xi / a$	1989	Mehran and Anderson [45]
3.	EPR line width	$\Delta(\xi/a)^3$	1990	Chakravarty and Orbach [46]
4.	NMR line width	$\Delta H(\xi/a)$	1990	Chakravarty and Orbach [46]
5.	NMR line width	$\Delta H_1(\xi_o/a)^2 + \Delta H_2 \ln(\xi/a)$	1991	Monien, Pines and Takigawa [53]

Here l is the mean free path, a is the lattice constant and ΔH_1 and ΔH_2 are two additive contributions to the line width.

Earlier than Mehran and Anderson, K.N. Shrivastava [4.50] has suggested that in the coherent state the lines are broadened according to the response function,

$$\chi'' = \frac{l\delta\Omega}{\Lambda\xi_o[(\omega - \omega_J)^2 - (\delta\Omega)^2]} \tag{4.6}$$

where l is the mean free path and $\Lambda = ne^2x/m$ for the electron with n as the charge density, e the charge and m the mass of an electron. Near the resonance $\omega = \omega_J$ so that

$$\chi_o'' = -\frac{l}{\Lambda \xi_o(\delta\Omega)} \tag{4.7}$$

which shows that the line width has widened to $\delta\Omega\xi_o/l$ due to coherence. Actually it is the value of Λ which multiplies the current which has been changed to $\Lambda\xi/l$ upon going to the coherent state [4.50]. However, the coherence length diverges near the T_c so that the linewidth (4.3) also diverges,

$$\Gamma_\perp^S \propto \left(1 - \frac{I}{T_c}\right)^{-3\nu} \tag{4.8}$$

For a three dimensional solid $\nu = 0.69$ so that the exponent of the line wdith is ~ 2.1. Thus the importance of the coherence length in determining the line width is quite clear.

4.2.Nuclear Magnetic Resonance

The nuclei usually relax by giving energy to the electrons. However, in the superconductors, the electrons are paired and there is a large gap in the dipersion relation so that the nuclear magnetic resonance which uses the radio frequency can not occur. What it means is that the nuclear magnetic resonance intensity changes upon cooling from the finite value in the normal state to zero in the superconducting state. The shift in nuclear magnetic resonance frequency of a nucleus of ^{63}Cu in going from an insulator, such as $CuCl_2$ to the copper metal, is known as the Knight shift. It is caused by the conduction electron contact hyperfine field at the site of the nucleus of the Cu atom. In the case of a superconductor, the Knight shift becomes zero at zero temperature.

Millis et al [4.52-4.57] have made an effort to describe the nuclear relaxation rates in terms of atomic structure factors which in turn can be written in terms of spectral moments which are functions of the coherence length. In this way the relaxation rates can be understood in terms of a Fermi liquid theory in the normal state. These theories suggest that the nuclear relaxation in the entire range of tempeatures belongs to the normal state. As the temperature is varied through the superconducting transition temperature there is no effect on the nuclear relaxation rate and the product of the relaxation time and that of the static susceptibility remains a constant. Using superconducting coherence length it may be concluded that there will be a peak in the relaxation time at the transition temperature. There are several distances in the problem, the short range antiferromagnetic coherence length, the superconducting coherence length and the London penetration depth. Theories at this time are not clear about the treatment of all these three distances.

The dynamical structure factor, in a one component model is given by

$$S(q,\omega) = \frac{1}{1 - e^{-\hbar\omega/k_BT}}\chi''(q,\omega) \tag{4.9}$$

In the small frequency approximation, $\hbar\omega << k_B T$,

$$S(q,\omega) = \frac{T}{\omega}\chi''(q,\omega) \ . \tag{4.10}$$

We consider a diatomic molecule, $Cu - O$ with the electronic configuration $3d^9$ for Cu^{2+} and $2s^2p^6$ for O^{--}. The spin orientations of $3d^9$ are five electrons with spin $\uparrow$ and four with spin $\downarrow$ so that there is a net magnetic moment $\uparrow$ on Cu^{2+} site. This $\uparrow$ moment interacts with $\uparrow$ moment at the O^{--} site so that out of $2p(\uparrow\downarrow)$ the energy of $\uparrow$ moment is reduced while that of O^{--} ($\downarrow$) remains unchanged. Accordingly, the probability of finding $\uparrow$ electron at the O^{--} site is slightly increased compared with that of the $\downarrow$ electron. So that there is a net spin density at the O^{--} site. Thus we have a self hyperfine field at the Cu^{2+} site and a transferred hyperfine field at the O^{--} site. Similarly in a system of three atoms $Cu^{2+} - O^{--} - Cu^{2+}$ there is a hyperfine coupling at the Cu^{2+} due to its unpaired electron in the $3d$ shell and a transferred hyperfine interaction due to another Cu^{2+} atom, and a transferred hyperfine interaction at the O^{--} site. In addition to the transferred hyperfine interaction there is a returning contribution due to the charge redistribution caused by the transfer [4.58-4.59]. Monien et al [4.53] have considered the hyperfine as well as the transferred hyperfine interactions at the ^{63}Cu site and the hyperfine interactions at ^{17}O and ^{89}Y sites. They have written [4.53,4. 60] the relaxation rates in terms of atomic structure factors, $S(q,\omega)$ as

$$^{63}W_\perp = \frac{3}{4\mu_B^2\hbar}\lim_{\omega\to 0}\sum_q \{A_\perp - 2B\left[\cos(q_x a) + \cos(q_y a)\right]\}^2 S(q,\omega) \tag{4.11}$$

where A is the hyperfine coupling constant at the Cu^{2+} site, B is the transferred hyperfine interaction constant at the Cu^{2+} site, q_x and q_y are the x- and y-components of the wave vector and a is the unit cell constant in the ab-plane. The spectral moments are calculated in terms of the coherence length as

$$S_o = \frac{\pi\bar{\chi}_o(k_B T)}{\hbar\Gamma}\left\{1 + \frac{\beta}{\pi^2}\left[\frac{\pi}{4}\left(\frac{\xi}{a}\right)^2 - \left(\frac{1}{8\pi} + \frac{1}{4\pi^2}\right)\right]\right\}$$

$$S_1 = \frac{\pi\bar{\chi}_o(k_B T)}{\hbar\Gamma}\left\{1 + \frac{\beta}{\pi^2}\left[\frac{\pi}{8}\ln\left(\frac{\xi}{a}\right)^2 + 0.1703\right]\right\}$$

$$S_2 = \frac{\pi\bar{\chi}_o(k_B T)}{\hbar\Gamma}\left(1 + 0.25\,\beta/\pi^2\right)$$

$$S_3 = \frac{\pi\bar{\chi}_o(k_B T)}{\hbar\Gamma}\left(1 + 0.19\,\beta/\pi^2\right) \tag{4.12}$$

where $\beta = (a/\xi_o)^4$. Thus S_o varies as ξ/a, S_1 as $\ln\xi$ and S_2 and S_3 are independent of ξ. The relaxation rate for ^{63}Cu is given by

$$W_{\parallel} = \frac{3}{4\mu_B^2 \hbar}\left[(A_\perp - 4B)^2 S_o + 8B(A_\perp - 4B)S_1 + 20B^2 S_2\right]$$

$$W_\perp = \frac{3}{8\mu_B^2 \hbar}\left[\left\{(A_\perp - 4B)^2 + (A_\parallel - 4B)^2\right\}S_o + 8B(A_\perp + A_\parallel - 8B)S_1 + 40B^2 S_2\right] \quad (4.13)$$

with $1/T_1 = (2/3)W$ for ^{63}Cu with similar results for other nuclei. For $\xi^2/a^2 >> 1$ antiferromagnetic correlations play a dominant role in determinign $T_1(^{63}Cu)$ with leading term proportional to ξ^{-2} and a logarithmic contribution playing more important role for $\xi/a \geq 1.5$.

The nuclear magnetic resonance of ^1H [61-64], ^2D [4.64], ^7Li [4.65], ^{17}O [4.66-4.77], ^{27}Al [4.78], ^{51}V [4.79], ^{63}Cu [4.80-4.93], ^{89}Y [4.94-4.103], ^{137}Ba [4.104], ^{139}La [4.105], ^{141}Pr [4.106], ^{169}Tm [4.106] nuclei in superconductors have been reported. There is considerable interest in the general studies on NMR [4.107-4.126], on the coherence peak [4.127], Knight shift [4.128] and the nuclear relaxation from the $t - J$ model hamiltonian [4.60, 4.129] and relaxation [4.130-4.153]. There are several NQR studies using ^{63}Cu [4.154-4.169], ^{137}Ba [4.170], ^{139}La [4.170-4.172] and ^{181}Ta [4.173] nuclei.

The nuclear magnetic resonance of protons in $YBa_2Cu_3O_{6.94}H_x$ ($x = 0.2, 0.5$) shows that H atoms are moving above 200K but trapped in sites below 130K. The nuclear relaxation rate upon cooling the sample shows an enhancement below T_c [4.128]. In $Sm_{1.85}Ce_{0.15}O_4H_x$ the hydrogen atoms are coupled with Sm^{3+} ions by dipolar interactions. Some of the hydrogen atoms are located in the interstitial sites in the Cu-O planes in the metallic area [4.129]. In the Fourier transform NMR there is a single peak above T_c which splits upon cooling below T_c. The splitting increases with decreasing temperature [4.130]. In earlier work, the line width was treated according to the condition of flux quantization $\Delta H = \phi_o/\lambda^2$. However, later work shows that short range antiferromagnetic correlations are important so that the lines are broadened due to coherence.

The ^{17}O NMR consists of five lines, a central line caused by all O atoms due to the $I_z = -\frac{1}{2} \leftrightarrow \frac{1}{2}$ transitions and satellite lines by O atoms in the CuO_2 planes due to the $I_z = \pm\frac{1}{2} \leftrightarrow \pm\frac{3}{2}$ and $\pm\frac{3}{2} \leftrightarrow \frac{5}{2}$ transitions. The relaxtion rate of ^{17}O in the CuO_2 planes of $YBa_2Cu_3O_{7-\delta}$ is proportional to the temperature in the range 90 to 250K in the normal state. It is field independent in the range 3.7 to 4.0T. In the superconducting state it has no enhancement just below T_c, indicating that the additional contribution associated with the spin diffusion to the vortex is small. In the mixed state there is an additional contribution to T_1 from the vortex cores,

$$\frac{1}{T_1} \simeq \frac{1}{T_{1N}}\frac{H\xi^2}{\phi_o} \quad (4.14)$$

The normal state relaxation time is obtained by extrapolation to 4.2K, $T_{1N} \simeq 1s$ and $H = 8T$, $\xi = 20\overset{\circ}{A}$ predicts $T_1 \sim 63s$ which is longer than observed at 4.2K.

The frequency shift approaches zero towards, $T = 0$ in both Bi as well as Tl containing superconductors consistent with the B.C.S. theory [4.69, 4.71]. In Fig. 4.1 we show [4.5] the Knight shift of ^{17}O which depends linearly on the static susceptibility. Similar result are found in $YBa_2Cu_4O_8$ for which $T_c \sim 74K$ also [4.74]. The effective field gradient shift as well as quadrupole interaction at O site for $Bi_2Sr_2Ca_{n-1}Cu_nO_{4+2n}$ $(n = 1, 2, 3)$ for structures 2201, 2212 and 2223 have been determined [4.68, 4.76]. In $n = 3$ sample [4.75] at T_c the spin component of the Knight shift, K_s, has dropped to half of its room temperature value. For outer planes the ^{17}O shift is almost temperature independent untill 120 K when it decrease sharply and the spin component of the Knight shift at T_c is 2/3 of its room temperature value. The temperature dependence of the ^{17}O shift of the outer two $Cu - O$ planes differs from that of the ^{17}O of the central plane showing that the coupling between planes is small. The relaxation behaviour for both sites is Korringa like ($K_s^2 T_1 T$ is constant) in the normal state.

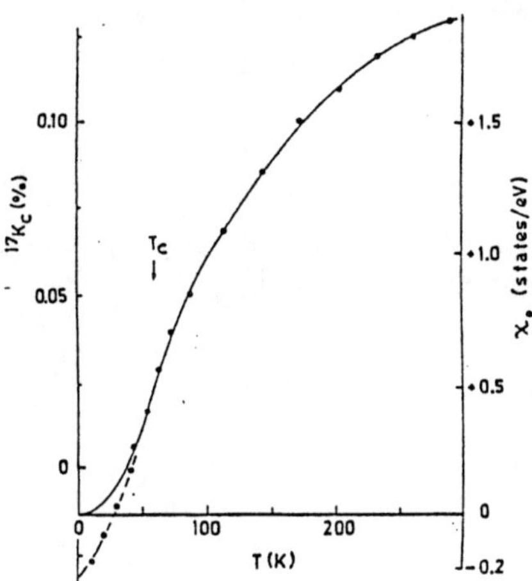

[Fig.4.1.] The experimental O Knight shift is plotted versus temperature showing linear dependence on static planar susceptibility. It may be noted that Knight shift of both the nuclei ^{63}Cu, and ^{17}O have the same functional dependence on temperature [4.53].

The value of $K_s^2 T_1 T$ is close to the theoretical value for simple s-type metals with no electron-electron interaction. The relaxation rate obeys a dynamical relation,

$$\frac{1}{T_1} = \frac{\gamma_N^2 k_B T}{2\mu_B^2} \sum_q |A_q|^2 Im\chi(q, \omega_o) \tag{4.15}$$

where $\chi(q, \omega_o)$ is the dynamical spin susceptibility, ω_o is the resonance frequency, γ_n is the nuclear gyromagnetic ratio and

$$A_q = \sum_i A_i \exp(iq\gamma_i) \tag{4.16}$$

with A_i the hyperfine coupling. The imaginary part of the susceptibility has two parts, a static part and an antiferromagnetic part correlated over short range,

$$Im\frac{\chi(q,\omega_o)}{\omega_o} = \frac{\pi\chi(0)}{\Gamma} + \frac{\pi\chi(0)}{\Gamma_{AF,q}} \tag{4.17}$$

$\chi(q)$ has a peak at $q = Q_{AF}(\pi,\pi)$ and A_q goes to zero at $q = Q_{AF}$ [4.97]. The relaxation rate then becomes

$$\frac{1}{T_1/T} = \sum_{q\simeq0} \frac{Im\chi(q,\omega_o)}{\omega_o} \simeq \frac{\pi\chi(0)}{\Gamma} \tag{4.18}$$

Γ is independent of $\chi(0)$ and K_sT_1T approaches a constant. In contrast for a Fermi liquid with weak magnetic correlations, Γ is inversely proportional to the density of states and thus $\chi(0)$ leads to the Korringa relation. The nuclear relaxation time may be expressed [4.146] as,

$$\frac{1}{T_1} = \frac{0.8\xi}{T_{1\infty}Z_c a}\left(\frac{T}{2\pi\rho_s}\right)^{3/2}\left[\frac{1}{1 + T/2\pi\rho_s}\right]^2 \tag{4.19}$$

at low temperatures whereas according to Moriya' calculation [4.174] for $S = 1/2$, for a square lattice at high temperatures,

$$T_1 = T_{1\infty}(1 + J/4T)^{1/2}\exp\left[(J/2T)^2(1 + \sigma/4T)\right] \tag{4.20}$$

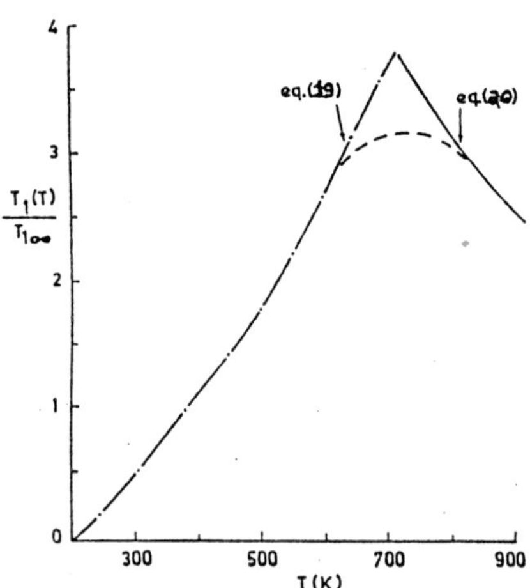

[Fig.4.2.] A plot of the NMR $(T_1/T_{1\infty})$ versus temperature for Cu^{2+} in normal La_2CuO from (4.19) and (4.20). The dashed connecting line is the predicted behaviour since a peak occurs in (4.19) [4.46,4.50]. Note that pure La_2CuO_4 does not have a superconducting phase.

Thus at low temperatures, T_1 increases with increasing temperatures while at high temperatures T_1 decreases with increasing temperature. Therefore, combining the two results we plot $T_1(T)$ in Fig.4.2 which shows that at some temperature $T_1(T)$ has a peak. This predicted behaviour of $T_1(T)$ is completely in agreement with the experiments [4.53, 4.175, 4.176]. The original idea [4.50-4.51] of using the coherence length in the line width thus works very well.

However, these discussions on the studies of ESR and NMR spectra are of very general nature and need not apply fully to the alkali doped C_{60} problems like K_3C_{60} etc. But still we beleive that since the super conducting C_{60} compounds show metallic nature, the above formulae should become applicable, with suitable modifications to these compounds also.

We have reviewed various properties of the carbon molecule C_{60} and have studied its energy levels. We have described the applications of Hubbard model to this molecule. In particular, we have studied the electron-phonon interaction and have calculated the transition temperature for going from normal to the superconducting state. We have also included a small discussion on the suggestion of Phillips using polarizability in addition to the electron-phonon interaction to understand the transition temperature of the alkali atom carbon superconductors.

We have made an effort to understand the effect of coherence on the EPR and NMR line widths as well as on the nuclear spin lattice relaxation rates. This idea is well developed ,particularly by Pines , to understand the relaxation rates and Knight shifts in copper oxide containing superconductors.It remains to be applied to the C_{60} superconductors. From this study our conclusion is that the recently discovered carbon superconductors form interesting materials for the application of theoretical ideas.Considerable amount of has been done in the past two years or so and a good deal of work is yet to be done.

References

1.1 N. Ashby and W.E. Brittin, Am. J. Phys. 54 (1986) 776.

1.2 P.W. Fowler, Chem. Phys. Lett. 131 (1986) 444.

1.3 A.J. Stone and D.J. Wales, Chem. Phys. Lett. 128 (1986) 501.

1.4 P.W. Fowler, J.C.S., Chem. Commun. (1987) 1403.

1.5 H.W. Kroto, Nature 322 (1986) 766; Astrophys. J. 314 (1987) 352.

1.6 D.J. Wales, Chem. Phys. Lett. 141 (1987) 478.

1.7 K. Balasubramanian, J. Comput. Chem. 9 (1988) 406.

1.8 A.T. Balaban, Comput. Mat. 17 (1989) 397.

1.9 H.W. Kroto, Comput. Mat. 17 (1989) 417.

1.10 D.R. Huffman, Phys. Today. 44, Nov. (1991) 22.

1.11 T. Erber and G.M. Hockney, Fermilab.-Pub.- 91/222-T; I. Stewart, Nature 351 (1991) 103.

1.12 O. Hahn, F. strassman, J. Mattauch and H. Ewald; Naturewiss 30 (1942) 541.

1.13 J. Mattauch, H. Ewald, O. Hahn and F. Strassman, Z. Phys. 20 (1943) 598.

1.14 H.W. Kroto, J.R. Heath, S.C. O'Brein, R.F. Curl and R.E. Smalley, Nature 318 (1985) 162.

1.15 J.R. Heath, S.C. O'Brein, Q. Zhang, Y. Liu, R.F. Curl, H.W. Kroto, F.K. Tittel and R.E. Smalley, J. Am. Chem. Soc. 107 (1985) 7779.

1.16 Y. Liu, S.C. O'Brein, Q. Zhang, J.R. Heath, F.K. Trittel, R.F. Curl, H.W. Kroto and R.E. Smalley, Chem. Phys. lett. 126 (1986) 215.

1.17 S.H. Yang, C.L. Pettiette, J. Conceicao, O. Chesnovsky, and R.E. Smalley, Chem. Phys. Lett. 139 (1987) 233.

1.18 B. Miller, J.M. Rosamilia, G. Dabbagh, R. Tycko, R.C. Haddon, A.J. Muller, W. Wilson, D.W. Murphy and A.F. Hebard, J. Am. Chem. Soc. 133 (1991) 6291.

1.19 P.A. Limbach, L. Schweikhard, K.A. Cowen, M.T. McDermott, A.G. Marshall and J.V. Coe, J. Am. Chem. Soc. 113 (1991) 6795.

1.20 R.F. Curl and R.E. Smalley, Science, 242 (1988) 1017.

1.21 M. Moskovit, Ultra Micros. 20 91986) 83.

1.22 S. Iijima, J. Phys. Chem. 91 (1987) 3466.

1.23 H.W. Kroto, J.E. Frommer, and P.C. Arnett, Nature 331 (1988) 328.

1.24 P. Ball, Nature 348 (1990) 583.

1.25 R.J. Wilson, C.T. Meijer, D.S. bethune, R.D. Johnson, D.D. Chambliss, M.S. Dvries, H.E. Hunziker and H.R. Wendt, Nature 348 (1990) 621.

1.26 J.L. Wragg, J.E. Chamberlain, H.W. White, W. Kratschmer, and D.R. Huffman, Nature 348 (1990) 623.

1.27 Y.Z. Li, J.C. Patrin, M. Chander, J.H. Weaves, L.P.F. Chibante and R.E. Smalley, Science 252 (1991) 547.

1.28 Y.Z. Li, M. Chander, J.C. Patrin, J.H. Weaver, L.P.F. Chibante and R.E. Smalley, Science 253 (1991) 429.

1.29 J.M. Wood, B. Kahr, S.H. Hoke II, L. Dejarme, R.G. Cooks and D. Ben-Amotz, J. Am. Chem. Soc. 113 (1991) 5907.

1.30 F. Dieberich, R. Ettl, Y. Rubin, R.L. Whetten, R. Beck, M. Alvarez, S. Anz, D. Sensharma, F. Wudl, K.C. Khemani, A. Koch, Science 252 (1991) 548.

1.31 P.J. Fagan, J.C. Calabrese and B. Malone, Science 252 (1991) 1160.

1.32 J.M. Hawkins, A. Meyer, T.A. Lewis, S. Loren and F.J. Hollander, Science 253 (1991) 312.

1.33 P.M. Allemand, K.C. Khemani, A. Koch, F. Wudl, K. Holczer, S. Donovan, G. Gruner and J.D. Thampson, Science 253 (1991) 301.

1.34 L.M. Roth. Y. Haung, J.T. Schwedler, C.J. Cassady, D. Ben-Amotz, B. Kahr and B.S. Freiser, J. Am. Chem. Soc. 113 (1991) 6298.

1.35 A. Penicaud, J. Hsu, C.A. Reed, A. Koch, K.C. Khemani, P.M. Allemand and F. Wudl, J. Am. Chem. Soc. 113 (1991) 6698.

1.36 S.J. Cyvin, E. Brendsdal, B.N. Cyvin and J. Brunvoll, J. Chem. Phys. Lett. 143 (1988) 377.

1.37 R.E. Haugler, J. Conceicao, L.P.F. Chibante, Y. Chai, N.F. Byrne, S. Flangan, M.M. Haley, S.C. O'Brein, C. Pan, Z. Xiao, W.E. Billups, M.A. Ciufolini, R.H. Hauge, J.L. Margrave, L.J. Wilson, R.F. Curl and R.E. Smalley, J. Phys. Chem. 94 (990) 8634.

1.38 Y. Wang and N. Herran, J. Phys. Chem. 91 (1987) 257.

1.39 N. Herron, Y. Wang, M.M. Eddy, G.D. Stucky, D.E. Cox, K. Moller and T. Bein, J. Am. Chem. Soc. 111 (1989) 530.

1.40 N.E. Blank, M.W. Haenel, C. Kruger, V.H. Tsay and H. Weintages, Angew. Chem. 27 (1988) 1064.

1.41 I.P. Buffey, W.B. Brown and H.A. Gebble, Chem. Phys. Lett. 148 (1988) 281.

1.42 F. Carnovale, J.B. Peel and R.G. Rothwell, J. Chem. Phys. 88 (1988) 642.

1.43 C.Y. Kung, R.A. kennedy, D.A. Dalson and T.A. Miller, Chem. Phys. Lett. 145 (1988) 455.

1.44 P.W. Fowler, P. Lazzeretti and R. Zanasi, Inorg. Chem. 27 (1988) 1298.

1.45 K.D. Kolenbrander and M.L. Mandich, J. Chem. Phys. 92 (1990) 4759.

1.46 L. Song and M.A. El-Sayed, J. Phys. Chem. 94 (1990) 7907.

1.47 U. Ray, M.F. jarrold, E. Bower, and J.S. Kraus, J. Chem. Phys. 91 (1989) 2912.

1.48 M.E. Geusic, R.R. Freeman, and M.A. Duncan, J. Chem. Phys. 88 (1988) 163.

1.49 A.M. Thayer, M.L. Steigerwald, T.M. Duncan and D.C. Douglas, Phys. Rev. Lett. 60 (1988) 2673.

1.50 S. Koka, A. Chatterjee and K.N. Shrivastava, Solid State Commun. 81 (1992) 509.

1.51 W. Kartschmer, L.D. Lamb, K. Fostiropoulos and D.R. Huffmann, Nature 347 (1990) 354.

1.52 D.H. Parker, P. Wurz, K. Chatterjee, K.R. Lykke, J.E. Hunt, M.J. Pellin, J.C. Hemminger, D.M. Gruen and L.M. Stock, J. Am. Chem. Soc. 113 (1991) 7499.

1.53 M.D. Newton, and R.E. Stanton, J. Am. Chem. Soc. 108 (1986) 2469.

1.54 K. Raghavachari, R.A. Whiteside and J.A. Pople, J. Chem. Phys. 85 (1986) 6623.

1.55 C.J. Sandroff, S.P. Kelty and D.M. Hwang, J. Chem. Phys. 55 (1986) 5337.

1.56 Q.L. Zhang, S.C. O'Brien, J.R. Heath, Y. Liu, R.F. Curl, H.W. Kroto and R.E Smalley, J. Phys. Chem. 90 (1986) 525.

1.57 C.E. Brown, J. Pol. Sc. Pl. 24 (1986) 519.

1.58 C.E. Brown, Synthetic Metals 15 (1986) 265.

1.59 D.M. Cox, D.J. Trevor, K.C. Reichmann and A. Kaldor, J. Am. Chem. Soc. 10 (1986) 2457.

1.60 J.R. Heath, Q. Zhang, S.C. O'Brein, R.F. Culr, H.W. Kroto and R.E. Smalley, J. Am. Chem. Soc. 109 (1987) 359.

1.61 J.R. Heath, R.F. Curl and R.E. Smalley, J. Chem. Phys. 87 (1987) 4236.

1.62 M. Van Thiel and F.H. Ree, J. Appl. Phys. 62 (1987) 1761.

1.63 H. Ajie, M.M. Alvarez, S.J. Anz, R.D. Beck, F. Diederich, K. Fostiropoulos, D.R. Huffman, W. Kratschmer, Y. Rubin, K.E. Schriver, D. Sensharma and R.L. Whetten, J. Phys. Chem. 94 (1990) 8630.

1.64 R.L. meng, D. Ramirez, X. Jiang, P.C. Chow, C. Diaz, K. Matsuishi, S.C. Mos, P.H. Hor and C.W. Chu, Appl. Phys. Lett (in press).

1.65 D. Schmicker, S. Schmitt, J.G. Skofronick, J.P. Teonnies and R. Vollmer, Phy Rev. B44 (1991) 10995.

1.66 J.B. Howard, T. McKinnon, y. Makarovsky, A.L. Lefleur and M.E. Hohnson, Nature 352 (1991) 139.

1.67 R. Taylor, J.P. Parsons, A.G. Avent, S.P. Rannard, T.J. Dennis, J.P. Hare, H.W. Kroto, D.R.M. Walton, Nature 351 (1991) 277.

1.68 A.L. MacKay, Nature (1990) 336.

1.69 O'Keefe, M.M. Ross and A.P. Barnavsky, Chem. Phys. Lett. 130 (1986) 17.

1.70 J. Bernholc and J.C. Phillips, J. Chem. Phys. 85 (1986) 3258.

1.71 S.W. McElvany, H.H. Nelson, A.P. Baranavski, C.H. Watson and J.R. Eyler, Chem. Phys. Lett. 134 (1987) 214.

1.72 J.T. Brenna, W.R. Creasy and W. Volksen, Chem. Phys. Lett. 163 (1989) 499.

1.73 W.R. Creasy and J.T. Brenna, J. Chem. Phys. 92 (1990) 2269.

1.74 W.R. Creasy, J. Chem. Phys. 92 (1990) 7223.

1.75 M. Kataoka, T. Nakajima, Tetrahedron 42 (1986) 6437.

1.76 D.J. Klein, T.G. Schmalz, G.E. Hite and W.A. Seitz, J. Am. Chem. Soc. 108 (1986) 1301.

1.77 D.S. Marynick and S. Estreicher, Chem. Phys. Lett. 132 (1986) 383.

1.78 S.C. Obrien, J.R. Heath, H.W. Kroto, R.L. Curl and R.E. Smalley, Chem. Phys. Lett. 132 (1986) 99.

1.79 M. Ozaki and A. Takahashi, Chem. Phys. Lett. 127 (1986) 242.

1.80 S. Satpathy, Chem. Phys. Lett 130 (1986) 545.

1.81 R.G. Schmalz, W.A. Seitz, D.J. Klein and G.E. Hite, Chem. Phys. Lett 130 (1986) 203.

1.82 N. Trinajst, Int. J. Quant. 699 (1986).

1.83 R.L. Disch and J.M. Schulman, Chem. Phys. Lett. 125 (1986) 465.

1.84 P.W. Fowler and J. Woolrich, J. Chem. Phys. Lett. 127 (1986) 78.

1.85 R.C. Haddon, L.E. Brus and K. Raghavachari, Chem. Phys. Lett. 131 (1986) 165: 125 (1986) 459.

1.86 P.D. Hale, J. Am. Chem. Soc. 108 (1986) 6087.

1.87 W.B. Brown, Chem. Phys. Lett. 136 (1987) 128.

1.88 V. Elser and R.C. Haddon, Nature 325 (1987) 792.

1.89 V. Elser and R.C. Haddon, Phys. Rev. A36 (1987) 4579.

1.90 P. Gerhardt, S. Laffler and K.H. Homann, Chem. Phys. Lett. 137 (1987) 306.

1.91 M.F. Guest, Theor. Chim. 71 (1987) 117.

1.92 R.C. Haddon, J. Am. Chem. Soc. 109 (1987 1676.

1.93 I. Laszlo and L. Udvardi, Chem. Phys. Lett. 136 (1987) 418.

1.94 Luthi H.P. and Almlof, J. Chem. Phys. Lett. 135 (1987) 357.

1.95 M.L. McKee, Theochem. 38 (1987) 75.

1.96 J.S. Miller and A.J. Epstein, J. Am. Chem. Soc. 109 (1987) 3850.

1.97 J.A. Northby, J. Chem. Phys. 87 (1987) 6166.

1.98 A. Otto, Z. Chem. 27 (1987) 380.

1.99 J.C. Phillips, J. Chem. Phys. 87 (1987) 1712.

1.100 C.T. Phillinge, Phil. Trans. Roy Soc. A323 (1987) 313.

1.101 K. Raghavachari, J. Chem. Phys. 87 (1987) 2191.

1.102 J.M. Schulman, R.L. Disch, M.A. Miller and R.C. Peck, Chem. Phys. Lett. 14 (1987) 45.

1.103 T.I. Shibuya and M. Yoshitani, Chem. Phys. Lett. 137 (1987) 13.

1.104 D. Tomanek and M.A. Schluter, Phys. Rev. B36 (1987) 1208; Phys. Rev. Lett 67 (1991) 2331.

1.105 P.W. Fowler, J.E. Cremona and J.I. steer, Theor. Chim. Acta 73 (1988) 1.

1.106 M. Frenklach and L.B. Ebert, J. Phys. Che,. 92 (1988) 561.

1.107 R.C Haddon, Acc. Chem. Res. 21 (1988) 243.

1.108 G.E. Hite, Zivkovic, T.P. and Klein, D.J. Theor. Chim. Acta 74 (1988) 349.

1.109 J.M. Rudzingk, Thermoc Acta 125 (1988) 155.

1.110 Z. Slanina, Chem. Phys. Lett. 142 (1987) 512.

1.111 Z. Slanina, Carbon 25 (1987) 747.

1.112 Z. Slanina, Coll. Czech 52 (1987) 2831.

1.113 Z. Slanina, Thermoc Acta 127 (1988) 237.

1.114 Z. Slanina, Thermoc Acta 128 (1988) 157.

1.115 E. Brendsda, Theochem 57 (1989) 55.

1.116 Y. Jiang and H. Zhang, Theor. Chim. Acta 75 (1989) 279.

1.117 I. Laszl, Theochem. 52 (1989) 271.

1.118 A. Rosen and B. Wastberg, J. Am. Chem. Soc. 110 (1988) 701.

1.119 K. Balasubramanian, Chem. Phys. Lett 175 (1990) 273.

1.120 P.P. Radi, M.T. Hsu, M.E. Rincon, P.R. Kemper and M.T. Bowers, Chem. Phys. Lett. 174 (1990) 223.

1.121 Z. Slanina, Chem. Phys. Lett. 173 (1990) 164.

1.122 Z. Slanina, Theochem. 61 (1989) 169.

1.123 Z. Slanina, Thermoc A C 156 (1989) 285.

1.124 Z. Slanina, Thermoc A C 168 (1990) 89.

1.125 W. Kratschmer, Chem. Phys. Lett. 170 (1990) 167.

1.126 T. Takai, C. Lee, T. Halicioglu and W.A. Tiller, J. Phys. Chem. 94 (1990) 4480.

1.127 E. Manousakis, Phys. Rev. B44 (1991) 10991.

1.128 W.G. Harter and D.E. Weeks, Chem. Phys. Lett. 132 (1986) 387.

1.129 Z.C. Wx, D.A. jelski and T.F. George, Chem. Phys. Lett 137 (1987) 291.

1.130 R.E. Stanton and M.D. Neuton, J. Phys. Chem. 92 (1988) 2141.

1.131 D.E. Weeks and W.G. Harter, Chem. Phys. Lett. 144 (1988) 366.

1.132 D. Bethune, G. Sm Meijer, W.C. Tang and H.J. Rosen, Chem. Phys. Lett. 174 (1990) 219.

1.133 A. Cheng and M.L. Klein, J. Phys. chem. 95 (1991) 6750.

1.134 G.W. Hayden and E.J. Mele, Phys. Rev. B36 (1987) 5010.

1.135 F. Negri, G. Orlandi and F. Zerbetto, Chem. Phys. Lett. 144 (1988) 31.

1.136 A. Ceulemans and P.W. Fowler, J. Chem. Phys. 93 (1990) 1221.

1.137 G. Stollhoff, Phys. Rev. B44 (1991) 10998.

1.138 P.A. Heiney, J.E. Fischer, A.R. McGhie, W.J. Romanow, A.M. Denenstein, Jr. J.P. McCauley, A.B. Smith III, and D.E. Cox, Phys. Rev. Lett. 66 (1991) 2911.

1.139 R.L. Garrel, T.M. Herne, C.A. Szafranski. F. Diederich, F. Ettl. and R.L. Wehtten, J. Am. Chem. Soc. 113 (1991) 6302.

1.140 A. Kaldor, Z. Phys. D3 (1986) 195.

1.141 T.P. Martin, Z. Phys. D3 (1986) 211.

1.142 M. Pettini,m Astrophys. J. 310 (1986) 700.

1.143 R. Rabilizi, Astrophys. Sp. Sc. 125 (1986) 331.

1.144 J. August, Astrophys. Sp. Sc. 128 (1986) 411.

1.145 M.E. Geusic, Z. Phys. D3 (1986) 309.

1.146 F. Hoyle and N. Wickramansinghe, Astrophys. Sp. Sc. 122 (1986) 181.

1.147 A. Leger, Astron. Astr. 203 (1988) 145.

1.148 A. Leger, Astron Astr. 216 (1989) 148.

1.149 T.P. Snow, Astron Astr. 213 (1989) 291.

1.150 W.B. Somerville, Month. Not. Roy Ast. 240 (1989) P41.

1.151 S.B. Trickey, Astrophys. J. 336 (1989) L37.

1.152 I.J. Allamand, Astrophys, J.S. 71 (1989) 733.

1.153 S.P. Balm, Month Not. Roy. Ast. 245 (1990) 193.

1.154 J.M. Perrin, Astrophys. J. 364 (1990) 146 R.

1.155 A. Webster, Nature 352 (1991) 412.

1.156 D. Heymann J. Geo. R, SE 91 (1986) E135.

1.157 E.A. Rohlfing, J. Chem. Phys. 89 (1988) 6103.

1.158 C. Witting, S. Sharpe and R.A. Beaudet, Acc. Chem. Res. 21 (1988) 341.

1.159 H. Feld, R. Zurmuhlen, A. Leute and A. Benninghoven, J. Phys. Chem. 94 (1990) 4595.

1.160 W.Y. Ching, M.Z. Huang, Y.N. Xu, W.G. Harter and F.T. Chan, Phys. Rev Lett. 67 (1991) 2045.

1.161 T.W. Ebbesen, K. Tanigaki, S. Kuroshina, Chem. Phys. Lett. 181 (1991) 501.

1.162 G.f. Bertsch, A. Bulgac, D. Tomanek and Y. Wang, Phys. Rev. Lett. 67 (1991) 2690.

1.163 E.A. Rohlfing, J. Opt. Soc. Am. B7 (1990) 1915.

1.164 E. Sohmen, J. Fink, R.H. Baughman and W.Z. Krätschmer, Z. Phys. B (1991); J. Fink, E. Sohmen and W. Krätschmer, Physica C (preprint).

1.165 R.R. Huang and J. Granpwski, J. Phys. Chem. 95 (1991) 6073.

1.166 D.L. Andrews, and K.P. Hopkins, Chem. Phys. Lett. 146 (1988) 37.

1.167 T.P. Martin, Angew Chem. 25 (1986) 197.

1.168 J.C. Phillips, Chem. Rev. 86 (1986) 619.

1.169 J. Bernholc and J.C. Phillips, Phys. Rev. B33 (1986) 7395.

1.170 L. Brus, J. Phys. Chem. 90 (1986) 2555.

1.171 R.E. Davies, Nature 324 (1986) 10.

1.172 M.Y. Hahn, E.C. Honea, A.J. Paguva, K.E. Schriver, A.M. Camarena and R.L. Whetten, Chem. Phys. Lett. 130 (1986) 12.

1.173 J. Almlof, A.C.S. Symp. Ser. 353 (1987) 35.

1.174 W.L. Brown, R.R. Freeman, K. Raghavachari and M. Schluter, Science 235 (1987) 860.

1.175 C. Coulombe, J. Chim. Phys. 84 (1987) 875.

1.176 C.O. Dietrich, C.O.D. Buchecker and J.P. Sauvage, Chem. Rev. 87 (1987) 795.

1.177 P. Garratt, Endeavour 11 (1987) 36.

1.178 M.E. Geusic, M.F. Jarrold, T.J. McLrath, R.R. Freeman and W.L. Brown, J. Chem. Phys. 86 (1987) 3862.

1.179 R. Hofmann, Am. Scientist 75 (1987) 619.

1.180 D.E.H. Jones, Chem. Brit. 23 (1987) 465.

1.181 A. Kaldor, Chem. Tech. U.S. 17 (1987) 300.

1.182 C. Keller, Chem. Zeitun, 111 (1987) 103.

1.183 K. Kovacevi, Int. J. Quant. 589 (1987).

1.184 H.W. Kroto, Nature 329 (1987) 529.

1.185 R.B. Mallion, Nature 325 (1987) 760.

1.186 K.M. Jr. Merz, R. Hoffman and A.T. Balaban, J. Am. Chem. Soc. 109 (198
6742.

1.187 M. Randic, S. Nikolic and N. Trinajstic, Gaz. Chim. Italiana 117 (1987) 69.

1.188 P. Scheier, Int. J. Mass. 76 (1987) R11.

1.189 M.S. Shaw and J.D. Johnson, J. Appl. Phys. 62 (1987) 2080.

1.190 Z. Slanina, Int. R. Phys. Chem. 6 (1987) 251.

1.191 P.L. Timms, Acc. Symp. S. 333 (1987) 1.

1.192 D.J. Trevor, Acc. Symp. S. 333 (1987) 43.

1.193 Y. Wang. A. Suna, W. Mahler and R. Kasowski, J. Chem. Phys. 87 (1987) 73

1.194 D.E. Williams, J. Chem. Phys. 87 (1987) 4207.

1.195 J. Aihara and H. Hosoya, Bull. Chem. Soc. Japan 61 (1988) 2657.

1.196 J. Brandmul, Croat. Chem. 61 (1988) 267.

1.197 E. Brendsda, Spect. Lett. 21 (1988) 313.

1.198 E. Brendsda, Spect. lett 21 (1988) 319.

1.199 C.E. Borwn, J. Pol. Sc. Pc. 26 (1988) 131.

1.200 O.S. Bruinsma, Fuel 67 (1988) 1190.

1.201 R.L. Cappelletti, J.R.D. Copley, W.A. Kamitakahara, Li. Fang, J.S. Lannin a
D. Ramage, Phys. Rev. Lett. 66 (1991) 3261.

1.202 Y. Chai, T. Guo, C. Jin, R.E. Haufler, L.P.F. Chibante, J. Fure, L. Wang, J.
Alford and R.E. Smalley, J. Phys. Chem. 95 (1991) 7564.

1.203 R.J. Van Zee, R.F. Ferrante, K.J. Zeringhe, W. Weltner Jr. and D.W. Ewing,
Chem. Phys. 88 (1988) 3465.

1.204 P.J. Krusic, E. Wasserman, B.A. Parkinson, B. Malone, E.R. Holler Jr. P.
Keizer, J.R. Morton and K.F. Preston, J. Am. Chem. Soc. 113 (1991) 6274.

1.205 P.N. Keizer, J.R. Morton, K.F. Preston and A.K. Sugden, J. Phys. Chem.
(1991) 7117.

2.1 A.F. Hebbard, M.J. Rosseinsky, R.C. Haddon, D.W. Murphy, S.H. Glarum, T.T.
Palstra, A.R. Ramirez and A.R. Kortan, Nature 350 (1991) 600.

2.2 K. Holczer, O. Klein, S.-M. Haung, R.B. Kaner, K.F. Fu, R.L. Whitten, F. Diederich, Sci. 252 (1991) 1154.

2.3 M.J. Rosseinsky, A.P. Ramirez, S.H. Glarum, D.W. Murphy, R.C. Haddon, A.F. Hebard, T.T.M. Palstra, A.R. Kortan, S.M. Zahuxk and A.V. Makhija, Phys. Rev. Lett. 66 (1991) 2830.

2.4 Z. Iqbal, R.H. Baughman, B.L. Ramakrishna, S. Khare, N.S. Murthy, H.J. Bornemann and D.E. Morris, Science 254 (1991) 826.

2.5 R.C. Haddon, Nature 350 (1991) 320.

2.6 K. Holczer, O. Klein, G. Grünes, J.D. Thompson, F. Diederich and R.L. Whetten, Phys. Rev. Lett. 67 (1991) 271.

2.7 Z. Zhang, C.C. Chen, S.P. Kelty, H. Dai and C.M. Lieber, Nature 353 (1991) 333.

2.8 J.E. Fischer, P.A. Heiney, A.R. McGhie, W.J. Romanow, A.M. Denenstein, J.P. McCauley Jr., A.B. Smith III, Science 252 (1991) 1288.

2.9 S.J. Duclos, K. Brister, R.C. Haddon, A.R. Kortan and F.A. Thiel, Nature 351 (1991) 380.

2.10 J.E. Schirber, D.L. Overmyer, H.H. Wang, J.M. Williams, K.D. Carlson, A.M. Kini, U. Welp, W.K. Kwok, Physica C178 (1991) 137.

2.11 G. Sparn, J.D. Thompson, S.M. Hang, R.B. Kaner, F. Diederich, R.L. Whetten, G. Gruner and Holczer, Science 252 (1991) 1829.

2.12 L. Chandran, L. Jansen and E. Lambardi, Physica C (in press); L. Jansen, R. Block and E. Lambardi, Physica C 182 (1991) 17.

2.13 L. Janesen and L. Chandran, Europhysics. Lett (in press).

2.14 P.W. Stephens, L. Mihaly, P.L. Lee, R.L. Whetten, S.M. Haung, R. Kaner, F. Deiderich and K. Holczer, Nature 351 (1991) 632.

2.15 K. Tanigaki, T.W. Ebbesen, S. Saito, J. Mizuki, J.S. Tsai, Y. Kubo and S. Kuroshima, Nature 352 (1991) 222.

2.16 S.P. Kelty, C.C. Chen and C.H. Lieber, Nature 352 (1991) 223.

2.17 R.M. Fleming, M.J. Rosseinsky, A.P. Ramirez, D.W. Murphy, J.C. Tully, R.C. Haddon, T. Siegrist, R. Tycko, S.A. Glarum, P. Marsh, G. Dabbagh, S.M. Zahurak, A.V. Makhija and C. Hampton, Nature 352 (1991) 701.

2.18 R.M. Fleming, A.P. Ramirez, M.J. Rosseinsky, D.W. Murphy, R.C. Haddon, S.M. Zahurak and A.V. Makhija, Nature 352 (1991) 787.

2.19 C.C. Chen, S.P. Ketty and C.M. Lieber, Science 252 (1991) 886.

2.20 M.J. Rosseinsky, A.P. Ramirez, S.H. Glarum, D.W. Murphy, R.C. Haddon, A.F
Hebard, T.T.M. Palstra, A.R. Kortan, S.M. Zahurak and A.V. Makhija, Phys. Rev
Lett. 66 (1991) 2830.

2.21 A.A. Zakhidov, A. Ugawa, K. Imaeda, K. Yukushi, H. Inokuchi, K. Kikuchi, I
Ikemoto, S. Suzuki and Y. Achiba, Solid State Communication 79 (1991) 939.

2.22 A.A. Zakhidov, K. Imaeda, A. Ugawa, K. Yukushi, H. Inokuchi, Z. Iqbal, R.H
Baughman, B.L. Ramakrishna and Y. Achiba, Physica C (in press).

2.23 A.A. Zakhidov, A. Ugawa, K. Yukushi, K. Imaeda, H. Inokuchi, I.I. Khairullin, P.K
Khabibullaev, Physica C (in press).

2.24 K. Kikuchi, S. Suzuki, K. Saito, H. Shiromaru, I. Ikemoto, Y. Achiba, A.A. Za
khidov, A. Ugawa, K. Imaeda, H. Inokuchi and K. Yukushi (Preprint, 1991); Y
Maruyama, T. Inabe, H. Agata, Y. Achiba, S. Suzuki and I. Ikemoto, Chemistr
Lett. (1991) 1849.

2.25 A.A. Zakhidov, K. Yukushi, K. Imaeda, H. Inokuchi, K. Kikuchi, S. Suzuki, I
Ikemoto and Y. Achiba, Molec. Cryst. Liq. Cryst. (preprint).

2.26 H.H. Wang, A.M. Kini, B.M. Savall, K.D. Carlson, J.M. Williams, K.R. Lykke
D.H. Parker, P. Wurz, M.J. Pellin, D.M. Gruen, U. Welp, W.K. Kwok, S. Fleshler
G.W. Crabtree, Inorg. Chem. 30 (1991) 2838.

2.27 H.H. Wang, A.M. Kini, B.M. Savall, K.D. Carlson, J.M. Williams, M.W. Lathrop
K.R. Lykke, D.H. Parker, P. Wurz, M.J. Pellin, D.M. Gruen, U. Welp, W.-K. Kwok
S. Fleshler, G.W. Crabtree, J.E. Schirber and D.L. Overmyer, Inorg. Chem. 3
(1991) 2962.

2.28 M. Kraus, J. Freytag, S. Garter, H.M. Vieth, W. Krätschmer and K. Luders, Z
Phys. B-condensed matter 85 (1991) 1.

2.29 R.C. Haddon, Acc. Chem. Res. 25 (1992) 127.

2.30 P.W. Stephens, L. Mihaly, P.L. Lee, R.L. Whetten, S.-M. Haung, R. Kaner, F
Diederich and K. Holczer, Nature 351 (1991) 632.

2.31 M.Z. Haung, Y.N. Xu and W.Y. Ching, J. Chem. Phys. (preprint).

2.32 Y.N. Xu, M.Z. Haung and W.Y. Ching, Phys. Rev. B (15 Dec. 1991).

2.33 E. Manousakis, Phys. Rev. B44 (1991) 10991.

2.34 S.C. Erwin and M.R. Pederson, Phys. Rev. Lett. 67 (1991) 1610.

2.35 M.R. Pederson, S.C. Erwin, W.E. Pickett, K.A. Jackson and L.L. Boyer (preprint

2.36 P.J. Benning, J.L. Martins, J.H. Weaver,, L.P.F. Chibante and R.E. Smalley, Science 252 (1991) 1417.

2.37 G.K. Werthein, J.E. Rowe, D.N.E. Buchanan, E.E. Chaban, A.F. Hebard, A.R. Kortan, A.V. Makhija and R.C. Haddon, Science 252 (1991) 1419.

2.38 Y. Fukuda, N. Sanada, M. Nagoshi, T. Takahashi, H. Katayama-Yoshida, K. Kikuchi. S. Suzuki, I. Ikemoto, Y. Achiba, Y. Syono and M. Tachiki, Physica C 181 (1991) 320.

2.39 H. Hoshi, N. Nakamura, Y. Maruyama, T. Nakagawa, S. Suzuki, H. Shiramaru, Y. Achiba, Jpn. J. Appl. Phys. 30 (1991) L1397.

2.40 C.T. Chen, L.H. Tjeng, P. Rudolf, G. Meigs, J.E. Rowe, J. Chen, J.P. McCauley Jr, A.B. Smith III, A.R. McGhie, W.J. Ramanow and E.W. Plummer, Nature 352 (1991) 603.

2.41 R.C. Haddon, S.H. Glarum, S.V. Chichester, A.P. Ramirez and N.H. Zimmerman, Phys. Rev. B43 (1991) 2646.

2.42 R. Tycko, G. Dabbagh, R.M. Fleming, R.C. Haddon, A.V. Makhija and S.M. Zahurak, Phys. Rev. Lett 67 (1991) 1886.

2.43 H. Schwartz, Angew. Chem. 30 (1991) 884.

2.44 P.M. Allemand, K.C. Khemani, A. Koch, F. Wudl, K. Holczer, S. Donovan, G. Gruner and J.D. Thompson, Science 253 (1991) 301.

2.45 M. Saunders, Science 253 (1991) 330.

2.46 D. Swinbanks, Nature 352 (1991) 749.

2.47 D. Swinbanks, Nature 353 (1991) 377.

2.48 Y. Wang, D. Tomanek and G.F. Bertsch, Phys. Rev. B44 (1991) 6562.

2.49 G. Gensterblum, J.J. Pireaux, P.A. Thiry, R. Caudano, J.P. Vigneron, Ph. Lambin, A.A. Lucas and W. Kratschmer, Phys. Rev. Lett. 67 (1991) 2171.

2.50 E.E.B. Campbell, G. Ulmer and Hertel IV, Phys. Rev. Lett. 67 (1991) 1986.

3.1 F.C. Zhang, H. Ogata and T.M. Rice, Phys. Rev. Lett. (in press).

3.2 W.P. Su, J.C. Schriffer and A.J. Heeger, Phys. Rev. Lett. 42 (1979) 1698.

3.3 S. Chakravarty and S. Kivelson (unpublished). These authors have argued a netative U Hubbard model for the doped fullerene from a different point of view.

3.4 Schluter, M. Lannoo, M. Needles, G.A. Baraff and D. Tomanek, Phys. Rev. Lett. (1991) (preprint).

3.5 J.L. Martins, N. Troullier and M. Schabel, Phys. Rev. B (in press).

3.6 S.C. Erwin and W.E. Pickett, Science (preprint).

3.7 Y. Takada, Phys. Rev. Lett. (1991) (preprint).

3.8 G. Stollhoff, Phys. Rev. B44 (1991) 10998.

3.9 N. D' Ambrumenil, Nature 352 (1991) 472; Y.J. Vemura, A. keren, L.P. Lee, G.M. Luke, B.J. Sternlieb, W.D. Wu, J.H. Brewer, R.L. Whetten, S.M. Haung, S. Lin, R.B. Karen, F. Diederich, S. Donovan, G. Gruner and K. Holczer, Nature 352 (1991) 605.

3.10 O. Zhou, J.E. Fischer, N. Coustel, S. Kycia, Q. Zhu, A.R. McGhie, W.J. Romanow, J.P. McCauley. Jr, A.B. Smith III and D.E. Cox, Nature 351 (1991) 462.

3.11 Y. Guo, N. Karasawa and W.A. Goddard III, Nature 351 (1991) 464.

3.12 J.C. Philips, Solid State Commun. 80 (7) (1991) 517.

3.13 R. Tycko, G. Dabbagh, M.J. Rosseinsky, D.W. Murphy, R.M. Fleming, A.P. Ramirez and J.C. Tully, Science 253 (1991) 884.

3.14 Q. Zhu, O. Zhou, N. Coustel, G.B.M. Vaughan, J.P. McCauley. Jr, W.J. Romanow, J.E. Fischer and A.B. Smith III, Science 254 (1991) 545.

3.15 S. Koka, A. Chatterjee, K.N. Shrivastava, Solid State Commun. 81(6) (1992) 509.

3.16 S. Koka and K.N. Shrivastava, Solid State Physics Symp. C34 (1991) held at Varanasi. India. Dec. 1991, Dept. of Atomic Energy of the Govt. of India, Bombay.

3.17 K.N. Shrivastava and K.P. Sinha, Phys. Rep. 115 (1984) 93.

3.18 K.N. Shrivastava, Phys. Rep. C200 (1991) 51.

3.19 R.M. Fleming, A.P. Ramirez, M.J. Rossimsky, D.W. Murphy, R.C. Haddon, S.M Zhahurak and A.V. Makhija, Nature 352 (1991) 787.

3.20 A.R. Kortan, N. Kopylov, S. Glarum, E.M. Gyorgy, A.P. Ramirez, R.M. Fleming, O. Zhou, F.A. Theil, P.L. Trevor and R.C. Haddon, Nature 360 (1992) 566.

4.1 H. Masuda, F. Mizuno, Y. Yamada, K. Sugawara, I. Hirabayashi, Y. Shiohara and S. Tanaka, Supercond. Sci. Technol. 4, S295 (1991).

4.2 V.Y. Nagy, T.A. Orlova, L.A. Matveeva, S.S. Grahulene and O.M. Petrukhin, Anal. Chem. USSR. Eng. Trans. 45, 875 (1990).

4.3 H. Masuda, F. Mizuno, I. Hirabayashi and S. Tanaka, Phys. Rev. B43, 7871 (1991).

4.4 N. Guskos, G.P. Triberis, M. Calamiotou, C. Trikalinos, A. Koufoudabris, C. Mitros, H. Gamariseale, D. Niarchos, Phys. Stat Solidi (B) 163, K89 (1991).

4.5 N. Guskos, M. Calamiotou, S.M. Paraskevas, A. Konfondakis, C. Mitros, H. Gamariseale, J. Kuriata, L. Sadlowski and M. Wabia, Phys. Stat. Solidi (B) 162, K101 (1990).

4.6 N. Guskos, C.A. Londos, C. Trikalinos, S.M. Paraskevas, A. Koufoudakis, C. Mitros, H. Gamariseale and D. Niarchos, Phys. Stat. Solidi (B)165, 249 (1990).

4.7 L. Kan, S. Elschner and B.Elschner, Solid State Commun. 79, 61 (1991).

4.8 P. Wang, S. Brandow, J.W. Zhan and D.B. Zhu, Synthet. Met. 43, 4077 (1991).

4.9 I.A. Garifullin, N.N. Garifyanov, N.E. Alekseevskii and S.F. Kim, Physica C179, 9 (1991).

4.10 H.C. Kim, H. So and H.K. Lee, Bull. Kor. Chem. Soc. 12, 499 (1991).

4.11 A. Janossy and R. Chicault, Physica, C192, 399 (1992).

4.12 V.F. Meshcheryakov and V.A. Murashov, Zh. Eksp. Teor. Fiz. 101, 241 (1992).

4.13 O.N. Bakharev, A.G. Volodin, A.V. Duglov, A.V. Egorov, M.S. Tagirov and M.A. Teplov, Fiz. Niz. Temp. 17, 1337 (1991).

4.14 J.M. Dance, A. Tressaud, B. Chevalier, C. Brisson and J. Etourneau, Physica C198. 185 (1992).

4.15 M.A.T. Dawoud, A.A. Elhamalawy and E.A. Ghali, J. Mat. Sc. 27, 4016 (1992).

4.16 V. Kataev, G. Winkel, D. Khomskii, D. Wohlleben, W. Crump, K.F. Tebbe and J. Hahn, Solid State Commun. 83, 435 (1992).

4.17 P. Delhaes, J. Amiell, L. Ducasse, B. Hilti, C.W. Mayer, and J. Zambounis, Physica B182, 99 (1992).

4.18 V.E. Kataev, Y.S. Greznev, E.F. Kukovitskii, G.B. Teitelbaum, M. Brener and N. Knauf, J.E.T.P. Lett. 56, 385 (1992).

4.19 N. Guskos, V. Lyvkodimos, C.A. Londos, A. Kondos, S.M. Paraskevas, A. Koufoudakis, C. Mitros, H. Gamariseale, D. Niarchos and I. Krak, J. Supercond. 5, 457 (1992).

4.20 P. Ganguly, N. Shah and F.C. Matacotta, Physica C206, 70 (1993).

4.21 J. Stankowski, W. Hilczer, J. Baszynskii, B. Czyzak and L. Szczepanska. Solid State State Commun. 77, 125 (1991).

4.22 A. Moto, A. Morimoto, M. Kumeda and T. Shimizu, Supercond. Sci. Technol. **3**, 579 (1990).

4.23 A. Janossy, A. Rockenbauer, S. Pekker, G. Oszlanyi, G. Faigel and L. Korecz, Physica C**171**, 457 (1990).

4.24 N. Guskos, G.P. Triberis, M. Calamiotou, C. Trikalinos, A. Koufoudakis, C. Mitros, H. Gamariseale and D. Niarchos, Phys. Stat. Solidi B**162**, 243 (1990).

4.25 N. Guskos, G.P. Triberis, U. Lykadimos, W. Windsch, H. Metz, A. Koufoudakis, C. Mitros, H. Gamariseale, and D. Niarchos, Phys. Stat. Solidi B**166**, 233 (1991).

4.26 F. Scheerlinck, I. Francois, J. Vanbockstal, P. Janssen, P. Degroot, F. Herlach and J. Witters, Physica B**177**, 101 (1992).

4.27 H. Shimizu, T. Kiyama and F. Nakamura, Physica C**107**, 225 (1993).

4.28 N. Guskos, J. Kuriata, V. Likodimos, L. Sadlowski, A. Koufoudakis, C. Mitros, H. Gamariseale, and D. Niarchos, Phys. Stat. Solidi B**175**, K61 (1993).

4.29 F.J. Owens, Z. Iqbal, A.A. Zakhidov and I.I. Khairullin, Physica C**174**, 309 (1991).

4.30 H. Maruyama and I. Shiozaki, Jpn. J. Appl. Phys. Pt. 2 Lett. **30**, L694 (1991).

4.31 R. Eremina, M. Falin, S. Gudenko, V. Kosynkin, S. Kucheyko, A. Mezhuev, D. Pishagin, A. Yakubovskii and E. Zhaeglov, Physica C**185**, 849 (1991).

4.32 S.K. Misra, M. Kahrizi, J. Kotlinski, M.O. Steinitz and F.S. Razavi, Solid State Commun. **81**, 503 (1992).

4.33 R.J. Barham and D.C. Doetschman, J. Mat. Res. **7**, 565 (1992).

4.34 F. Bensebaa, B.S. Xiang and L. Kevan, J. Phys. Chem. **96**, 6118 (1992).

4.35 N. Kinoshita, Y. Tanaka, M. Tokumoto and S. Matsumiya, Solid State Commun. **83**, 883 (1992).

4.36 F. Mehran, A.J. Schellsorokin and C.A. Brown, Phys. Rev. B**46**, 8579 (1992).

4.37 P. Byszewski, R. Jablonski and S. Kolesnik, Solid State Commun. **84**, 1111 (1992).

4.38 N. Kinoshita, Y. Tanaka, M. Tokumoto and S. Matsumiya, Int. J. Mod. Phys. B**6**, 4025 (1992).

4.39 M. Bonvalot, M. Puri and L. Kevan, J. Chem. Soc. Faraday Trans. **88**, 238 (1992).

4.40 M. Bonvalot and L. Kevan, J. Phys. Chem. **96**, 9992 (1992).

4.41 Y.N. Shvachko, A.A. Koshta, A.A. Romanyukha, V.V. Ustinov and A.I. Akimov, Physica C174, 447 (1991).

4.42 N. Bontemps, D. Davidov, P. Monod and R. Even, Phys. Rev. B43, 11512 (1991).

4.43 J.T. Masiakowski, M. Puri and L. Kevan, J. Phys. Chem. 95, 8968 (1991).

4.44 Y.N. Shvachko, A.A. Koshta, A.A. Romanyukha, S.V. Naumov and V.V. Ustinov, Physica C197, 27 (1992).

4.45 F. Mehran and P.W. Anderson, Solid State Commun. 71, 29 (1989). The result given have is the same as that of ref. 184.

4.46 S. Chakravarty and R. Orbach, Phys. Rev. Lett. 64, 224 (1990).

4.47 R. Janes, K.K. Singh, S.D. Burnside and P.P. Edwards, Solid State Commun. 79, 241 (1991).

4.48 A. Deville, L. Bejjit, B. Gaillard, J.P. Sorbier, O. Monnereau. H. Noel and M. Potel, Phys. Rev. B47, 2840 (1993).

4.49 K.N. Shrivastava, Bull. Mat. Sci. (India) 14, 625 (1991).

4.50 K.N. Shrivastava, J. Phys. (Paris) 49, C8-2239 (1988).

4.51 K.N. Shrivastava, Solid State Commun. 68, 1019 (1988).

4.52 A.J. Millis, H. Monien and D. Pines, Phys. Rev. B42, 167 (1990).

4.53 H. Monien, D. Pines and M. Takigawa, Phys. Rev. B43, 258 (1991).

4.54 H. Monien, P. Monthoux and D. Pines, Phys. Rev. B43, 275 (1991).

4.55 A.J. Mills and H. Monien, Phys. Rev. B45, 3059 (1992).

4.56 D. Thelen, D. Pines and J.P. Liu, Phys. Rev. B47, 9151 (1993).

4.57 N. Balut and D.J. Scalapino, Phys. Rev. Lett. 68, 706 (1992); Phys. Rev. B47 3419 (1993).

4.58 K.N. Shrivastava, J. Phys. C. 3, 535 (1970).

4.59 K.N. Shrivastava, Phys. Rev. B21, 2702 (1980).

4.60 B.S. Shastry, Phys. Rev. lett. 63, 1288 (1989).

4.61 H. Niki, T. Higa, S. Tomiyoshi, M. Omori, T. Kajitani, T. Sato, T. Shinohara, T. Suzuki, K. Yagasaki and R. Igei, J. Mag. Mag. Mat. 90, 672 (1990).

4.62 N.M. Suleimanov, A.D. Shengelaya, R.G. Mustafin, E.F. Kukovitskii, P.W. Klamut, G.W. Chadzynski, H. Drulis and J. Janczak, Physica C185, 759 (1991).

4.63 H. Niki, H. Kyan, T. Shinohara, S. Tomiyoshi, M. Omori, T. Kajitani, T. Sato and R. Igei, Physica C185, 1133 (1991).

4.64 S.D. Goren, C. Korn, V. Volterra, H. Riesmeier, E. Rossier, H.M. Vieth and K. Luders, Phys. Rev. B46, 14142 (1992).

4.65 M. Nicolas, C.N.V. Huong, A. Dubon, G. Vetter and J. Conard, J. Phys. III (Paris) 3, 13 (1993).

4.66 M. Takigawa, A.P. Reyes, P.C. Hammel, J.D. Thompson, R.H. Heffner, Z. Fisk and K.C. Ott, Phys. Rev. B43, 247 (1991).

4.67 Y. Kohori, T. Sugata, T. Kohara and Y. Oda, J. Mag. Mag. Mat. 90, 667 (1990).

4.68 R. Dupree, Z. P. Han, A.P. Howes, D.M. Paul, M.E. Smith and S. Male, Physica C175, 269 (1991).

4.69 L. Reven, J. Shore, S. Yang, T. Duncan, D. Schwartz, J. Chung and E. Oldfield, Phys. Rev. B43, 10466 (1991).

4.70 F. Barriquand, P. Odier and D. Jerome, Physica C177, 230 (1991).

4.71 A. Trokiner, L. LeNoc, J. Schneck, A.M. Pougnet, R. Mellet, J. Primot, H. Savary Y.M. Gao and S. Aubry, Phys. Rev. B44, 2426 (1991).

4.72 A. Trokiner, L. Lenoc, R. Mellet, A.M. Pougnet, D. Morin, Y.M. Gao, J. Primot and J. Schneck, Phase Trans. 30, 147 (1991).

4.73 K. Ishida, Y. Kitaoka, G.Q. Zheng and K. Asayama, J. Phys. Soc. Japan 60, 3516 (1991).

4.74 G.Q. Zheng, Y. Kitaoka, K. Asayama, Y. Kodama and Y. Yamada, Physica C193 154 (1992).

4.75 A.P. Howes, R. Dupree, D.M. Paul and S. Male, Physica C193, 189 (1992).

4.76 I. Mangelschots, M. Mali, J. Roos, D. Brinkman, S. Rusiecki, J. Karpinski and E Kaldis, Physica 194, 277 (1992).

4.77 N.H. Hur, Y.K. Park, J.C.Park, Y.H. Ko and H.C.. Lee, Solid State Commun. 82 547 (1992).

4.78 M. Kyogaku, Y. Kitaoka, K. Asayama, C. Geibel, C. Schank, and F. Steglich, J Phys. Soc. Japan 61, 2660 (1992).

4.79 Y. Kishimoto, T. Ohno, T. Kanashiro, Y. Michihiro, Y. Kitaoka and K. Asayama Physica C185, 2731 (1991).

4.80 P. Mendels, H. Alloul, J.F. Marucco, J. Arabski and G. Collin, Physica C171, 42 (1990).

4.81 T. Kohara, K. Ueda, Y. Kohori and Y. Oda, J. Mag. Mag. Mat. **90**, 669 (1990).

4.82 K. Kumagai, M. Abe, S. Tanaka, Y. Maeno, and T. Fujita, J. Mag. Mag. Mat. **90**, 675 (1990).

4.83 D.P. Tunstall and W.J. Webster, Supercond. Sci. Technol. **4**, S406 (1991).

4.84 A.P. Reyes, D.E. MacLaughlin, T. Takigawa, P.C. Hammel, R.H. Heffner, J.D. Thompson and J.E. Crow, Phys. Rev. **B43**, 2989 (1991).

4.85 M. Mali, I. Mangelschots, H. Zimmermann, D. Brinkman, Physica **C75**, 581 (1991).

4.86 N.Q. Fan and J. Clarke, Rev. Sci. Instru. **62**, 1453 (1991).

4.87 K. Ishida, Y. Kitaoka, T. Yoshitomi, N. Ogata, T. Kamino and K. Asayama, Physica **C179**, 29 (1991).

4.88 T. Goto, K. Miyagawa and T. Fukase, Physica **C185**, 1081 (1991).

4.89 A.K. Rajarajan, L.C. Gupta and R. Vijayaraghavan, Physica **C193**, 413 (1992).

4.90 M. Bankay, M. Mali, J. Roos, I. Mangelschots and D. Brinkaman, Phys. Rev. **B46**, 11228 (1992).

4.91 M. Takigawa, J. Phys. Soc. Japan **53**, 1651 (1992).

4.92 V.S. Kasperovich and E.V. Charnaya, Fiz. Tverd. Tela **34**, 2040 (1992).

4.93 M. Horvatic, T. Auler, C. Berthier, Y. Berthier, P. Butaud, W.G. Clark, J.A. Gillet, P. Segransan and J.Y. Henry, Phys. Rev. **B47**, 3461 (1993).

4.94 H. Alloul, T. Ohno, H. Casalta, J.H. Marucco, P. Mendels, J. Arabski, G. Collin and M. Mehbod, Physica C **171**, 419 (1990).

4.95 T. Ohno, H. Alloul, P. Mendels, G. Collin and J.F. Marucco, J. Mag. Mag. Mat **90**, 657 (1990).

4.96 H.B. Brom and H. Alloul, Physica **C177**, 297 (1991).

4.97 T. Ohno, T. Kanashiro and K. Mizuno, J. Phys. Soc. Japan **60**, 2040 (1991).

4.98 Z.P. Han, R. Dupree, D.M. Paul, A.P. Howes and L.W.J. Caves, Physica **C181**, 355 (1991).

4.99 T. Ohno, K. Mizuno, T. Kanashiro and H. Alloul, Physica **C185**, 1067 (1991).

4.100 W.J. Webster, D.P. Tunstall, P.F. Freeman and J.R. Cooper, Physica **C185**, 1079 (1991).

4.101 P. Carretta and M. Corti, Phys. Rev. Lett. **68**, 1236 (1992).

4.102 D. Brinkmann, Z. Naturf. A47, 1 (1992).

4.103 R. Dupree, A. Geneten and D.M. Paul, Physica C193, 81 (1992).

4.104 J. Shore, S.T. Yang, J. Haase, D. Schwartz and E. Oldfield, Phys. Rev. B46, 595 (1992).

4.105 P.C. Hammel, E.T. Ahrens, A.P. Reyes, R.H. Heffner, P.C. Canfield, S.W. Cheong, Z. Fisk and J.E. Schirber, Physica C185, 1095 (1991).

4.106 M.A. Teplov, O.N. Bakharev, A.V. Dooglav, A.V. Egorov, M.V. Eremin, M.S. Tagirov, A.G. Volodin and R.S. Zhdanov, Physica C185, 1107 (1991).

4.107 Y. Kitaoka, K. Fujiwara, K. Ishida, T. Kondo and K. Asayama, J. Mag. Mag. Mat. 90, 619 (1990).

4.108 Y. Maniwa, T. Mituhashi, K. Mizoguchi and K. Kume, Physica C175, 401 (1991).

4.109 H. Alloul, Physica B169, 51 (1991).

4.110 A.A. Gippius, E.A. Kravchenko, V.V. Moshchalkov, Zh. Neorg. Khim SSSR 36 568 (1991).

4.111 M.C. Nuss, P.M. Mankiewich, M.L. Omalley, E.H. Westerwick, and P.B. Littlewood, Phys. Rev. lett. 66, 3305 (1991).

4.112 K. Asayama, G.Q. Zheng, Y. Kitaoka, K. Ishida and K. Fujiwara, Physica C178 281 (1991).

4.113 R. Dupree, Z.P. Han, D.M. Paul, T.G.N. Babu and C. Greaves, Physica C179 311 (1991).

4.114 K. Fujiwara, Y. Kitaoka, K. Ishida, K. Asayama, Y. Shimakawa, T. Manako an Y. Kubo, Physica C184, 207 (1991).

4.115 J.A. Martindale, S.E. Barrett, C.A. Klug, K.E. Ohara, S.M. Desoto, C.P. Slichte T.A. Friedmann and D.M. Ginsberg, Physica C185, 93 (1991).

4.116 T. Takahashi, K. Kanoda and G. Saito, Physica C185, 366 (1991).

4.117 S. Sasaki, K. Kinoshita and A. Matsuda, Physica C185, 1083 (1991).

4.118 K. Ishida, Y. Kitaoka, N. Ogata, T. Kamino and K. Asayama, Physica C185, 11 (1991).

4.119 M.W. Pieper, Physica C190, 261 (1992).

4.120 P. Carretta, M. Corti, A. Rigamonti, R. Derenzi, F. Licci, C. Paris, L. Bonol M. Sparpaglione and L. Zini, Physica C191, 97 (1990).

4.121 F.J. Ohkawa, J. Phys. Soc. Japan **61**, 1157 (1992).

4.122 Z.P. Han, R. Dupree, A. Geneten, R.S. Liu and P.P. Edwards, Phys. Rev. Lett. **69**, 1256 (1992).

4.123 D.R. Grempel and M. Lavagna, Solid State Commun. **83**, 595 (1992).

4.124 A.A. Lundin, Zh. Eksperim. Teor. Fiz. **102**, 352 (1992).

4.125 A. Trokiner, L. Lenoc, J. Schneck, A.M. Pougnet, R. Mellet, J. Primot and H. Savary, Ferroelectrics **128**, 149 (1992).

4.126 C. Bandte, Phys. Rev. **B47**, 5473 (1993).

4.127 P.B. Allen and D. Rainer, Nature **349**, 396 (1991).

4.128 T. Machi, I. Tomeno, T. Miyatake, N. Koshizuka, S. Tanaka, T. Imai and H. Yasuoka, Physica **C173**, 32 (1991).

4.129 P. Kalinay, Physica **C177**, 145 (1991).

4.130 S. Chakravarty, M.P. Gelfand, P. Kopietz, R. Orbach and M. Wollensak, Phys. Rev. **B43**, 2976 (1991).

4.131 T. Kostyrko, Phys. Stat. Solidi **B164**, K45 (1991).

4.132 T. Kostyrko, Physica **C178**, 182 (1991).

4.133 R.E. Walstedt, R.F. Bell and D.B. Mitzi, Phys. Rev. **B44**, 7760 (1991).

4.134 S. Wermbter and L. Tewordt, Phys. Rev. **B44**, 9524 (1991).

4.135 M.A. Teplov, Usp. Fiz. Nauk **161**, 195 (1991).

4.136 A.S. Alexandrov, Physica **C182**, 327 (1991).

4.137 A.B. Nazarenko and S.E. Shafranyuk, Phys. Stat. Solidi **B167**, K93 (1991).

4.138 S. Wermbter and L. Tewordt, Physica **C183**, 365 (1991).

4.139 J.A. Martindale, S.E. Barrett, C.A. Klug, K.E. O'Hara, S.M. DeSoto, C.P. Slichter, T.A. Friedmann and D.M. Ginsberg, Phys. Rev. Lett. **68**, 702 (1992).

4.140 M. Takigawa, J.L. Smith and W.L. Hults, Physica **C185**, 1105 (1991).

4.141 R. Combescot and G. Varelogiannis, Europhys. Lett. **17**, 625 (1992).

4.142 V. Hardy, D. Groult, J. Provost and B. Ravean, Physica **C190**, 289 (1992).

4.143 S. Fujimoto, J. Phys. Soc. Japan **61**, 765 (1992).

4.144 T. Ohno, K. Mizuno, T. Kanashiro, Y. Kishimoto and H. Alloul, J. Alloy Comp. **193**, 135 (1993).

4.145 Y. Itoh, H. Yasuoka, Y. Fujiwara, Y. Ueda, T. Machi, I. Tomeno, K. Tai, N. Koshizuka and S. Tanaka, J. Phys. Soc. Japan **61**, 1287 (1992).

4.146 H. Bahlouli, Phys. Lett. A**164**, 206 (1992).

4.147 R. Friedberg, T.D. Lee and H.C. Ren, Phys. Rev. B**45**, 10732 (1992).

4.148 A.V. Bondar, A.A. Motuz, S.M. Ryabchenko and Y.V. Fedotov, Fiz. Niz. Temp. **17**, 1272 (1991).

4.149 V.V. Naletov, A.V. Egorov, R.S. Zhdanov, M.S. Tagirov and M.A. Teplov, Fiz. Niz. Temp. **17**, 1341 (1991).

4.150 T. Ohno, H. Yasuoka and Y. Ueda, J. Phys. Soc. Japan **61**, 3869 (1992).

4.151 T.P. Devereaux, Z. Physik B**90**, 65 (1993).

4.152 Q.P. Li and R. Joynt, Phys. Rev. B**47**, 530 (1993).

4.153 J.A. Martindale, S.E. Barrett, K.E. O'Hara, C.P. Slichter, W.C. lee and D.M Ginsberg, Phys. Rev. B**47**, 9156 (1993).

4.154 I. Mangelschots, M. Mali, J. Roos, H. Zimmermann, D. Brinkman, S. Rusiecki, J karpinski, E. Kalids, and E. Jilek, Physica C**172**, 57 (1990).

4.155 R.G. Graham, P.C. Riedi and B.M. Wanklyn, J. Phys. Condens. Mat. **3**, 13 (1991).

4.156 G.Q. Zheng, E. Yanase, K. Ishida, Y. Kitaoka, K. Asyama, Y. Kodama, R. Tanaka S. Nakamichi, and S. Endo, Solid State Commun. **79**, 51 (1991).

4.157 S. Ohsugi, Y. Kitaoka, K. Ishida and K. Asayama, J. Phys. Soc. Japan **60**, 235 (1991).

4.158 V.L. Matukhin, I.A. Safin, V.N. Anashkin, O.V. Zarikov and Y.A. Ossipyan, Sol State Commun. **79**, 1063 (1991).

4.159 B.W. Stat and L.M. Song, Physica C**183**, 372 (1991).

4.160 M. Zoli, J. Phys. Soc. Japan **60**, 3837 (1991).

4.161 O. Okada, T. Oashi, K. Kumagai, T. Noji, Y. Koike and Y. Saito, Physica C**18** 1075 (1991).

4.162 S. Ohsugi, Y. Kitaoka, K. Ishida and K. Asayama, Physica C**185**, 1099 (1991).

4.163 H. Yamagata, K. Inada and M. Matsumura, Physica C**185**. 1101 (1991).

4.164 H. Tou, M. Matsumura and H. Yamagata, J. Phys. Soc. Japan **61**, 1477 (1992).

4.165 Y.A. Osipyan, O.V. Zharikov, V.N. Anashkin, V.L. Matukhin, A.L. Pogoreltsev. I.A. Safin and G.B. Teitelbaum. Fiz. Niz. Temp. **17**, 1331 (1991).

4.166 G. Markandeyulu, K.V. Gopalkrishnan, A.K. Rajarajan, L.C. Gupta, R. Vijayaraghavan, A.S. Tamhane, K.I. Gnanasekar and R. Pinto, Phys. Rev. B**47**, 1123 (1993).

4.167 T. Imai, C.P. Slichter, K. Yoshimura and K. Kosuge, Phys. Rev. Lett. **70**, 1002 (1993).

4.168 S.D. Goren, C. Korn, H. Riesemier and K. Luders, Phys. Rev. B**7**, 2821 (1993).

4.169 T. Shimizu, J. Phys. Soc. Japan **62**, 779 (1993).

4.170 S.B. Sulaiman, N. Sahoo, T.P. Das and O. Donzelli, Phys. Rev. B**45**, 7383 (1992).

4.171 V.N. Anashkin, V.L. Matukhin, E.F. Kukovitskii, A.I. Pogoreltsev and I.A. Sajin, Fiz. Niz. Temp. **17**, 1334 (1991).

4.172 J.H. Cho, F. Borsa, D.C. Johnston and D.R. Torgeson, Phys. Rev. B**46**, 3179 (1992).

4.173 A.A. Sorkin, E.N. Shirani, L.G. Shpinkova, Z.Z. Akselrod, B.A. Komissarova, G.K. Ryasny, S.I. Semyonov, G.A. Denisenko, I.P. Zibrov and A.R. Buev, Hyp. Int. **73**, 337 (1992).

4.174 T. Moriya, Prog. Theoret. Phys. **28**, 371 (1962).

4.175 Y.Q. Song, M.A. Kennard, K.R. Poeppelmeier and W.P. Halperin, Phys. Rev. Lett. **70**, 3131 (1993).

4.176 G.Q. Zheng, T. Kuse, Y. Kitaoka, K. Ishida, S. Ohsugi, K. Asayama and Y. Yamada, Physica C**208**, 339 (1993).

Low Temperature Specific Heat and Related Studies on Pure and Substituted Phases of High T$_C$ Cuprates

Deepak Varandani, A.K. Bandyopadhyay and A.V. Narlikar

National Physical Laboratory
Dr. K.S. Krishnan Road
New Delhi 110012, India

ABSTRACT:

Specific heat measurements on various high Tc cuprate superconductors have been carried out. In the present review, we discuss briefly a few of the important results obtained in our studies. The results on the single phase samples of Bi- and Tl- based cuprates have been discussed first /1/. Subsequently, from the specific heat and other measurements, we discuss the possible effect of lead substitution in Bi-based cuprate system where both low-Tc and high-Tc phases are present /2/. Finally, we summarize our work on the excess specific heat and the fluctuation induced effects in conjunction with substitution in RE-1-2-3 systems /3/. A brief description of the instrumentation has also been given.

I.   INTRODUCTION

The specific heat measurement at low temperature is recognized as an important tool in many areas of solid state physics. It yields various important parameters and provides an insight into phenomena, such as, the contribution of lattice vibrations [Debye Temperature (θ_D)], the density of states, Sommerfeld constant (γ), energy levels in magnetic materials, order-disorder transitions etc.. Thermodynamically,the specific heat data are considered to be the most reliable signature of any bulk phase transition - the specific heat anomaly associated with superconducting transition may be cited as an example. It is thus quite appropriate that since the discovery of these high Tc superconductors, an extensive specific heat measurements on different cuprate systems have been reported in literature /4/. Among them, in the most studied Y-based 1-2-3 $(Y_1Ba_2Cu_3O_{7-y})$ system, the specific heat results show the characteristic jump at Tc. Contrarily, the limited published data on Bi-Ca-Sr-Cu-O and Tl-Ca-Ba-Cu-O based systems signify two types of behaviours - (i) a distinc anomaly and (II) a broad hump in the vicinity of Tc. I is noteworthy to mention here that both these Bi- and Tl based systems can yield what are commonly known as low-T and high-Tc phases respectively represented by th stoichiometric compositions of 2-1-2-2 and 2-2-2-3 of Bi Ca-Sr-Cu-O. The Tc values of the 2-1-2-2 for Bi- and Tl based systems are around 85 K and 105 K, while of the 2-2 2-3 they are around 107 K and 125 K. In contrast to this the Y-based 1-2-3 system with Tc of about 90 K can be syn thesized fairly easily in homogeneous and single phas form suitable for investigation of their physica properties. It is, therefore, conjectured that the reaso of different behaviour of Bi- and Tl- based systems partly linked with the difficulties involved in realizir good quality samples which would be single phase or a least contain a substantial volume fraction of the desire phase. However, all these studies clearly show that th observed specific heat anomaly at Tc, both in the pure and substituted samples of Y-1-2-3 or in the other Bi-based and Tl-based systems, is weak in comparison to conventional low-Tc superconductors. This indicates that although inhomogeneous broadening due to the sample imperfections might partially cause weak anomaly or a broad hump in the specific heat data at Tc, the contribution of the intrinsic fluctuations which are found to be inherently present in the system, cannot be neglected. As the fluctuations in the superconducting order parameter, just above Tc, have long since been considered as an important effect responsible for a host of interesting phenomena /5/. Moreover, owing to the quasi two-dimensional nature, coupled with their intrinsically very small coherence length ξ (10 A), the fluctuation effects tend to get significantly pronounced in high-Tc cuprates than in conventional low-Tc superconductors which form three-dimensional systems of large ξ of about 1000 A /6/. The low dimensionality of cuprate superconductors essentially stems from their unusual layered structure which comprises one or more planar CuO_2 layers intervened by other cationic-oxide (or pure cationic) layers /7/.

It is in this context that the specific heat measurements on various high Tc superconductors have been pursued in the last five years in our laboratory. In the present review, we will (A) first discuss briefly the instrumentation and then (B) summarize the results in the following sequence: [B(1)] single phase samples of Bi- and Tl- based cuprates, [B(2)] effect of lead substitution in Bi-based cuprate containing both low-Tc and high-Tc phases, and [B(3)] the excess specific heat and the fluctuation induced effects in conjunction with the site dependent substitutions in RE-1- 2-3 system.

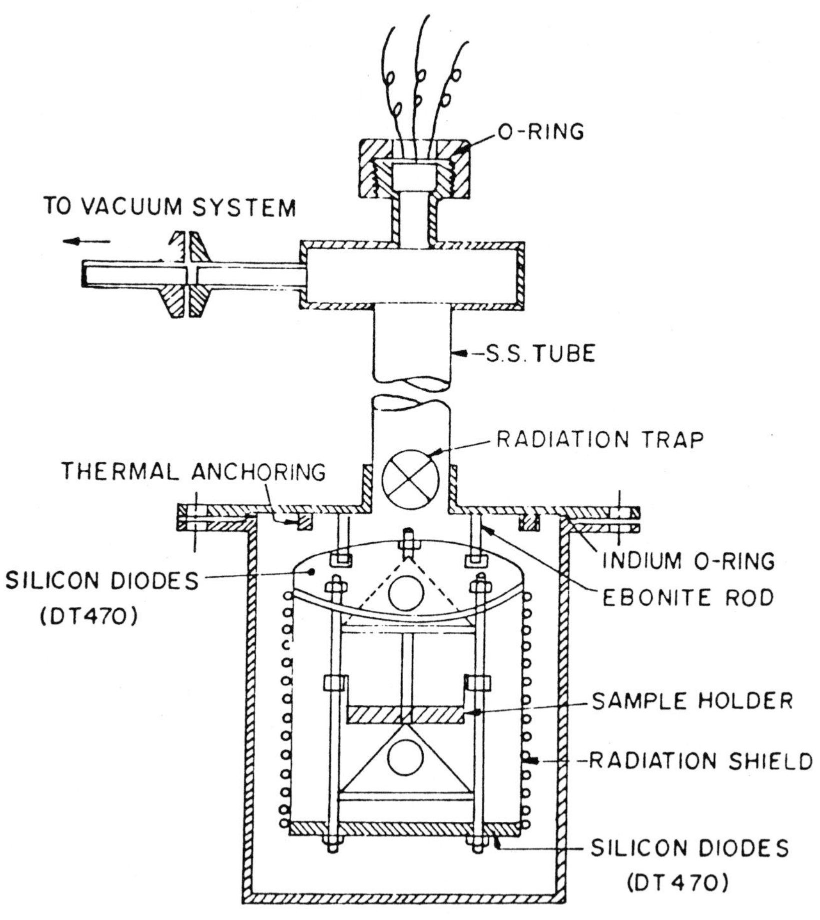

Figure 1 : Schematic representation
of the calorimeter

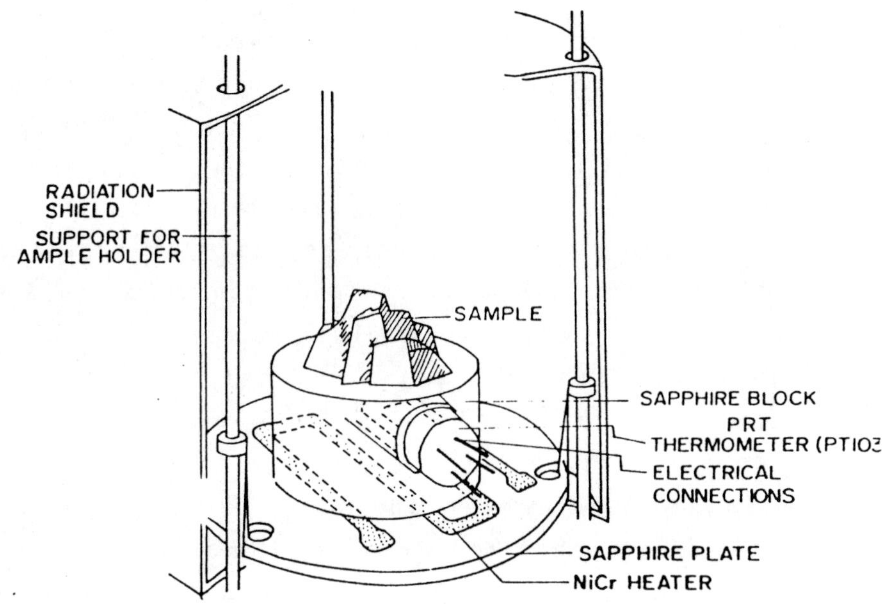

Figure 2 : Sample holder supported
by tripod stand with
nylon threads

II.   EXPERIMENTAL SET-UP:

The schematic representation of the calorimeter is
depicted in Fig. 1. As is clear that the calorimeter is
essentially comprised of (i) the sample chamber and (ii)
the outer jacket. The sample chamber is suspended by using
ebonite rods, the heating element [25 Ohm] of the radiation shield is wound around the copper jacket with GE var-
nish. Two silicon diodes [Lakeshore cryotronics DT-470]
have been used for the measurement and control of the temperature of the shield. The sample holder (Fig. 2) con-
sisted of a sapphire plate embedded in a heater and a sapphire block. The bottom side of the sapphire plate (14 mm

in dia. and 1 mm thick) is evaporated with a thin film of NiCr of 1430 ohms and is used as the heater. The sapphire block (8 mm in dia. and 5 mm in thickness) has the provision of inserting the thermometer. A thin coating of Apiezon-N grease is applied between the sapphire heater plate and the block; and the sapphire block and the sample (1 gm). The entire sample holder assembly is supported by three cotton threads from the tripod-stand. Figure 3 shows a schematic diagram of a complete measuring cycle. The sample is heated step wise; and the heat capacity is obtained upon division of the applied energy ΔQ by the increase in temperature ΔT. The ΔQ is estimated by measuring heating time, current and voltage drop across the heater. The ΔT is obtained from the temperature drift of the sample before and after heating and this is extrapolated towards the mid point of the heating time. In the isothermal mode of operation the temperature of the radiation shield is kept constant during such a cycle. The instrument controlling and data acquisition were carried out through a Hewlett-Packard (HP) 9332 model computer. The

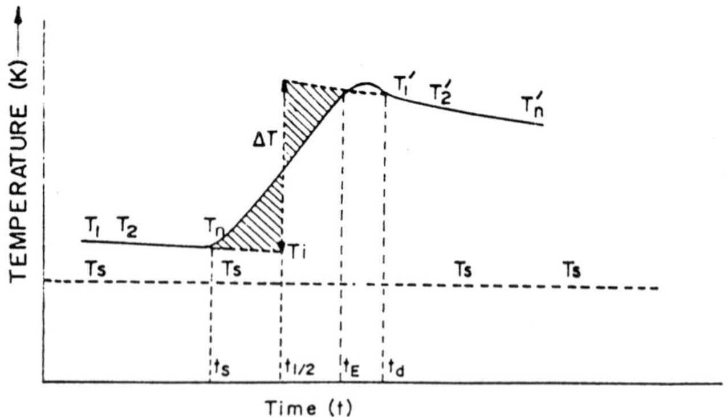

Figure 3 : Schematic representation
of a measurement cycle

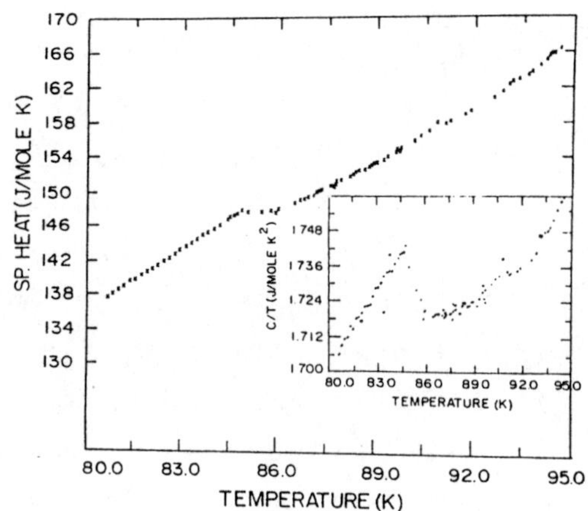

Figure 4 : C vs T curves for
Bı 2-1-2-2. Inset: C/T vs T.

specific heat of the sample was obtained by subtracting
the heat capacity of the addenda consisting of a sapphire
block, a heating element, and apiezon N grease, etc.; OFHC
(Oxygen Free High Conductivity) copper was used for
calibration, and it turns out that the absolute error in
the specific-heat measurement is estimated to be less than
1%. The details of the set up and the calibration procedure of the instrument will be published elsewhere /8/.

III. RESULTS AND DISCUSSIONS:

It may be mentioned here that although we have
studied in the temperature range 300-77 K, we concentrate
here on the data over a narrow temperature range close to
Tc.

[B(1)] Single phase samples of Bı- and Tl- based cuprates:

We first report the of specific heat measurements
carried out on single phase samples of Bı 2-1-2-2 and Tl

2-2-2-3 compounds. The data are compared with those obtained on Y 1-2-3 system. The samples were simultaneously characterized for superconductivity using ac susceptibility and four-probe resistivity techniques and for their structural aspects using XRD. Figure 4 shows the specific heat data of Bi 2-1-2-2 samples as a function of temperature, while Fig.5 depicts the same for Tl 2-2-2-3 samples. As may be seen, there is a distinct anomaly in specific heat at a temperature of 85 K for Bi 2-1-2-2 and at 110 K for Tl 2-2-2-3 samples. These values closely match with the bulk Tc of the samples determined from the susceptibility data. In the insets of Figs 4 and 5, Cp/T has been plotted as a function of T. The specific heat jump (ΔCp) at the Tc for Bi 2-1-2-2 is found to be 2.2 J/mole K, while for the Tl 2-2-2-3 phase the value is 5.6 J/mole K. The specific heat at Tc, Cp(Tc) for these samples is respectively 147.7 J/mole K and 244.2 J/mole K.

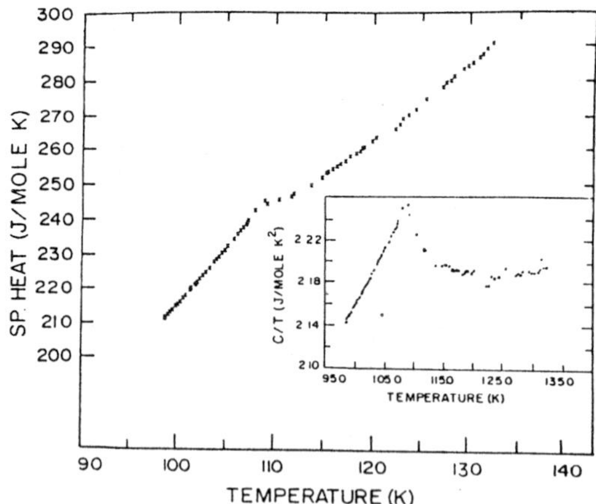

Figure 5 : C vs T curves for
Tl 2-2-2-3. Inset: C/T vs T.

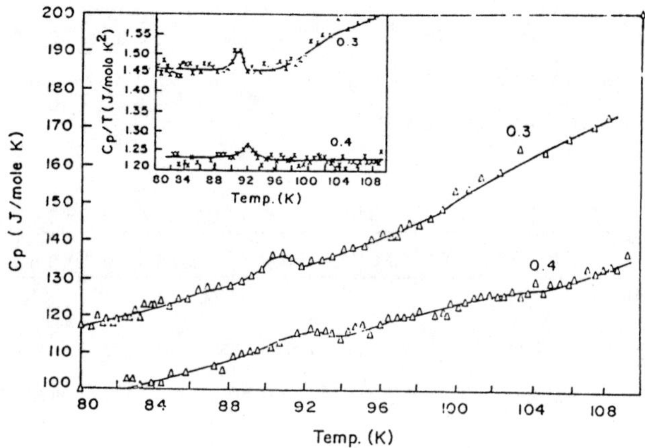

Figure 6 : Molar specific heat (Cp)
           and temperature (T) curves
           for BSCCO-3 AND BSCCO-4;
           the inset shows Cp/T
           with T for these samples.

Thus [ΔCp/Cp(Tc)] for the former is 1.48% while for the latter it is 2.29% which essentially indicates that the fraction of bulk superconductivity for the former is less than for the latter. This is substantiated from the d.c. susceptibility studies where it has been observed that the volume fraction of the superconducting phase is 80% of the Bi 2-1-2-2 sample with Tc of 85 K while about 90% of the Tl 2-2-2-3 with a Tc of 110 K. However, the resistive transition shows that Tc(R=0) for Bi 2-1-2-2 and Tl 2-2-2-3 samples is 91 K and 126 K, respectively. Thus Tc obtained from specific heat and susceptibility is lower than Tc(R=0) and this has been explained from the fact the former two measurements show the Tc of the bulk sample which in resistive measurements may get short-circuited by a much smaller volume fraction of continuous percolative paths of locally enhanced Tc. The local Tc enhancement may be the consequence of higher ordering of the regions forming the percolative path.

[B(2)] Effect of lead substitution in Bi-based cuprate containing both low-Tc and high-Tc phases

We report the results of resistivity, ac-susceptibility, and specific-heat measurements carried out on samples of $Bi_{2-x}Pb_xSr_2Ca_2Cu_3O_{10-y}$ (BSCC-X, X=0-7) that were characterized by x-ray diffraction and high-resolution scanning tunneling microscopy. The zero resistance temperature, Tc(R=0), shows a peak at x = 0.3 and the susceptibility data indicate that although a high-Tc

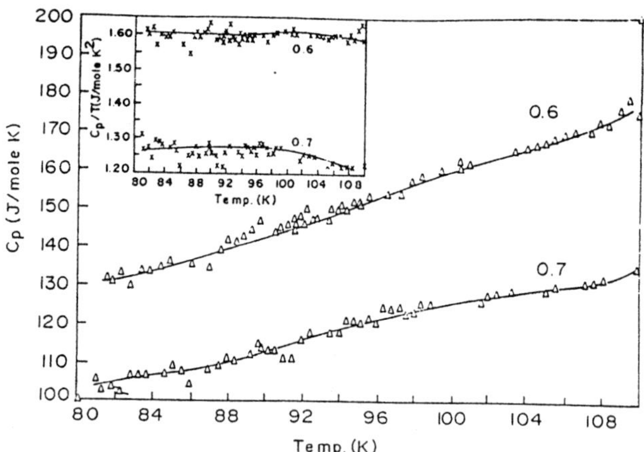

Figure 7 : Molar specific heat (Cp) and temperature (T) curves for BSCCO-6 AND BSCCO-7; the inset shows Cp/T with T for these samples.

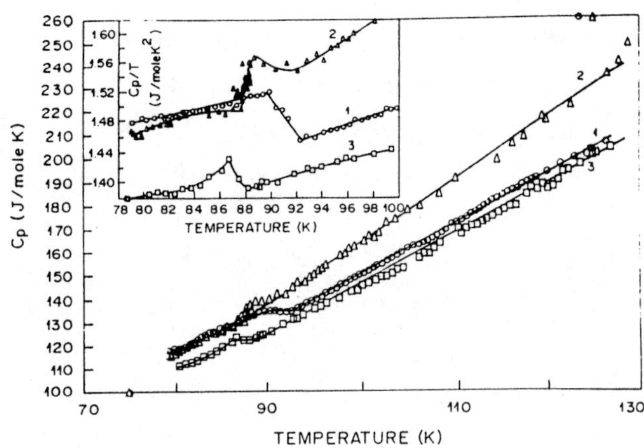

Figure 8 : Molar specific heat (Cp)
            vs. temperature (T) curves
            for (1) Er-1-2-3 (pure),
            (2) Er-1-2-3 (Ni), and
            (3) Er-1-2-3 (Zn); the inset shows
            Cp/T vs T for these samples.

2:2:2:3 phase is present at 110 K, there is a second transition at 90 K which seems to indicate the intergrowth of
an additional low-Tc 2:1:2:2 phase. Figure 6 shows the
calculated molar specific heat of the samples BSCCO-3 and
BSCCO-4. A distinct and clear specific heat anomaly can be
seen at 92 K for these samples. Figure 7 shows the molar
specific heat data of the samples BSCCO-6 and BSCCO-7. It
can be seen that the specific heat anomaly, which is observed for the samples BSCCO-3 and BSCCO-4, is not observed in BSCCO-6 and BSCCO-7. A little fluctuation in
data was observed for these samples [BSCCO-6 and BSCCO-7]
in the vicinity of the transition temperature although all
these experiments were carried out under identical conditions. This can be further cleared by plotting Cp/T versus

T for these samples, which are shown in the inset of Figs.6 and 7 and the observed transition at 89 and 92 K for the samples BSCCO-3 and BSCCO-4 is in agreement with others. The specific heat jump, which is associated with bulk superconducting properties, turns out to be for the samples BSCCO-3 and BSCCO-4, 4.8 and 3.2 J/mole K, respectively. The estimated $\Delta Cp/Tc$ for these samples is 54 and 35 mJ/mole K^2. The results from these combinations of experiments indicate two distinct types of behaviors, (i) in the region of Pb content x < 0.4 and (ii) for x > 0.4. Tc(R=0) from the resistivity measurement, the flux exclusion onset temperature from susceptibility and the anomaly observed in the specific-heat indicate that a Pb substitution of x around 0.4 stabilizes the superconducting phase consisting of both high-Tc and low-Tc fractions. However, when x is beyond 0.4, the absence of the second transition in X and also the specific-heat anomaly at 90 K suggest that the overall proportion of the superconducting phase in the sample diminishes as a result of the decrease of both 2:2:2:3 phases in the system. As mentioned, trivalent Bi^{3+} is having an ionic radius 1.03 A , while the divalent Pb^{2+} is 1.19 A . Hence, the substitution of Bi^{3} with larger ionic radius by Pb^{2+} results in the increase in c-parameter as is substantiated through XRD measurements. Therefore, it has been conjectured that when Pb content >0.4 , there are two possibilities (i) without being substituted as Pb^{2+} at Bi^{3+} site, Pb remains unreacted in the system and (ii) Pb may substitute in tetravalent state Pb^{4+} [where the ionic radius is 0.84 A at Ca^{2+} site the ionic radius 0.99A). As we do not find any line corresponding to the unreacted Pb in the XRD pattern, the first possibility may be ruled out. The other possibilit seems to suggest that c should decrease with x when the Pb content of the sample is more than 0.4 which is also not observed. Therefore, it is suggested that when Pb is more than 0.4, it is partially substituted at the Bi^{3+} as Pb^{2+} and partially at the Ca^{2+} site in Pb^{4+} state. Since the Pb substitution at these two different sites have opposite effects on c, there is no significant change in the value of c of the samples when Pb content increases above 0.4. A similar possibility of Pb getting substituted as Pb^{4+} at Ca^{2+} site in Bi 2-2-2-3 has been discussed from neutron diffraction experiments by Sastri et al /9/.

[B(3)] <u>The excess specific heat and the fluctuation in effects   conjunction with substitution in (rare earth) 1-2-3 systems.</u>

Finally, we now describe a systematic study of the resistivity, magnetic susceptibility and specific heat anomaly in Er-123 system, observed both in the pure phase, as well as when Cu is partly (0.5%) replaced by Ni, Zn, Fe, Co and Ga. These substitutions are particularly interesting as the first two members tend to preferentially occupy the Cu(2) site of the CuO_2 planes (to be referred to as in-plane disorder), while the latter three prefer the Cu(1) site of the CuO chains (i.e., out-of-plane disorder). Purely from superconductivity considerations, the main difference between these two substitutions is that the former adulterates CuO_2 planes where superconductivity is primarily supposed to reside, while the latter adulterates the chains and thereby influences the interlayer coupling between the successive stacks of CuO_2 planes of the neighboring unit cells along the c-direction. The former has the effect of lowering the Tc while the latter tends to broaden the resistive transition. The role of interlayer coupling in broadening the resistive transition of Bi-2122 system has been elaborated by Samanta et al. /10/. A comparative study of ensuing effects of the in-plane and out-of-plane substitutions in Er-123 system on the resistivity, magnetic susceptibility and specific heat anomaly should provide a useful insight into the possible role of fluctuations in conjunction with the site dependent disorder in Er-123 system. Figure 8 shows distinct and clear Cp anomaly at 90, 88, and 84 K respectively, for the samples Er-1-2-3 (Pure), Er-1-2-3 (Ni) and Er-1-2-3 (Zn). The inset of Fig.8 shows the plot of Cp/T against temperature for the same samples. Figure 9 shows the specific heat data of the sample Er-1-2-3 (Fe), Er-1-2-3 (Co) and Er-1-2-3 (Ga) and the inset depicts the plot of Cp/T with temperature. The specific heat anomaly is seen for these samples also, but it is relatively broad and smeared. The specific heat jump turns out to be within 4.5 J/mol K for the pure, Ni and Zn substituted samples, while for the samples with Fe,Co and Ga, it is less than 3 J/mol K. The estimated ΔCp/Tc for these samples is 20-48 mJ/mol K^2 which agrees well with the published data /4/. We have analyzed also the specific heat data of all samples in the MFT to determine the excess Cp due to fluctuation. It may be emphasized here that XRD and SEM studies of the above series of samples have given no indication of any macroscopic inhomogeneties and thus the above results may be ascribed primarily to the substitutions. In this context, it is interesting to mention that bulk Y-124 samples had showed that the transition at Tc to be noticeably broadened for Fe, Co and Ga substitution than for Ni and

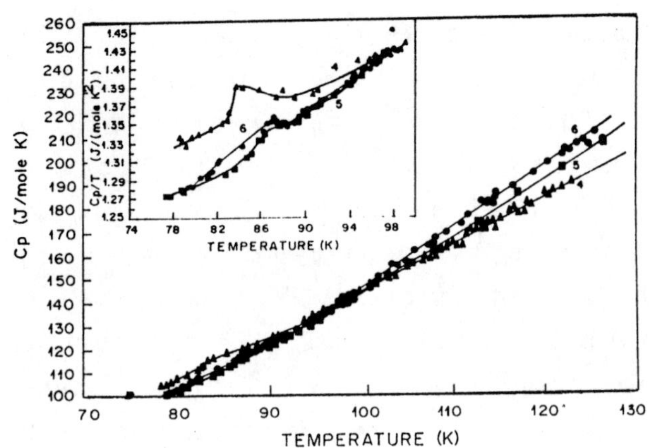

Figure 9 : Molar specific heat (Cp) vs.
temperature (T) curves for
(4) Er-1-2-3 (Fe), (5) Er-1-2-3
(Co), and (6) Er-1-2-3 (Ga);
the inset shows Cp/T vs T
for these samples.

Zn substitutions. Following resistivity, a smooth background of phononic contribution $Cp^B/T = a_0 + bT$ is constructed both above and below Tc and this is shown in
Fig.10 for the pure sample. The excess specific heat (ΔCp)
which is essentially (Cp $-Cp^B$) is shown (1) in Fig.11 for
the samples Er-1-2-3 (pure), Er-1-2-3(Ni) and Er-1-2-3
(Zn), respectively, and (11) in Fig.12 for Er-1-2-3 (Fe),
Er-1-2-3(Co) and Er-1-2-3 (Ga), respectively. A plot obtained between ln(ΔCp) and ln((T-Tc)/Tc) revealed (not
shown) that there is no single linear region indicating
the absence of any single exponent power law dependence of
the fluctuations. However, when the fit was carried out in
the region within 5 K of Tc, the slope of the linear
region turns out to within 1.0 $\pm$ 0.1 which is indicative

of two-dimensional nature of fluctuation. Further, we see that the observed Cp anomaly for Ni and Zn substituted samples to be as distinct as for the pure sample. On the other hand, Cp anomaly observed for Fe, Co and Ga substituted samples is significantly broad, illustrating that the effect of fluctuations in Fe, Co and Ga substituted samples is comparatively much larger than in Ni and Zn substituted samples. This is again indicative of a more pronounced fluctuation effect occurring when Fe, Co and Ga are incorporated at Cu(1) site. These findings, we suggest, are the direct manifestation of in-plane and out-of-plane disorder which the above substituents create in 123 structure. Zn and Ni, as they occupy CuO_2 plane, mainly contribute in decrease of Tc while the interlayer

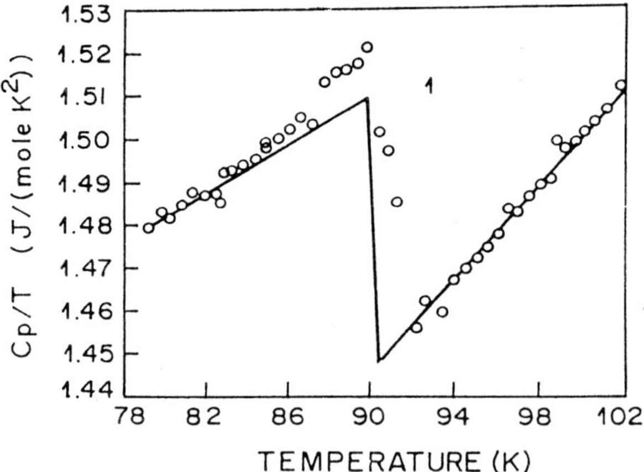

Figure 10: Cp/T vs temperature for (1) Er-1-2-3 (pure); the continuous line shows the fitting of the data with $Cp^B/T = a_0 + bT$ both above and below Tc.

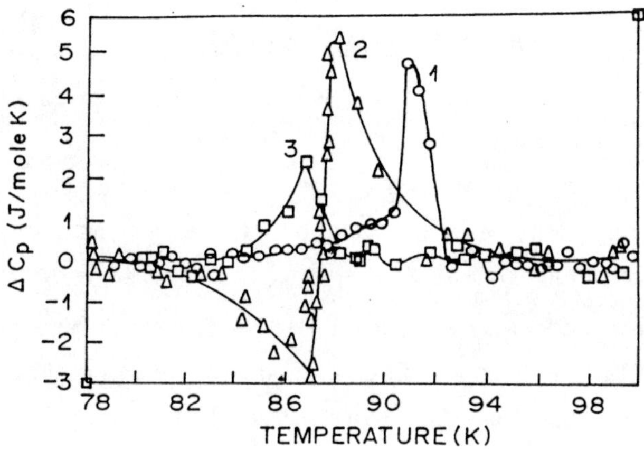

Figure 11: Excess specific heat (ΔCp)
vs. temperature for
(1) Er-1-2-3 (pure),
(2) Er-1-2-3 (Ni),
and (3) Er-1-2-3 (Zn).

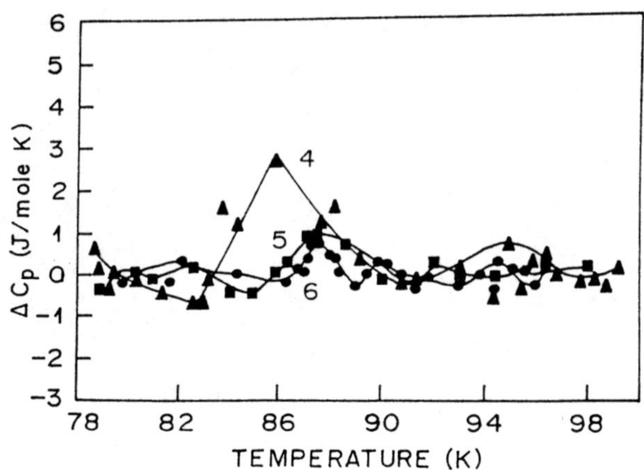

Figure 12: Excess specific heat (ΔCp)
vs. temperature for (4)
Er-1-2-3 (Fe), (5) Er-1-2-3
(Co), and (6) Er-1-2-3 (Ga).

coupling between CuO_2 across the neighboring unit cells is little influenced by them. On the other hand, Fe, Co and Ga preferring the chainer Cu(1) site create out-of-plane disorder between the CuO_2 layers of the adjoining unit cells and thereby weaken the coupling between them in the c-direction, where the range of coherence is already low. In effect, this possibly results in transforming the system towards more two dimensional, leading to an increased fluctuation effects, observed in the specific heat data of Fe, Co and Ga doped samples.

It is to be mentioned that the characteristic parameters, like the Sommerfeld constant or Debye temperature, are difficult to estimate in these oxide cuprates because the electronic contribution to the specific-heat anomaly is strongly affected by the phononic contribution. The difficulty is that at high-temperatures, phononic contribution cannot be measured independently through suppression of superconductivity by using a high magnetic field or creating enormous disorder by irradiation /11/. The multiphase compound complicates phonon density of states and thus the application of the mean-field approximation (MFA) is difficult. However, in order to indicate the possible trend in the nature of coupling, Sommerfeld constant was estimated /1/. After correcting the temperature-independent susceptibility for diamagnetic core and Landau contributions in the usual way, we obtain in these systems which are lying in the range 20-5 $mJ/mole\ K^2$ /4/. Similarly, the Debye temperature is calculated from the specific-heat data at temperature using the theoretical Debye function assuming the three-dimensional Debye model is valid. An iterative method has been developed to obtain the best-fitted value. The values of the Debye temperatures are found to be temperature dependent, and the average value around the transition temperature is in accord with the reported data /4/.

Acknowledgments:

The authors thank Prof. S.K. Joshi, Director General, CSIR, for his keen interest and they are grateful to Prof. Dr.E.Gmelin for collaborative help. They acknowledge experimental assistance given by Mr.B.V. Kumarswamy, Dr. S. P.Pandy, Mr. V. P. S. Awana and Mr. V. N. Moorthy in sample preparation and some of the measurements.

V.   REFERENCES :

/1/   A. K. Bandyopadhyay, P. Maruthikumar, G.L. Bhalla, S.K.Agarwal, and A.V. Narlikar, Physica C 165, 29 (1990).

/2/   A. K. Bandyopadhyay, E. Gmelin, B. V. Kumaraswamy, V. P. S. Awana, Deepak Varandani, Nirupa Sen, and A.V.Narlikar, Phys. Rev B. 48, 6470 (1993).

/3/   A. K. Bandyopadhyay, E. Gmelin, Deepak Varandani, and A. V. Narlikar, Phys. Rev B. 50, 462 (1994).

/4/   D. R. Harshman, and A. P. Mills, Jr., Phys. Rev.B 45, 10684 (1992).

/5/   Y.Iye, "Studies of High Temperature Superconductors" vol. 1 (Ed. A. Narlikar, Nova Science Publishers, New York, 1989), p.166.

/6/   M. Akinaga, "Studies of High Temperature Supercon
      ductors " vol. 8 (Ed. A. Narlikar, Nova Science
      Publishers, New York, 1989), p.297.

/7/   A. V. Narlikar, S.K. Agarwal and C. V. N. Rao, "
      Studies of High Temperature Superconductors" vol. 1
      (Nova Science Publishers, New York, 1989) edited by
      A.V. Narlikar, p.341.

/8/   A. K. Bandyopadhyay, E. Gmelin, V.S. Yadav and A. V.
      Narlikar (to be published).

/9/   S. B. Samanta, P.K. Dutta, V.P.S Awana, E. Gmelin,
      and A.V. Narlikar, Physica C 178, 171 (1991).

/10/ A. Sequera, J. V. Yakhmi, R. M. Iyer, and H.
      Rajagopal, Physica C 167, 291 (1990).

/11/ N. Sen, A.K. Bandyopadhyay, P. Sen, U. Tiwari,
      Deepak Varandani, V.P.S.Awana and A.V. Narlikar,
      Sol. State Comm. 82, 555 (1992).

Microwave Surface Resistance of HTSC Thin Films

R. Pinto

Tata Institute of Fundamental Research
Homi Bhabha Road
Bombay – 400 005

Abstract

The surface resistance, R_s, at microwave frequencies has become an important qualification parameter of high temperature superconductor (HTSC) thin films. In this talk, the measurement of R_s of $YBa_2Cu_3O_{7-x}$ (YBCO) thin films in situ grown on <100> MgO and $LaAlO_3$ substrates by pulsed laser deposition will be presented. R_s has been measured at various frequencies in the range 1-12 GHz by patterning the films into microstrip resonators.

It has been observed that the value of R_s at a given frequency and temperature critically depends upon the granularity and the epitaxial quality of the films. For example, YBCO films in situ grown on MgO have been found to be weaklink-limited with a significant microwave power dependence of R_s. YBCO films on $LaAlO_3$ have shown much better results. The best results, however, have been obtained on Ag-doped YBCO films which have not only shown the lowest R_s values of 210 $\mu\Omega$ at 77K at 10GHz but also have not shown any microwave power dependence upto 13dBm input power. This has been found to be due to the improved epitaxy of Ag-doped YBCO films.

However, due to the presence of Ag at the grain boundaries, Ag-doped films have shown higher residual resistance, R_{res}, at 15K. Nevertheless, Ag-doped YBCO films have been found to be superior to undoped films due to their lower R_s at 77K, a temparature at which HTSC devices are expected to operate.

1. Introduction

The electromagnetic response and surface resistance, R_s of high temperature superconductors (HTSC) at microwave frequencies has been a subject of great importance both from fundamental and technological considerations.[1-6] R_s depends on a number of physical parameters such as coherence length, energy gap, mean free path and penetration depth all of which, in turn, depend on the intrinsic as well as the extrinsic properties of the material. Obviously therefore, R_s has become an important characterization parameter for HTSC materials, especially for HTSC thin films. Furthermore, work of Halbritter[7] and others[8-10] has shown that the origin of the observed residual surface resistance, R_{res} (as T ⇒ 0, where T is temperature), and the temperature dependence of R_s (for $T < T_c/2$, where T_c is the superconducting transition temperature), both of which are in distinct disagreement with the predictions of Bardeen-Cooper-Schrieffer (BCS) theory, can be attributed to locally weak superconductivity of weak links in the granular superconductors.

Granularity of HTSC materials has been the most serious problem ever since their discovery.[11-12] While the single crystal grains of HTSC are intrinsically capable of carrying very high supercurrents with critical current density, J_c, $\geq 10^7 Acm^{-2}$ @ 77K,[13] most practical HTSC materials have shown J_c values in the range $10^2 - 10^4$ Acm^{-2} at 77K. The main cause of these low values of J_c are the grain boundaries separating the crystalline grains. The grain boundaries can be 10-200Å in thickness and may consist of a simple boundary between the misoriented grains on the one extreme and insulating second phases or impurity phases on the other. Hence, depending upon their nature the grain boundaries exhibit superconductor-normal metal-superconductor (S-N-S) or superconductor-insulator-superconductor (S-I-S) type of weak-link Josephson junction (JJ) behaviour.[11] These weak-link JJs not only govern the current transport in ceramic HTSC materials but also affect its dependence on magnetic field.[14]

Significant progress has been made over the last few years on the improvement of the quality of superconducting $YBa_2Cu_3O_{7-\delta}$ (YBCO) thin films. The realization of high quality YBCO thin films with low R_s was primarily due to the improved epitaxy and reduced weak link effects. We have shown recently that doping of YBCO films with Ag during in situ growth by pulsed laser deposition (PLD) not only improves critical current density, J_c, by a factor of 2 to 3 (highest J_c being $1.4 \times 10^7 Acm^{-2}$ @ 77K on $SrTiO_3$)[15] but also reduces R_s, the lowest reported being 210 $\mu\Omega$ at 77K at 10 GHz for patterned films on $LaAlO_3$.[16] This has been shown to be due to the improved epitaxy and reduced weak link effects in Ag-doped films.

However, while Ag-doping improves the microstructure and reduces weak link effects thereby reducing R_s at 77K, the Ag atoms segregating at the grain boundaries have been found to increase R_{res}, thus introducing the metallic contribution to R_{res}. Although the residual losses in HTSC have been primarily attributed to extrinsic weak link effects, the possibility of losses arising from the inclusion of normal metallic material in HTSC has been conjectured by others earlier.[9,10] Here, we show that while Ag-doping of YBCO thin films reduces R_s at 77K due to improved epitaxy, the segregation of Ag atoms at the grain boundaries increases R_{res}, and makes the observed R_s vs T behaviour deviate further from the $R_s \propto \exp(-\Delta/kT)$ relation predicted by the BCS theory, where Δ is the gap parameter.

2. Experiment

The microwave measurements were carried out at X-band frequencies on microstrip resonators fabricated on 10mm x 10mm x 0.5mm MgO and $LaAlO_3$ substrates using 5000Å thick YBCO films in situ grown by pulsed laser deposition process. Both undoped and 5 wt% Ag-doped YBCO pellets having 15mm diameter and 3mm thickness were used for the growth of films. The detailed procedure for in-situ growth of both undoped and Ag-doped YBCO films is reported elsewhere.[15,17] In order to study the effect of film microstructure and grain boundary weak links on microwave transmission, undoped and Ag-doped YBCO films were grown on <100> MgO and <100> $LaAlO_3$. The films characterized by X-ray diffraction were found to be highly oriented with c-axis normal to the substrate.   The microstructure of partially etched films was also studied using scanning electron microscope. The results showed that the films on $LaAlO_3$ had larger grains than those grown on MgO. Ag-doped films on $LaAlO_3$ showed a further improvement in grain size. Furthermore, they also showed a distinct alignment in the a-b plane. Transport measurement performed with 2000Å thick and 10 μm wide microbridges showed J_c of 1.2×10^6 Acm^{-2} and 2.5×10^6 Acm^{-2} at 77K on MgO and $LaAlO_3$, respectively. The Ag-doped films on $LaAlO_3$, however, showed J_c $\sim 8 \times 10^6$ Acm^{-2} at 77K.[15]

The design and fabrication of X-band mcirostrip resonator is reported elsewhere.[18] A schematic of transmission type microstrip resonato

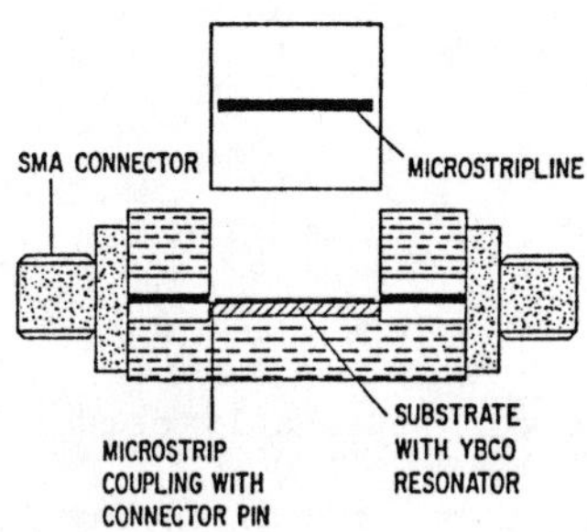

Fig.1   Schematic of transmission type microstrip resonator at X-band.

at X-band frequencies is shown in Fig.1.   It is a half-wavelength resonator with input/output   couplings   through   capacitive   gaps   between   the   YBCO transmission line and the SMA connector pins.   Both the device ports are loosely coupled such that the measured Q-factor is approximately equal to the unloaded Q-factor, Q_o, of the resonator.   The microwave measurements were carried out using HP83620A synthesized sweeper and HP8757C scalar network analyser with the associated reflectometry set up.   The input power level applied was in the range -20dBm to 13dBm.   The low temperature measurements were performed by mounting the resonator on the cold head of a closed cycle He cryocooler with a temperature range of 10-300K.

3. Results and Discussion

The Q-factor of resonators was measured from the HP8757C scalar network analyser at various resonant frequencies (fundamental and higher harmonics) and temperatures by calculating the ratio $f/\Delta f$, where f is the centre frequency and Δf is the 3dB bandwidth of resonant curves.   Since the devices were loosely coupled with 18-20dB insertion loss the measured

Q-factor is nearly equal to Q_o, the unloaded Q-factor. The Q-factor due t

conductor loss, Q_c, is given by

$$1/Q_c = 1/Q_o - 1/Q_d - 1/Q_r \qquad\qquad (1)$$

where Q_d and Q_r are the Q-factors due to dielectric and radiation losses

respectively. Here, one can reasonably assume $Q_d \sim (\tan\delta)^{-1} \sim 10^4$ fc

$LaAlO_3$. Since the losses due to radiation have been minimized by providir

effective shielding at $\lambda/2$ spacing around the device, they can be neglecte

as compared to Q_d. The values of R_s corresponding to values of Q_c were the

calculated using the expressions given by Pucel et al.[19]

Shown in Fig.2 and Fig.3 are the variations of Q_c and R_s wit

input power at 77K obtained for undoped and Ag-doped YBCO films on MgO an

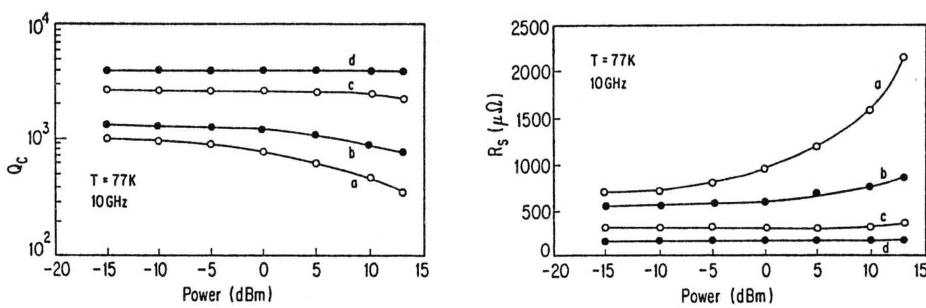

Fig.2 Variation of Q_c with input power at 77K and 10 GHz for YBCO

thin film microstrip resonators: (a) undoped film on MgO, (b)

5 wt% Ag-doped film on MgO, (c) undoped film on $LaAlO_3$, and

(d) 5 wt% Ag-doped film on $LaAlO_3$.

Fig.3 Variation of R_s with input power at 77K and 10 GHz for YBCO

thin films: (a) undoped film on MgO, (b) Ag-doped film on MgO,

(c) undoped film on $LaAlO_3$, and (d) Ag-doped film on $LaAlO_3$.

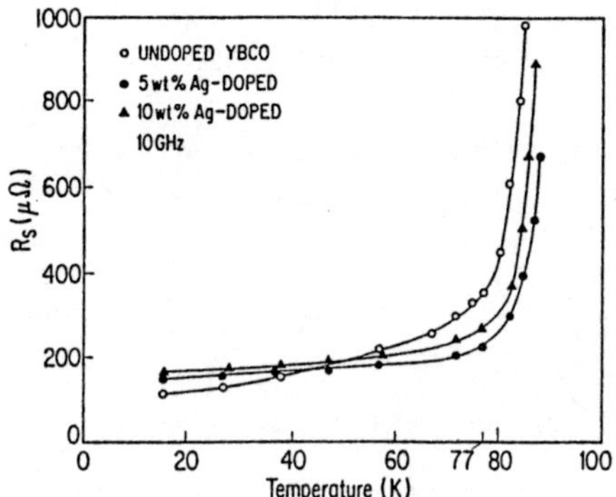

Fig.4 R_s vs T plot for undoped and Ag-doped YBCO films on $LaAlO_3$ at 10 GHz.

$LaAlO_3$.   It is seen in Fig.2 that Q_c of undoped YBCO film on MgO decreases considerably with increasing input power from -10dBm, whereas Q_c of undoped film on $LaAlO_3$ decreases only after 10dBm.   The Ag-doped films on both MgO and $LaAlO_3$ however, show significant improvement over the undoped films. Similarly, as seen in Fig.3, R_s of undoped film on MgO increases sharply with input power right from -10dBm; whereas, R_s of undoped film on $LaAlO_3$ shows a dependence on input power only from 10dBm onwards. R_s of Ag-doped film on $LaAlO_3$, however, shows no dependence on input power.

     The variation of R_s with temperature, T, for undoped and Ag-doped films, measured at 10 GHz is shown in Fig.4.   Two observations are evident from Fig.4.   First, the R_s value of 5 wt% Ag-doped film at 77K is lower than the corresponding value of the undoped film.   Second, the R_s vs T curve of undoped film crosses that of 5wt% Ag-doped film at ~45K indicating thereby that R_s of undoped film is lower than that of 5wt% Ag-doped film below ~45K. The R_s vs T plot of 10wt% Ag-doped film is similar to that of 5wt% Ag-doped

                    R. Pinto

film except that it has higher values of R_s and crosses R_s vs T plot of undoped film at a higher temperature of ~50K.

The temperature dependence of R_s of superconductors based on BCS theory can be expressed as[8]

$$R_s = R_{res} + A \exp(-\Delta/kT) \qquad (2)$$

where R_{res} is the non-BCS residual resistance and Δ is the gap parameter. As expected, it is difficult to fit the R_s vs T curves in Fig.4 with Eq.(2) for the full temperature range to find a unique value of Δ. However, what is interesting is that R_{res} increases with Ag-doping. Therefore, while Ag-doping reduces R_s at 77K through improved epitaxy and reduced weak link effects, it increases R_{res} implying an additional contribution to R_{res}. We ascribe this additional contribution to R_{res} to the presence of Ag as agglomorates at the grain boundaries. It may be mentioned here, that our value of $R_{res} = 110\mu\Omega$ at 15K for undoped film is higher than the lowest value reported.[4] But this is due to the fact that our R_{res} value was measured on patterned resonators having contribution from dielectric and radiation losses which becomes a significant fraction of total R_{res} at low temperatures. What is important however, is that the lower limit of contribution to R_{res} is 30 $\mu\Omega$ with 5wt% Ag doping and $40\mu\Omega$ with 10wt% Ag-doping.

4. Conclusion

In conclusion, we have shown that R_s of HTSC thin films is ver sensitive to the grain boundary weak links in the films. YBCO films grow on MgO have been found to show highest values of R_s due to the granula nature of the films. Ag-doped films on MgO have shown lower values of R_s

Granularity of YBCO films has also been found to determine the microwave power dependence of R_s. YBCO films on <100> $LaAlO_3$ have shown lower values of R_s, ~ 350 $\mu\Omega$ at 10 GHz at 77K. However, 5 wt% Ag-doped films have shown the lowest R_s of 210 $\mu\Omega$ at 10GHz at 77K and no power dependence upto 13 dBm. The value of R_{res}, however, has been found to be higher in Ag-doped films, presumably due to the presence of Ag in the grain boundaries of the films.

Acknowledgement

The author would like to thank Prof. R. Vijayaraghavan and Prof. L.C. Gupta for discussion. Thanks are also due to Dr. P.R. Apte and Dr. D. Kumar for some of the experimental work and S.P. Pai for growth of the YBCO films.

References

[1] S. Sridhar, D.H. Wu, and W. Kennedy, Phys.Rev.Lett. **63**, 1873 (1989).

[2] A. Inam, X.D. Wu, L. Nazar, M.S. Hegde, C.T. Rogers, T. Venkatesan, R.W. Simon, K. Daly, H. Padamsee, J. Kirchgessner, D. Moffat, D. Rubin, Q.S. Shu, D. Kalokitis, A. Fathy, V. Pendrick, R. Brown, B. Brycki, E. Belohoubek, L. Drabeck, G. Grüner, R. Hammond, F. Gamble, B.M. Lairson, and J.C. Bravman, Appl. Phys. Lett. **56**, 1178 (1990).

[3] A.M. Portis, D.W. Cooke, E.R. Gray, P.N. Arendt, C.L. Bohn, J.R. Delayen, C.T. Roche, M. Hein, N. Klein, G. Muller, S. Orbach, and H. Piel, Appl. Phys. Lett. **58**, 307 (1991).

[4] G. Muller, N. Klein, A. Brust, H. Chaloupka, M. Hein, S. Orbach, H. Piel, and D. Reschke, J. Superconduct. **3**, 235 (1990).

[5] D. Kalokitis, A. Fathy, V. Pendrick, E. Belohoubek, A. Findikoglu, A. Inam, X.X. Xi, T. Venkatesan, and J.B. Barner, Appl. Phys. Lett. **58**, 537 (1991).

[6] D.E. Oates and A.C. Anderson, IEEE Trans. Magn. **MAG-27**, 867 (1991).

[7] J. Halbritter, J. Appl. Phys. **68**, 6315 (1990).

[8] W.L. Kennedy and S. Sridhar, Solid State Commun. **68**, 71 (1988).

[9] T.L. Hylton, A. Kapitulnik, M.R. Beasley, J.P. Carini, L. Drabeck, and G. Grüner, Appl. Phys. Lett. **53**, 1343 (1988).

[10] J.P. Carini, A.M. Awasthi, W. Beyermann, G. Grüner, T. Hylton, K.Char, M.R. Beasley, and A. Kapitulnik, Phys. Rev. **B37**, 9726 (1988).

[11] S. Zhen-peng, Z. Yong, S. Shi-fang, C. Zu-Yao, C. xian-hui, Z. Qi-Yui, Solid State Commun. **69**, 1067 (1989).

[12] Y. Matsumoto, J. Hombo, Y. Yamaguchi, M. Nishida, and A. Chiba, Appl. phys. lett. **56**, 1585 (1990).

[13] R. Kromann, J.B. Bilde-Sorensen, R. de Reus, N.H. Andersen, P. Vase, and T. Frelotoft, J. Appl. Phys. **71**, 3419 (1992).

[14] Y.H. Kao, Y.D. Yao, L.Y. Jang, F. Xu, A. Krol, L.W. Song, C.J. Sher, A. Darovsky,, J.C. Phillips, J.J. Simmins, and R.L. Snyder, J. Appl. Phys. **67**, 353 (1990).

[15] D. Kumar, M. Sharon, R. Pinto, P.R. Apte, S.P. Pai, S.C. Purandare, L.C. Gupta, and R. Vijayaraghavan, Appl. Phys. Lett. **62**, 3522 (1993).

[16] R. Pinto, N. Goyal, S.P. Pai, P.R. Apte, L.C. Gupta, and R. Vijayaraghavan, J. Appl. Phys. **73**, 5105 (1993).

17 R. Pinto, S.P. Pai, C.P. D'Souza, L.C. Gupta, R. Vijayaraghavan, D. Kumar, and M. Sharon, Physica **C196**, 264 (1992).

18 R. Pinto, P.R. Apte, L.C. Gupta, R. Vijayaraghavan, K. Easwar and B.K. Sarkar, Supercond. Sci. Technol. 4, 577 (1991).

19 R.A. Pucel, D.J. Masse, and C.P. Hartwig, IEEE Trans. Microwave Theory Tech. MTT-16, 342 (1968).

Observation of Hysteretic to Flux–flow Transition in x_{AC} of HTSCs

G. Rajaram and T. Sobha Rani

School of Physics
University of Hyderabad
Hyderabad 500 046, India

Abstract

The occurance of inter and intra-grain dissipation peaks in the imaginary part of the AC magnetic susceptibility, $\chi_{AC}(T)$, is distinct regions in the AC amplitude – Temperature diagram in which χ''_{AC} is amplitude independent and the other in which χ''_{AC} is amplitude dependent– is present. A possible explanation that this arises from a hysteretic to flux-flow regimes is discussed.

Introduction

AC susceptibility, χ, has been routinely used to characterise the superconducting transition for the transition temperature, transition width, screened fraction. In the relatively more complex ceramic high temperature superconductors (HTSCs), the real and imaginary parts of χ (χ' and χ'') as functions of temperature, AC excitation amplitude, and DC bias fields show rich variety of structures obviously related to the complex microstructure of these materials. Since the experiment is sensitve, non-destructive and easily performed, much effort has gone into understanding this relationship for use in the characterisation of HTSCs during the past few years, with a good deal of success[1-6]. This article deals with some features of the intra-grain $\chi''(T)$ which have hitherto received little attention.

The experiment is typically performed as follows: The sample is subjected to an AC magnetic field, h(t), using a solenoid primary coil. A secondary coil with N_s turns is wound round the sample such that (a) the coil surrounds a small region of the sample or (b) the coil volume is much larger than that of the sample. In case (a) if there is no air-gap between the coil and sample, then the secondary output is proportional to the time-derivative of the induction, dB/dt, spatially averaged over the sample cross-section of area A. The voltage induced at the secondary coil is detected using a Dual-phase lock-in amplifier (LIA). A voltage in phase with the exciting AC field is used to provide a reference phase for the LIA, and this is typically derived from the primary coil current. The implicit assumption that the excitation field is in phase with the primary current needs to be explitly verified, under actual conditions, if there are good conductors, magnetic materials or other coils in the neighbourhood of the primary or secondary coils. The LIA typically measures the root-mean-square voltages which are in-phase and quadrature of the 1st harmonic of the reference frequency. The quadrature output is proportional to the real part of the permeability, μ', and the in-phase output is proportional to the imaginary part of the permeability, μ''. If B vs H relationship is non linear or if B has explicit time dependences (non-exponential relaxation effects) then the secondary output may have higher order harmonics. Irrespective of the presence of the harmonics, the area enclosed by the B-H curve (dissipation per cycle) is $1/2\mu''h_0^2$ [1].

A second coil may be wound in the opposite sense in series with the secondary coil containing the sample, so that in the absence of the sample, the voltage from the secondary pair is zero or as close to zero as possible. This reduces the 'background' signal close to zero and considerably enhances sensitivity. A center-tap of the secondary pair is usually drawn out and connected to circuit ground. In this case, the voltage across the 'compensating' coil can be used to 'sense' the excitation field and serves as reference for its phase if there are difficulties in using the primary current for this purpose. If the second coil is subjected to the same excitation field as the first, then the output from the secondary set varies as $d(B-H)/dt \sim dM/dt$, where M is the Magnetisation averaged across the cross-section of the sample. The LIA outputs are now proportional to the imaginary and real parts of the susceptibility. This is true even if there is an air gap between the sample and the first secondary, but for demagnetisation effects. If the sample has a demagnetising factor D, then the real and imaginary χ can be written as

$$\chi_r = K\, V_r/H_m(1-D)$$

$$\chi_i = K\, V_i/H_m(1-D)$$
$$\chi' = \chi_r - D\,(\chi_r^2 + \chi_i^2)/(1-\chi_r D)^2 + D^2\chi_i^2$$
$$\chi'' = \chi_i/(1-\chi_r D)^2 + D^2\chi_i^2$$

where $K = \sqrt{2}/\omega A N_s \mu_0$, V_i and V_r are the imaginary and real parts of the induced secondary voltage [7].

In configuration (b), if the sample volume is small relative to that enclosed by the secondary coil, the sample may be approximated by a dipole of moment M fixed at the center of the coil. The secondary voltage is proportional to dM/dt. The compensating secondary is used to null background. Larger sample volumes may be dealt with rigourously by taking into consideration higher order moments of the current distribution in the superconducting sample. Close coupling between the sample and the coil is desirable in order to have a large signal to noise ratio. In our system, the secondaries, with bore ~3mm, are a pair of coils with 1200 turns each, nulled to within 0.1% of the voltage in each coil in the absence of the sample. The samples are usually of dimension 1x1x6mm. The LIA outputs were logged as a function of temperature as the latter was swept at a rate of about 0.3 K/min. They were normalised by the excitation field amplitude.

The hysteretic contribution to χ'' has two origins in high T_c superconductors: one related to the granularity of high T_c superconductors, and the other due to flux vortex pinning, by defects in the mixed state. In the former, "weak links" typically larger than the coherence length, ξ of the superconductor, eg. impurity phases at grain boundaries in sintered superconductors, stacking faults, cracks etc.,restrict the screening currents to less than some "inter-grain" critical current J_{cj}, which has an effect on the local gradient of the magnetic field. In the latter, in the mixed state, meta-stable screening currents decrease to a value J_{cg} related to the pinning force density for flux vortices, to give a final static current distribution. Either situation is well described by the 'critical state model' which specifies the static current distribution to be the one that is the relevant critical current everywhere in the sample. One then needs to know explicit dependences of of J_c on H and the boundary conditions to solve for the current and field distributions and so obtain the hysteresis loops . A large J_c leads to large hysteresis between the applied field and local field. The final χ'' increases with J_c magnitude and decrases with the increase of shielded volume. Typically J_c decreases with T. The low temperature situation is that of a large J_c , large shielded volume and low χ'', while the high temperature situation is that of a sample completely penetrated but low J_c and negligible hysteresis. Therefore a peak occurs in $\chi''(T)$ around at intermediate temperatures e.g.the temperature where J_{cm} is such that the field variation due to H_{ac} first reaches the center of the sample.

The Bean critical state model [8] had been used in the past to describe the magnetic properties of hard type 11 conventional superconductors. In this model, J_c is assumed to be independent of H, the field profiles have constatnt gradients, dH/dx ~ $\pm J_c$. for a sample with cylindrical geomtery of radius a, $\chi'(T_p)=-5/16$ and $\chi''(T_p)=$ .212 (i.e. peak height is constant for different amplitudes and $J_c(T_p)=H_m/a$, where T_p is the temperature at which penetration of the field to the center of the sample takes place [7], hence a peak in χ'' occurs. The effect of a non-zero H_{c1} has been investigated by Lofland et al [9] assuming that for local fields less than H_{c1}, the field gradient is determined by Meissner shielding with the penetration depth lambda negligible relative to sample size. The peak in $\chi''(T)$ then occurs when

$$(H_m - 1.2\ H_{c1})/J_c a = 1.$$

Kim's model [10] assumes a critical current density dependent o field as $J_c(H_i)=k/(H_0+H_i)$, where k and H_0 are positive constants. This dependence is better suited to describe the inter-granular Jc's upto large field amplitudes. Chen and Goldfarb have calculated the magnetisation due to the shielding critical currents for the rectangular parellelopipedgeometry for the case of Kim's model for $J_c(H)$ [11]. Once the M(H) is known or modelled, the induced secondary voltage, V, proportional to dM/dt = (dM/dH)dH/dt+ dM/dt can be calculated. For the case of hysteresis losses, the assumption is that as H is changed, the final metastable states corresponding to the new H are established on a time scale much smaller than the experimental time scale which is $\sim 10^{-2}$ seconds. The second term is neglected and

$$\chi' = \left[1\ /\Pi\ H_m^2\right] \int_{-H_m}^{H_m} (dM/dH)\,H\ dH$$

$$\chi'' = \left[1/\Pi\ H_m^2\right] \int_{-H_m}^{H_m} (dM/dH)\sqrt{(H_m^2 - H^2)}\ dH$$

It has been extensively observed that $\chi''(T)$ of sintered samples consists of two peaks,(Fig 4) with degrees of overlap and relative peak heights. Typically, for the peak occuring at low temperatures, the peak position is sensitive to the excitation field amplitude, the high temperature peak less so. Furthermore, the low temperature peak loses its well defined maximum, when the sample is powdered; the former is assigned to hysteresis in the magnetisation due to inter-grain shielding currents and the second peak to intra-grain shielding currents. The latter correspond to smaller shielded or coherance areas and larger critical current densities. This 'intra-grain shielding currents' may, of course, be due to either shielding currents flowing across extended defects structures within grains that are not quite so weak, e.g. stacking faults, twinning boundaries, micro-cracks or due flux vortex pinning at smaller defects: oxygen vacancies, impurity inclusions. Explicit observations using decoration or scanning microscopy techniques are required to distinguish between the two. When the grain sizes or the coherance areas within the grains are of the order of λ at the elevated temperatures of this peak, the distinction between the two origins becomes even more blurred.The intra-grain dissipation is now also dependent on the grain position in the sample since the flux profile is determined by inter- and intra-grain shielding together. The combined effect of inter and intra-grain shielding, calculated within the Bean limit of the results of Ref. [11] for both inter and intra-grain shielding critical currents, and assuming that the grains see a field determined independently by the inter-grain shielding currents, is given in figs [1] and [2].Fig. [1]illustrates the effect of grain volume fraction, fand fig.[2] the effect of differences in magnitudes of inter- and intra-grain critical currents.

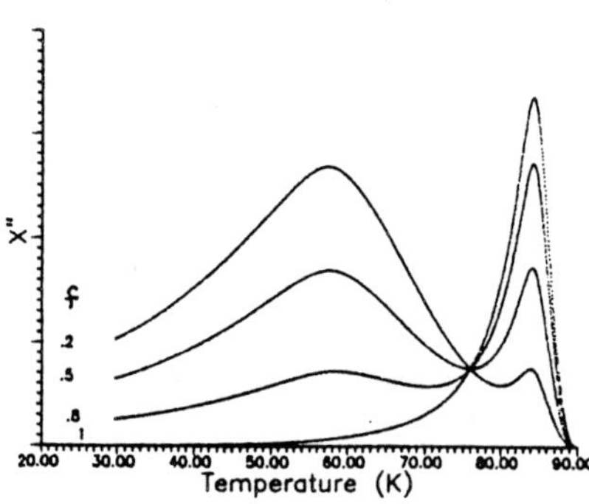

Fig.1.χ"(T) for various f

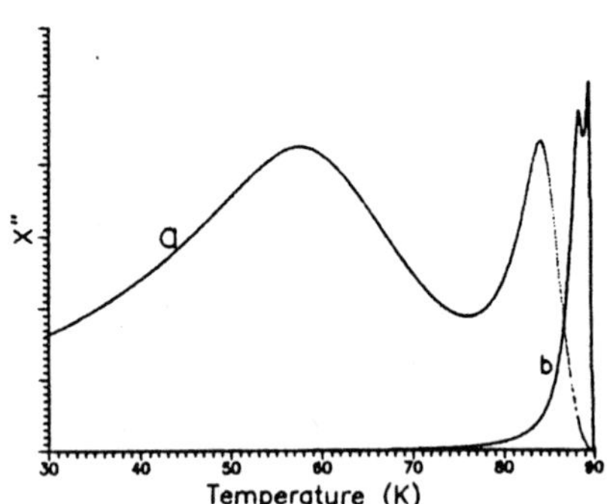

Fig.2.χ"(T) for different J_{cg} & J_{cj}
a. $J_{cj} \ll J_{cg}$ b. $J_{cj} = J_{cg}/10$

In H-T diagram a line separates the magnetically reversibl region from the magnetically irreversible regime or the hystereti regime. This line is defined as the irreversibility line becaus it separates the high temperature magnetically reversible region where the J_c is zero, from the low temperature magnetic hystereti region. This line can be determined from χ'' by using a large DC field with a small AC field superimposed on it. For various AC fields χ'' will be independent of field above the irreversible temperature, T_{irr} [12]. The irreversibility line can also be determined from the disappearance of third harmonic at T_{irr} for a particular field. The transition from hysteretic to flux flow, where pinning is absent, hence zero critical current density results.

The interpretation of this irreversibility line is a subject of controversy. This is explained as a transition of vortex glass to vortex liquid from the fit of T_{irr} Vs H to a power law, same as for the spin glass transition [13]. But Yeshuran et al [14] were able to explain the same results by using conventional flux creep model. The conventional flux creep model gives the dependence of B as logarithm of time. Hence, by using this they were able to explain the relaxation effects in magnetisation of 123 single crystals as

$$dM/dlnt=(aJ_c/4c)(kT/U_0)$$

The low U_0 (pinning potential) combined with short coherence lengths are the cause for the appreciable flux creep observed in the high T_c superconductors. The decrease of U_0 with temperature leads to the drop of J_c below the measurement threshold above a critical temperature by thermal activation. The condition for zero J_c is,

$$1-t=(8\Pi f^2 BkT_c ln(Bd\Omega/E_c)/2.56H_{c0}^2\phi_0\xi_0)^{2/3}$$
$$\text{where } t=T/T_c$$

In the absence of flux pinning , flux creep increases and vortices will be free to move , hence flux flow results. In this region a linear relationship between E and J exists.region. In this linear region χ'' don't depend on the applied field , but depends on frequency.

Fig.3 shows the $\chi'(T)$ and $\chi''T)$ for various field amplitudes for Bi-2223/Bi-2212 polycrystalline sample. $\chi''(T)$ shows the typical two peak structure described above. The peaks near T_c are all occuring above 90K, the transition temperature of Bi-2212 phase. Hence they are attributed to the dissipation in the Bi-2223

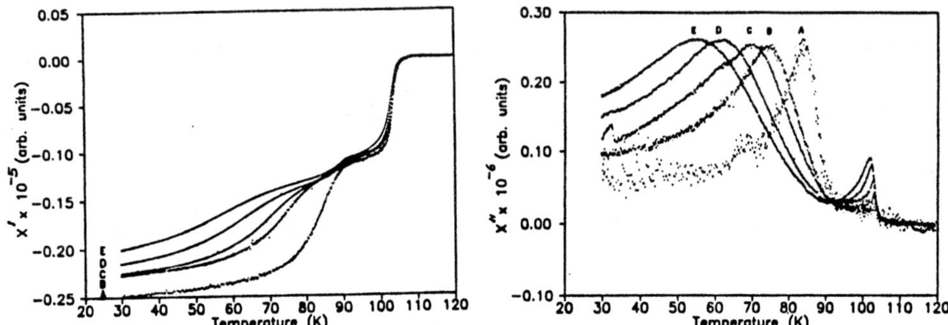

Fig.3 $\chi'(T)$ and $\chi''(T)$ for Bi-2223 sample for various field
A:0.5 Oe, B:1.4 Oe, C:2.4 Oe, D: 4.5 Oe, E: 6.4 Oe

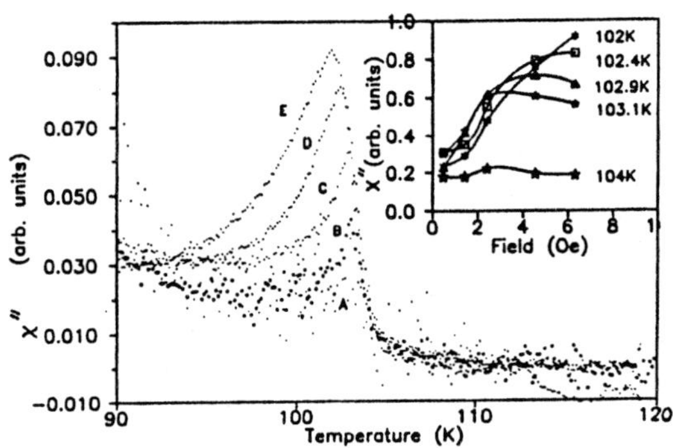

Fig4.The intra-grain peaks in $\chi''(T)$, that occur between 90K and T_c(onset) at different excitation fields, H_m as in Fig.3. The inset shows χ'' as a function of H_m at different temperatures.

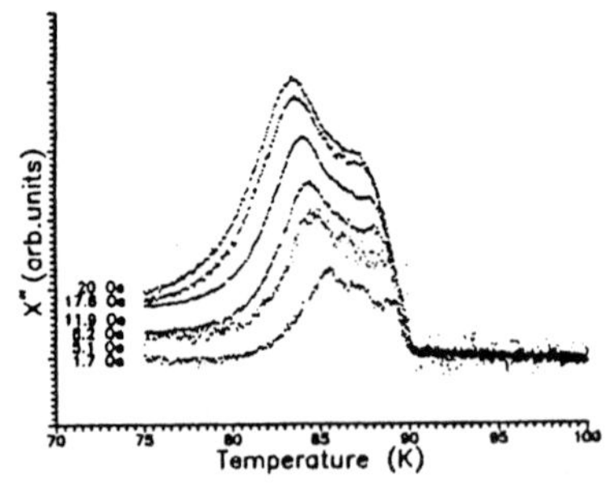

Fig.5.$\chi''(T)$ for 123 sample with silver addition

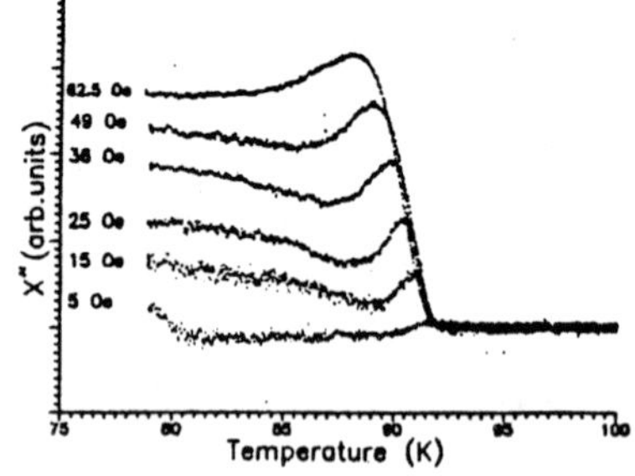

Fig.6.$\chi''(T)$ for 123 sample with 1% Pr substitution at Y site

grains. The inset of fig [4] shows the feature, above a certan temperature T_p, χ'' has the same value as those for higher amplitudes. In other words , in the (H_m,T) diagram $T_p(H_m)$ defines a line above which χ'' is independent of field, while below that line χ'' is field dependent. Fig.5 & 6 shows the χ'' formelt-grown 123 sample with silver addition and 123 sample with 1% Pr addition at the Y site. All these curves exhibit the feature above a certain temperature, where χ'' is independent of field. This kind of feature is already reported for the melt-grown sample [12]. But, it is for the first time such a feature is reported for the Bi 2223 system.The inset of Fig [4] illustrates the typical feature observed in all samples. As discussed above since , χ'' is independent of field , this feature is attributed to the flux flow. The novelty of this feature is that it is occuring for such low AC fields, without a DC bias field.

References:
1.J.R. Clem, in Magnetic Susceptibility of superconductors and other spin sysrems, edited by R.A. Hein, T. Francavilla and D. Liebenburg (Plenum, New York, 1992) pp 177-211
2.R.B. Goldfarb, M. Lelental, C.A. Thompson, in Magnetic Susceptibility of superconductors and other spin sysrems, edited by R.A. Hein, T. Francavilla and D. Liebenburg (Plenum, New York, 1992)
3.K.H. Muller, Physica C 159, 717 (1989)
4.R.B. Flippen, T.R. Askew and M.S. Osofsky, Physica C 201, 391 (1992)
5. R.B. Flippen, T.R. Askew, J.A. Fendrich and B.M. Vlcek, Physica C 228, 85 (1994)
6. H.B. Sun, K.N.R. Taylor and G.J. Russel, Physica C 227, 55 (1994)
7.S.D. Murphy, K. Renouard, R. Crittenden and S.M. Bhagat, Solid. State. commun., 69, 367 (1989)
8. C.P. Bean , Phys. Rev. Lett 6, 250 (1962)
9.S. Lofland, M.X. Huang and S.M. Bhagat, Physica C 203, 271 (1992)

10. Y.B. Kim, C.F. Hempstead and A.R. Strnad, Phys. Rev. Lett 7, 306 (1962)

11.D.X. Chen and R.B. Goldfarb, J. Appl. Phys 66, 2489 (1989)

12.X.C. Jin , Y.Y. Xue, Z.J. Huang, J. Bechtold, P.H. Hor and C.W. Chu, Phys. Rev B **47**, 6082 (1993).

13. K.A. Muller, M. Takashige and J.G. Bednorz, Phys. rev. Lett 58, 1143 (1987)

14.Y. Yeshurun and A.P. Molezemoff, Phys. Rev. Lett 60, 2202 (1988)

Specific Heat of Superconductors

Udayan De[1] and B.D. Bhattacharyya[2]

[1]Variable Energy Cyclotron Centre
Department of Atomic Energy
I/AF, Bidhan Nagar
Calcutta – 700 064, India

[2]Baidyabati – 712 222
Hooghly
West Bengal, India

Abstract :

Important observations of various specific heat measurements on high Tc oxide superconductors, including the fluctuation effects and the effect of radiation damage are reviewed. The main emphasis has been on possible theoretical explanations.

1.   Introduction

Specific heat in conventional superconductors has been covered in excellent texts like one by E. S. Rajagopal [1]. Although this review starts with some overview of the BCS concepts of specific heat, the experimental results on conventional or the confirmed BCS superconductors have been mostly excluded. Experimental and theoretical work on high Tc oxides has been covered, with some additional theoretical formulations to present a coherent explanation of the various effects.

2.   Basic concepts and observations

Specific heat anomalies of high Tc superconductors elucidate the nature of superconducting condensate in terms of the parameter; r = d/ξ, where ξ is the coherence length, or the average diameter of a

                    Udayan De and B.D. Bhattacharyya

pair of electrons (or change carriers to include hole doped supercond
uctors) in the condensate, and d the average distance between tw
pairs. The Bardeen-Cooper-Schrieffer (BCS) theory gives $\xi \propto V/\Delta$
where $V\Phi$ is the average Fermi velocity in the normal state and 2Δ th
binding energy of a Cooper pair. Since $d^{-3} \propto N(O)\Delta$, the density o
states at the Fermi level, $N(O)$ can the obtained from specific hea
measurements above $T\chi$. The specific heat jump $\Delta C\Pi$ at $T\chi$ gives $N(O$
from the BCS relationship in the weak coupling limit $(r \ll 1)$:

$$\Delta C\Pi = 1.43 \; \gamma \; T\chi,$$

with $T\chi \approx \theta\Delta \exp(-1/\lambda),$ (1)

and $\lambda = N(O)V$

Here γ is the electronic specific heat coefficient (Sommerfeild const
ant) :

$$C_P = \gamma \; T + \beta \; T^3,$$ (2)

where β is a constant, $\theta\Delta$ the Debye temperature, γ the phonon mediate
electron-electron coupling constant and λ the electron phonon (el-ph
interaction parameter. From a number of data [2, 3] it is found tha
the magnitude of the jump $(\Delta C\pi/\gamma T\chi)$ is consistent with the prediction
of the weak coupling BCS theory (1.43) to within 30% in lanthanum an
yttrium based cuprate superconductors.

The parameter r depends on both Δ and $N(O)$ in the BCS theor
and is a sensitive function of infrared (IR), Raman and specific hea
measurements. In the limit of $r \ll 1$, a mean-field (mf) treatment o
the condensate as in the BCS or Eliashberg theory appears to be appro
priate. In this limit $2\Delta/kT\chi \geq 3.5$, stable pairs are not present abov
$T\chi$, and they exist below $T\chi$. The IR studies [4-8] gives reduced energ
gap in the range : $2\Delta/kT\chi \approx 2 - 3.5$, smaller than the BCS value (
3.5kTc). Schleisinger et al. [9] reported $2\Delta\chi \approx 3$ kT$\leq$ and $2\Delta\alpha \; \beta \approx$
kT$\leq$ from IR studies of YBa$\angle$Cu$\neg$O$\propto\mp_\delta$. A comparison to IR data fro
Br Sr CaCu O ϑ also shows $2\Delta\alpha\beta \approx 8$ kT$\leq$. The $r \gg 1$ limit correspon
to Bose condensation. For an ideal Bose gas, the condensation tempera
ture $T_c \propto n^{2/3}(m)^{-1}$, where n is the density and m the mass of th
bosons. $r \gg 1$ reflects electron pairs which would also exist abov

T$\leq$. The pair binding energy, E$\leq$, is not related to T$\leq$ in this r >> 1 limit, and, therefore, it is reasonable to assume a large binding energy 2Δ >> kTχ. Raman studies [10] show that the 90K phase of YBa2Cu3O7-$_\delta$ have broad features at ~ 350 and ~ 500 cm^{-1} in the A1g and B γ symmetries, respectively, which exhibit behavior qualitatively similar to the 'gap' redistribution expected from r << 1 (BCS like superconductor). With decreasing oxygen concentration the untwinned Tχ = 60K single crystal of YBCO reveals non BCS features. The 500 cm^{-1} B1g peak is found to persist well into the normal state and is insensitive to the change in carrier concentration. The characteristic energy, E$_c$ $\approx$ 500 cm^{-1} cannot be assigned to a band structure feature of a straightforward BCS gap. Raman studies [10] thus suggest that in a coupling regime descriptive of the high T$\leq$ cuprates where the coherence length ξ is comparable to the interparticle spacing k$\bar{\Phi}^{1}$, bound pairs could exist at temperatures T > Tχ; and for T < Tχ these particles undergo condensation to bulk superconductive state when T is slightly larger than the mf transition temperature T$_c^{mf}$. Conversely, small regions fluctuate into the normal state when T is slightly lower than T$_c^{mf}$. The material containing a superconducting region and an adjacent normal (nonsuperconducting) region must have an intermediate region of size ξ in which the mf value of the superconducting order parameter goes gradually to zero. The size of the fluctuating region determined by ξ is temperature dependent, becoming very large near T$_\chi^{mf}$ at a rate which depends on the dimensionality of the system. Length ξ is much smaller in high T$_ç$ cuprates like $\xi\alpha\beta$ (0) = 13 Å [11], $\xi\chi$ (0) = 3 Å [12]. So only a small coherence volume, containing a small number of electrons, need to fluctuate into the superconducting or normal state. Fluctuations involving a small number of particles occur with greater probability than those involving many particles as is the case in conventional superconductors.

The superconductors La ξSrξCuO , YBa Cu O $_\delta$, DyBa Cu O$_\delta$ [13] show an anomalously large jump $\Delta C\pi$ in the region around Tχ. An anomalously large change in slope in Cπ is observed for

           Udayan De and B.D. Bhattacharyya

BiPb   SrCaCu O   and Tl Ba CaCu O   [14]. If one assumes that this change of slope in these latter materials is due to a BCS transition buried in the noise due to a distribution of $T\chi$'s of non-single phased materials, one would obtain a value roughly 100 times that of a classical BCS superconductor [14]. Concerning the jump in the specific heat observed in the above materials it could be due either to a coinciding lattice transition and hence have nothing to do at all with an electronic origin, or - as is now considered to be much more likely - to a strong coupling of the phonon system with the electrons Okazoki et al. [11] reported that an accurate evaluation of $\Delta C\pi/T\chi$ can be made only after separating electronic specific heat from lattice specific heat, although the former contribution is by far smaller (due to small value of $E\Phi$) than the latter in this high temperature range.

        Weber and Mattheiss [15] made a realistic tight- binding approach based on the energy-band results of Mattheiss and Hamann [16]. The strong coupling of the phonon system with electrons in YBa Cu O could give a correct evaluation of specific heat anomaly near $T\chi$. Their calculation, focused at first on the el-ph interaction within the three pdσ bands, gives a value of λ in the range 0.5 to 1 and corresponding $T\chi \sim$ 3 to 10K. Their second calculation incorporated the pdπ band giving λ ranging from 1.3 to 2.3 and corresponding $T\chi$ in the range 19 to 30K. Their final calculation assumed a formal Cu valence of +2 in order to accommodate one more electron per unit cell. This calculation yielded approximately the same values of λ and $T\chi$ as the first calculation. So Weber and Mattheiss [15] concluded from their calculations that conventional el-ph interaction cannot account for the high $T\chi$ in cuprates like YBa Cu O . Thus we need an additional or an alternative mechanism for the electronic pairing that could lead to no global lattice instability, but rather to a strong mixing of electronic and phonon degrees of freedom, common to small Jahn-Teller polaronic carriers. This leads to the kink like increase of the entropy [14, 18]. In this connection, let us mention some studies on field dependence of C near $T\chi$ [17, 18]. The behavior differs from that of conventional type II superconductors in that the anomaly is diminished with increasing field but neither the temperature of the maximum in C/T nor the temperature of the onset of superconductivity is substantially affected. For YBa Cu O $_\delta$ single crystals the strong anisotropic magnetic field dependence jump in C is suppressed for fields above ~ 6T perpendicular to the CuO_2 layers [18]. These features reflect the importance of fluctuation effects. The anomaly centered at about 107K in the reduced specific heat C(T)/T vs. T plot for $Bi_2Sr_2Ca_2Cu_3O_x$ [19] has very little resemblance with the usual BCS or two fluid type anomaly. This reflects the role of superconducting fluctuations.

Ginsberg et al [20] have measured the specific heat of a single crystal of $YBa_2Cu_3O_{7-\delta}$ and observed superconducting fluctuations as a BCS like step [20]. This is well described by 3-d Gaussian fluctuations. Okazaki [11] observed a broadened transition width and a sharp peak at $T\chi$ of the (Bi,Pb)-Sr-Ca-Cu-O superconductor and suggested them to be due to 2-d superconducting fluctuation effects. A relatively strong 2-d fluctuation of the superconductivity in Bi Sr CaCu O has been reported in this material [21]. The fluctuation effect is observed as a large scale crossover near $T\chi$, limiting the applicability of Hao-Clem's model to the temperature region much lower than $T\chi$.

Braun et al. [22] reported measurements of specific heat and thermal expansion of Bi-2212, Bi-2223, Tl-2212 and Tl-2223 superconductors and suggested that the sharp anomalies near $T\chi$ reveal clear evidence of strong fluctuations of superconducting order parameter. For temperatures more than ~ 5K away from T_c, the anomalies of all Bi and Tl phases can be explained with 2-d Gaussian fluctuations, while within ± 5K of T_c the fluctuation contribution is due to critical fluctuations. The critical fluctuations observable for high-T_c materials only involve interactions among the fluctuations and arises through the nonlinear term in Ginzburg-Landau (G-L) free energy. Analysis of polarized copper K-edge extended x-ray absorption fine structure (EXAFS) studies of $YBa_2Cu_3O_{7-\delta}$ and $TlBa_2Ca_3Cu_4O_{11}$ superconductors [23,

24] reflect structural fluctuations of apex oxygen which moves in a double-well potential described by G-L free energy :

$$V(\psi) = A\ \psi^2 + B\ \psi^4 \qquad (3)$$

This double well potential softens in the critical fluctuation regime : $(T_\chi - \delta\lambda T) < T < (T_\chi + \delta\upsilon T)$ where the double well collapses into effectively a structure with a single anharmonic minimum, the upper bound $\delta\upsilon T$ is 4 to 12K above the halfway point of T_χ. This change suggests nonlinear coupling between anharmonic phonons of the potential and superconducting order parameter.

Zhang et al [25] reported that fluctuation region in (Bi,Pb) $Sr\ CaCu\ O$ extends over temperatures $T > 0.9\ T_\chi$

A combination of not totally unrelated high T_χ parameters, namely : the high transition temperature (T_χ), the extremely short coherence length (ξ), the large value of the anisotropy ratio, $\Gamma\alpha = \lambda_\chi/\lambda\alpha\beta = \xi\alpha\beta/\xi_\chi = (m_\chi/m\alpha\beta)^{1/2}$, and the reduced dimensionality enhances the fluctuation effects and makes the critical fluctuation region experimentally accessible. Here $m\alpha\beta$ and m_χ are the effective mass tensor, $\xi\alpha\beta$ $(\lambda\alpha\beta)$ and $\xi_\chi(\lambda_\chi)$ are the coherence lengths (penetration depths) parallel and perpendicular to the superconducting CuO -layers. The width of the critical region $[\Delta T]_\chi$:

$$[\Delta T]_\chi \approx 2\ (2\pi T_\chi/\phi_0)^3\ (\chi^2\ \Gamma\alpha)^2\ [H_\chi\ (0)]^{-1} \qquad (4)$$

around the superconducting transition, T_χ, where mf Gaussian fluctuations break down due to the growing interaction between fluctuations. Here $\chi = \lambda\alpha\beta/\xi\alpha\beta$, H_χ (0) is the upper critical field at zero kelvin and $\phi = 2.07 \times 10^{-7}$ Gcm^2, the flux quantum. For YBa Cu O$_\delta$, th Ginzburg number is Gi $\sim 10^{-3}$. The fluctuation contributions to the specific heat and diamagnetic susceptibility above Tc in the Eliashberg model are enhanced by a factor $(1 + 3\lambda/2)$ for weak coupling $(\lambda << 1)$ and $3(1 + \lambda)/2$ for $\lambda >> 1$ in comparison with the corresponding BCS and GL results. The fluctuation contribution of specific heat Cπ and diamagnetic susceptibility χ in 3-d and 2-d are [26]

$$C_\square = \frac{1}{8} \left[\frac{\pi T \chi}{\phi}(-dH_{C2}/dT)_{T=T_c}\right]^{1/2} \left[(T-T_C)/T_C\right]^{3/2} \lambda' , \qquad (3\text{-}d)$$

$$(5)$$

$$\chi = - \frac{\pi T \chi}{48} \left(\frac{\pi}{\phi}\right)^{3/2} (T-T_C)^{1/2} \left[(-dH_{C2}/dT)_{T=T_c}\right]^{1/2} \lambda' \qquad (3\text{-}d)$$

and

$$C_\square = \frac{\pi T \chi^2}{4\phi(T-T\chi)} \left[(-dH_{C2}/dT)_{T=T_c}\right] \lambda' \qquad (2\text{-}d)$$

$$(6)$$

$$\chi = - \frac{\pi^2 T \chi}{24(T-T\chi)\phi} \left[(dH_{C2}/dT)_{T=T_c}\right]^1 \lambda' \qquad (2\text{-}d)$$

with $\lambda' = 1 + 3\lambda/2$ for weak el-ph interaction, and $\lambda' = 3(1 + \lambda)/2$ for strong el-ph interaction.

The slope, $dH\chi(T)/dT$, can be obtained from the BCS fit of $H_c(T)$ which yields $dH_{c2}/dT \simeq -3.68$ TK^{-1} near T_c for Bi$_2$Sr$_2$Ca$_2$Cu$_3$O$_{10}$, where $\chi\chi = 170.4$ is used [27] and $H\chi = \chi\chi\sqrt{2H}(T)$. The above equations (5, 6) may be applied to find C_Π for YBCO. But for Bi-family we have to modify the equations as discussed below.

For the instant electron-electron interaction as considered in Eq.(1) with $\lambda \ll 1$ (BCS model) one gets the renormalization of the Fermi velocity UΦ and as a result this renormalization is included in $-(dH\chi/dT)_{T_c}$. The relation [28]

$$\frac{H_c(T)}{H_C(0)} = 1.7367\left[1 - \frac{T}{T_C}\right]\left[1-0.2730\left(1 - \frac{T}{T_C}\right)-0.0949\left(1 - \frac{T}{T_C}\right)^2\right] \qquad (7)$$

keeps BCS fit to the $H_C(T)$ vs T behavior between 74 and 86K. It is found [27] that for temperatures above 100K, it shows deviation between experimental data and BCS correlation. This was attributed to the large fluctuation effect near T_C for the Bi-family superconductors. The large fluctuation near T_C results in the diverging behavior of $\chi\chi$, the G-L parameter which shows a limitation of Hao-Clem theory. This unusual temperature dependence of $\chi\chi$, increasing as temperature increases [27] near $T\chi$ is opposite to what we might expect for a conventional type II superconductor. This behavior could come from either the extended fluctuation below $T\chi$ or perhaps the quasi-2d nature of the bismuth based cuprates. It should be pointed out that the Hao-Clem model is based on a 3-d anisotropic mf theory which does not include the characteristic nature of a quasi-2d superconductor.

The angle resolved photo emission studies [29] for Bi-2212 ($T\chi$ = 80K) show that the Cu-O and the Bi-O layers were metallic and superconducting below $T\chi$, while for a sample with $T\chi$ = 89K only the Cu-O layers showed superconductivity. The latter could be responsible for the enhancement of the fluctuations with increasing $T\chi$: the structure with insulating Bi-O layers is obviously near to 2-d than the structure with superconducting Bi-O layers. The enhancement of the fluctuations in Bi-family with growing $T\chi$ is possibly due to a gradual transition of the Bi-O planes from metallic (oxygen rich samples) to insulating behavior [29] (deoxygenated samples).

3. Gaussian and critical fluctuations to the specific heat

The fundamental concept of conventional superconductivity is the existence of single valued order parameter (OP), ψ, which locally describes the density of charge carriers of superconducting current :

$$\rho\sigma\chi = \psi^{*}(\underset{\sim}{r})\,\psi(\underset{\sim}{r}) = <|\psi|^{2}> \qquad (8)$$

The OP ψ is a complex scalar field representing the quantum mechanical wave function of the paired electrons and can be characterized by its amplitude ψ and phase φ :

$$\psi(\underset{\sim}{r}) = \psi(\underset{\sim}{r})\,\exp\,[i\varphi(\underset{\sim}{r})] \qquad (9)$$

The superconducting state of a material is characterized by one or more OP(s). The nature of the electronic band structure, the pairing matrix depending on the symmetry of the crystalline lattice, time reversal and guage group put constraints on the superconducting OP. In conventional superconductors only the gauge symmetry is broken while all other symmetries are retained. On the other hand, in high temperature superconductors one or more of the other symmetries are also broken. The OP at the superconducting transition transforms according to an irreducible representation of the relevant symmetry group.

A careful analysis of specific heat measurements, particularly the anomaly near T_c in high T_c cuprates, reveal symmetry information which would be relevant for understanding the pairing and hence the microscopic mechanism. In absence of a clear microscopic picture, the G-L theory provides a rigorous approach to understand the broken symmetries of the OP. The G-L free energy is expanded as a power series in the OP in the vicinity of Tc and by using variational method the G-L equations are obtained. We note that the conventional G-L theory invokes one complex (n = 2) OP. The OP function used in the Hao-Clem model is a trial function for a variational approach to the G-L theory.

The solution of the G-L equations depends on the approximations and the associated boundary conditions. The simplest one is the mf approximation. The free energy F is calculated and the specific heat is given by

$$C_p = -T \, (\delta^2 F/\delta T^2) \tag{10}$$

Mean-field approximation is adequate for conventional superconductors which have an extremely small fluctuating region. For high T_C materials the critical region is experimentally accessible. Because the coherence length, ξ is large for the conventional superconductors, mf theory is valid to very small values of the reduced temperature t = $(T/T_\chi - 1)$, and the BCS specific heat jump is observed for these materials. Thouless [30] and then Aslamazov and Larkin [31] predicted that the specific heat has a Gaussian fluctuation contribution in the mf regime, so that

$$\Delta C = C_{\pm} \, |t|^{-1/2} \tag{11}$$

Here $C_{\pm}$ is the amplitude of the fluctuation contribution above and below $T_\leq$.

For quadratic (i.e., Gaussian) fluctuations about mf theory in a O(n) model, the amplitudes are in the ratio

$$C_+/C_- = n/2^{d/2} \tag{12}$$

where n is the number of components of the OP and d the dimensionality [32]. For $YBa_2Cu_3O_{7-\delta}$ with small ξ, the Ginzburg criterion for n = 2, is

$$|t| \gg (1/32\pi^2) \, (k_B/\Delta C \xi^3)^2.$$

This predicts that the critical effects from interactions between fluctuations which should occur close to $|t| = 0$, should not be observable for $|t| > 10^{-3}$.

In addition to the contribution of the Gaussian fluctuations to the specific heat, there is the usual BCS like contribution, so that following Ma [32] the electronic heat :

$$C(T) = C^{mf}(T) + \Delta C^{G}(T) \tag{13}$$

where
$$C^{mf}(T) = C^{BCS} = 1.43\ \gamma\varepsilon\phi\tau\ (1 + 1.83t) \tag{14}$$

or $C^{BCS}/T = (\Delta C\pi/T\chi)\ (1 + bt)$

where $\gamma\varepsilon\phi\tau$, the electronic specific heat coefficient includes possible strong coupling corrections.

Inderhees et. al [33] have observed inverse-square-root, behavior of specific heat of YBa Cu O$_{\delta}$ near $T\chi$ corresponding to Gaussian fluctuations. They found the excess specific heat above (+) and below (−) $T\chi$ in addition to the BCS type specific hear due to the fluctuation effect :

$$C(T) = C^{mf}(T) + \Delta C^{G}(T) \tag{15}$$

where $C^{mf}(T)$ is the mf contribution exhibiting a discontinuity,

and
$$\Delta C^{G}(T) = C^{\pm}|t|^{-2+d/2} \tag{16}$$

with
$$C^{+} = k_B/[8\pi\xi(0)^3] \tag{17}$$

As $H_{c2}(0)_{11} = \phi/2\pi\xi\alpha\beta^2(0)$, C^{+} is thus depending on $H\chi(0)$. The contribution of fluctuations to the specific heat in the presence of magnetic fields parallel to the C-axis is of the form [34]

$$C_H' = [k_B/4\pi C\xi\alpha\beta(0)^2]\ \tilde{C}_H$$

with
$$\tilde{C}_h = \sum_{N=0}^{\infty} 2h\ (\varepsilon\nu + 2hN + 1/2\gamma)\ (\varepsilon\eta + 2h)^{-3/2}\ (\varepsilon\eta + 2hN + \gamma)^{-3/2} \tag{18}$$

where $\varepsilon\eta = \varepsilon + h$, $h = \xi\alpha\beta(0)^2\ 2eH$, and $\gamma = \Delta[\xi\chi(0)/C]^2$

In the limit $h \to 0$ we recover the zerofield result. The parameter γ gives the temperature for crossover form 2-d to 3-d.

Lee and Shenoy [35] calculated the field dependence of the excess specific heat CH caused by Gaussian fluctuations in both clear and dirty limits :

$$C_H = h^{-1/2} f (t_H/h) \qquad (19)$$

with $h = \xi^2(O)H/\phi$, $t_H = [T/T\chi(H) - 1]$

For $\chi >> 1$, $f(x) \to \chi^{-1/2}$, leading to Eq. (16)

The magnetic field broadens the width of the critical fluctuation region [17] significantly. We therefore perform a scaling analysis, assuming that the magnetic field enters into the singular part of the free energy through the term $[(-ih\nabla - 2dA/C)\psi]^2$, where ψ is the OP and A the vector potential. The free-energy density in dimensionless units is expressed as

$$F_{SCH} = F_{SCO} + \frac{1}{2} |\psi|^2 + \frac{1}{2}|\psi|^4 + [(\frac{1}{ik}\Delta - A) \psi]^2 + h^2 \qquad (20)$$

where F_{SCH} and F_{SCO} are the free energy density of superconducting phase in a magnetic field and zero magnetic field, respectively. Assuming A scales as an inverse length we find the fluctuation contribution to the free-energy in scaling form :

$$\Delta F = h^{(d/2)} g(t_H/h^{(1/2\nu)}),$$

where ν is the exponent governing the divergence of the G-L parameter $\kappa\chi = \lambda\chi/\xi\chi = \frac{\phi}{2\pi\xi(O)^2} [\frac{1}{8\pi TX} \frac{}{\Delta C}]$. The specific heat now becomes :

$$CH = h^{(d/2-1/\nu)} g'' (t_H / h^{1/2\nu}) \qquad (22)$$

The mf result $\nu = 1/2$ leads to Eq. (19).

If we measure C in zero field and in a field at temperatures which are equidistant from a critical point on a t plot, we have

$$(C - CH)h^{(1/\nu-d/2)} = f (t/h^{1/2\nu}) \qquad (23)$$

where $f(x)$ is a scaling function.

Expanding the OP ψ in terms of Landau orbitals with gauge A = B $x\hat{y}$ Ikeda et. al. [36] obtained G-L functional for layered superconductors under a magnetic field $B||\hat{z}$:

$$H = \sum_{k,n} [\mu + 2nh + \frac{\lambda}{2} (1 - \cos(qs))] |\phi\nu\kappa|^2 + \frac{\tilde{g}o}{2}$$

$$\sum_{\{n_iK_i\}} [\delta K_1 K_2, K_3 K_4 \times U\Pi_1 \ \Pi_3 U\Pi_2 \Pi_3].$$

$$W(n_i,P_i)\phi\nu^*_{\kappa_1 1} \phi\nu_{\kappa_2 2} \phi\nu_{\kappa_3 3} \phi\nu_{\kappa_4 4} \qquad (24)$$

with

$$h = 2\pi\xi^2(0)B/\phi \; ,$$

$$\lambda = (2\xi\chi(0)/S)^2,$$

$$\mu = \varepsilon + h \simeq (-1 + T/T\chi o) + h,$$

$$U_p = \exp(-P^2/2h),$$

$$W(n\iota, P\iota) = \prod_{i=1}^{4} \lim_{Si \to D} (\frac{1}{2^n n!})^{1/2} \frac{\delta^{ni}}{\delta S\iota^{ni}} \exp [-Si (Si + 2 \frac{pi}{\sqrt{h}})$$

$$+ \frac{1}{2} \sum_{j,k=1}^{4} SJ (S\phi + \frac{p\phi}{\sqrt{h}}] \qquad (25)$$

K denotes the pair of momenta (p.q) and the parameter $\tilde{g}o$ is expressed in terms of specific heat jump ΔC or of the GL parameter $\kappa = \dfrac{\phi}{2\pi\xi(0)^2}$

$$(\frac{1}{8\pi T\chi \; \Delta C})^{1/2} \qquad \tilde{g}0 = g3 (\frac{2\pi}{h})^{1/2} \frac{1}{\tilde{A}\vartheta \; \tilde{A}\zeta},$$

$$g3 = \frac{k_B}{\Delta C} \frac{B}{\phi \; \xi\chi(0)} = \frac{4\kappa\tilde{\chi}^2 kBT\chi o}{\xi^2(0)\xi\chi(0)} (\frac{2\pi\xi^2(D)}{\phi o})^3 B.$$

Here $\tilde{A}y$ and $\tilde{A}z$ are system sizes as we have said before that the vector potential A scales as an inverse length, $\tilde{A}y$, $\tilde{A}z$ are the normalized coherence lengths in y and z directions, respectively. In the specific heat calculations Ikeda et al. [36] pointed out that the contributions of higher Landau levels (LL) fluctuations produce a nontrivial deviation from the scaling behavior :

$$T - T_{CH} \sim H^{2/3}, \text{ in } T > T_{CH} \qquad (26)$$

T_{CH} is the mf transition temperature in a field H, One can notice that the sign of the derivative $\delta C/\delta H$ of each of the YBCO curve [36,37] changes at an intermediate temperature (below TXH), so that the curves cross with one another.

In contrast, in BSCCO curves no such behavior is seen in the temperature range of interest. In the lowest LL fluctuation theory, one can more easily understand these features dependent on the sample dimensionality and the anomalous behavior of $\kappa\chi$ for the Bi-family. According to the Maxwell relation :

$$\frac{T\partial^2 M}{\partial T^2} = \frac{\delta C}{\delta H} , \ldots\ldots \qquad (27)$$

the sign change [36,37] corresponds to the broad maximum of $\delta M/\delta T$ which could probably be explained in the low temperature ($T < T_\chi$) by a naive extension of the Prange's Gaussian fluctuation theory to the renormalized one.

The magnetization taken as a function of temperature in various fixed magnetic fields in Bi Sr Ca Cu Oξ [19,Fig.3] reveal a broad maximum of $\delta M/\delta T$ at intermediate temperatures ($T < TXH$) in the low field case – but this maximum vanishes as the field increases. This field variation suggests field induced dimensional crossover from 3-d to 2-d i.e. the structure with superconducting Bi-O layers changes to insulating Bi-O layers.

4.   Theoretical Model and Discussions

The work of Wohlleben et al. [39] shows an overview of C/T vs T of five high Tc cuprates with specific heat anomalies near Tc. The largest anomaly in Y-123 is at a first glance reminiscent of the step like BCS type anomaly found in the conventional superconductors. The Bi- and Tl- families on the other hand show much smaller anomalies. All the five anomalies show one common feature :

Above the general downward curvature of the background, there are two points of inflection located below and above each Tc, where an upward curature start or ends. For the Bi- and Tl- family the interval between the two points of inflection which extends over 15-20 K. It is due to the fact that both these Bi- and Tl- based cuprates have low-Tc and high-Tc phases, respectively represented by the stoichicometric compositions of 2 : 1 : 2 : 2 and 2 : 2 : 2 : 3 of Bi-Ca-Sr-Cu-O. The Tc values of Bi-2122 and Tl-2122 are around 85K and 105K, while that of Bi-2223, Tl-2223 are 107K and 125K respectively.

Ishikawa et al. [40] reported two distinct peaks of specific heat around 90K suggesting the existence of another superconducting phase in YBa2Cu3O7. But to explain the two distinct peaks or points of inflection in specific heat studies of high-Tc cuprates we consider the two-way Jahn-Teller (JT) model [41,42]. We assume that the Hubbard model alone is almost superconducting and the additional el-ph coupli-

Udayan De and B.D. Bhattacharyya

ng [15] in the form of JT interaction just triggers the system to the superconducting phase. The existence of JT like interactions interactions in Cu-O2 planes and Cu-O chains have been reported by many workers, especially by Alex Miller [43], Brûesch and Bûhrer [44], Bishop et al. [45], Conradson and Raistrick [46], Xiong Shijie [47], Oguri [48] and Liu et al. [49]. The Hamiltonian of our problem may be conveniently expressed in a condensed form [41, 42] :

$$H = -t_\Pi \sum_{1i\sigma} n_{1i\sigma} - t\rangle \exp(-\lambda\Psi T) \sum_{i\sigma} (n_{\iota\uparrow} + n_{\iota\uparrow}) + (U\varepsilon\phi\phi)$$
$$\sum_{1i\sigma} n_{i1\uparrow} n_{i1\downarrow} + \sum_{iq\sigma} [\Omega(q)/t\chi \,\varepsilon\xi\pi \,\lambda\Psi T]^{1/2} (n_{1\uparrow} - n_{2\downarrow})$$
$$(b_q^+ + b_q) + \sum_q K\,\omega\theta\,b_\theta^+\,b_\theta^+ \tag{28}$$

with $n_{1i\sigma} = C_{1i\sigma}^+ C_{1i\delta}$, $\lambda\Psi T$ the JT el-ph interaction term, t_π and t_χ are intralayer and interlayer transfer integrals.

$U\varepsilon\phi\phi$ is JT renormalized term. The last term describes the phonon continuum with b_q^+ (b_q), the phonon creation (annihilation) operators with wave vector q

$$\langle\frac{1}{\sqrt{2}} (a_{Li\uparrow}^+ a_{Lj}^{+\downarrow} - a_{T\downarrow}^+ a_{T\uparrow}^+)\rangle \text{ and } \langle\frac{1}{\sqrt{2}} (a_{Li\uparrow}^+ a_{2j}^{+\downarrow} - a_{T\downarrow}^+ a_{T\uparrow}^+)\rangle \tag{29}$$

are the three singlet Cooper pairs, the first two (1 = 1,2) are intra band Cooper pairs and the second term in (29) is the inter band Cooper pairs, where

$$a_{1i\uparrow}^+ = \frac{1}{\sqrt{2}} (c_{1i\uparrow}^+ + c_{2i\uparrow}^+) \tag{30}$$
$$a_{2i\uparrow}^+ = \frac{1}{\sqrt{2}} (c_{12\uparrow}^+ - c_{2i\uparrow}^+)$$

The intra band Cooper pairs ($\psi^{+\uparrow}$, ψ^+) are not independent. Introducing $H_{\pi\eta}$ in the form :

$$H_{ph} = (P^2/2m + (1/2)\,k\chi^2), \tag{31}$$

we obtain the effective Hamiltonian for CuO plane

$$H_p^{eff} = (P^2/2m + (1/2)\,k\chi^2) - \frac{1}{g}(\chi^2 + \chi^2) + \frac{2}{g}\chi\chi \tag{32}$$

where $\chi = \psi^+\psi$ and $\chi = \psi^+\psi$

If $\chi = 1$, $\chi = 0$, the holes are close to the bottom of one potentia well with minima of the effective potential displaced to Q $\sqrt{[2E_\pi/k]}$. If $\chi = 0$, $\chi = 1$, the holes move in the opposite direction

the minimum of the effective potential is displaced to q , with $|Q^J.|$
$= |Q^{J-}|$. EJT is the JT stabilization energy and k the elastic constan-
t.

The expression of G.L free energy obtained from (28) is :

$$F(T) = \frac{1}{2}A(T-T_{co}^{mf} - \Delta T)\psi^2 + \frac{1}{2}A(T-T_{\chi o}^{mf} + \Delta T)\psi^2 + \frac{1}{4}B(\psi^4 + \psi^4)$$
$$+ \frac{1}{2}C\;\psi^2\psi^2 - g\psi\;\psi\;x + \frac{1}{2}\;kx^2 - hx, \qquad (33)$$

where $x = (2m\omega/h)^{1/2}\;Q$.

As x describes an odd parity JT distortion, it makes the OP(s) $\psi\;\psi$
coupled - this is evident also in eqs. (30, 32)

The jump of specific heat [40] at the first superconducting transition
$T_{c1} = (T_{co}^{mf} + \Delta T)$ is given by

$$(\Delta C\Pi)1 = \frac{1}{2}\;(A^2/B)\;T_{C1} \qquad (34)$$

The peak corresponding to T_{C1} as propounded by Ishikawa et al. yields

$$\frac{(\Delta C\Pi)}{T_{C1}} = \frac{1}{2}\;(\frac{A}{B}^2) = 62\;mJ/mol\;k^2 \qquad [40]$$

which gives a small value of $\gamma = 4.3$ mJ/mol-f.u.K^2 for YBa Cu O . For
BiPb-2212 samples Braun et.al. [22] reported that samples with lower
$T\chi$ show a relatively large $\gamma BX\Sigma$. The samples with higher $T\chi$ show a
much smaller $\gamma BX\Sigma$ - i.e. a large fluctuation contribution and a change
over to critical behavior.

Double superconducting transitions showing specific heat jumps in high
$T\chi$ cuprates [39, 40] is associated with a two step transition in resistivity, $\rho(T)$. YBa Cu $O_{\delta+x}$ superconductors with x < 0.9 are characterized be highly diffusive transitions. The temperature at which the
transition begins T_c^+ and the temperature at which it ends, can be
clearly identified. T_c^+, the temperature at which a superconducting
pairing begins, and can be defined experimentally as the temperature
below which the resistivity in strong magnetic fields is appreciably
higher than the resistivity at H = 0. T_c^- is the temperature at which
the coherence in the Cooper pairs is established and hence the resist-
ivity vanishes. The transition width is $\Delta T\chi = T_\chi^+ - T_\chi^-$. The zero resistance transition temperatures for YBCO layer thickness of one, two and

three unit cells isolated in a relatively thick PrBCO matrix are : $T\chi o$ = $T\bar{c}$ ~ 19, 54 and 70 K, respectively [50].

Shcherbakov et al. [51] observed a correlation between the broadening of superconducting transitions, $\Delta T\chi$, and the microwave field resonance absorption showing ferroelectric anomalies. The correlation determined by the effect of Cooper pairs on the establishment of coherence in the system, suggests a possible link between specific heat anomalies and ferroelectric anomalies [52] :

$$G(t) = \int_{-\infty}^{\infty} d\Delta T\chi \exp\left[-\left(\frac{\Delta T\chi}{\sqrt{2}\sigma}\right)^2 - \frac{16\ B}{\pi h A^2}(\Delta C\Pi)\ t\right] (2\pi\sigma)^{-1/2}, \tag{35}$$

where $(\Delta C\Pi)$ is given by eq. (34). Here $G(t)$ is the order parameter correlation function and is given by

$$G(t) = \langle\ \psi(t)\ \psi(0)\ \rangle \tag{36}$$

If we set $\sigma = \pi|e|V\rangle/2$, eq. (35) gives a consistent broadening of the superconducting transition by an amount ~ $|e|V\rangle$, where V is a Gaussian stray field with a correlation function

$$\langle\ V(t)\ V(0)\ \rangle = V_c^2\ K\ (t/\tau\chi)\ K(0) = 1 \tag{37}$$

where $V\chi$ is the fluctuation amplitude – located mainly above the temperature where one finds R = 0 i.e. at a temperature where coherence in Cooper pairs is established and ferroelectric anomaly appears. The broadening of the specific heat fluctuations ~ ± 10K [19, 22] gives us a measure of $\Delta T\chi \approx$ 10K in Bi∠Sr∠Ca∠Cu¬O⊖. If one takes the carrier density in Bi∠Sr∠Ca∠Cu¬O⊖ from Hall coefficient [19] at 120K (0.08 holes/Cu atom) one obtains $\Delta T\chi \approx$ 30K at T⊆. If one uses about one carrier per unit cell, as suggested by the band structure calculations [53], one finds $\Delta T\chi \approx$ 10K. Schnelle et al. [54] reported a flux expulsion of only about 30% at fields H smaller than the lower critical field, H⊆⊥ which is all equilibrium property of the fully superconducting polycrystaline samples of the layered HTSC [55]. A complete flux expulsion did not take place at fields of order 1 Gauss or even less. The critical question is whether the grains are in contact with each other by a sufficiently dense network of weak links. Very important in this connection are the transition widths of the

Meissner curves : these widths reach values of $\approx$ 1.0K for the Y-sample and $\approx$ 1.5 for the Dy samples at 5 Gauss. The transition widths at 5 gauss, therefore, provide an upper limit for the width of the distribution of T_c within the grains averaged over the pellet i.e., for the width of the specific heat anomaly, due to smearing by sample inhomogeneities. The link between ferroelectric anomalies and specific heat anomalies is probably determined by the effect of Cooper pairs on the establishment of coherence in the system, rather than by the effect of individual pairs on the energy characteristic.

The condition under which two successive resistive (specific heat) transitions may be seen are enumerated by Rosenblatt et al., [56]

- the grain size is larger than the critical size,
- the strength J of the Josephsen coupling between adjacent grain is smaller than the superconducting condensation energy

of the grain.

The chain superconductivity in $YBa_2Cu_3O_x$ single crystals [57] reveal that the mf value of the specific heat jump at T_c increases by about 25% between 6.94 and 7.0. In contrast, T_c decreases by about 4%.

It is also observed that the fluctuation contributions to the specific heat are largely reduced near n = 7.0. This can be explained from (28) and G-L functional for the chain which gives specific heat jump

$$\Delta C = \frac{10 \; \xi \chi(0) \xi}{L \; T_C} \left(\frac{A}{B}\right)^2 \tag{38}$$

like eq. (34) where L = 1500 Å, 800 Å and 300 Å as average twin spacings [58] for oxygen contents of 6.98, 6.62 and 6.55, respectively. Breit et al. [57] found

$\frac{\Delta C}{T_c}$	55 (mJ/mol K^2)	53 (mJ/mol K^2)
χ	6.97	6.99
T_c^{mf}	91.1 K	89.2 K

which agrees with eq. (38) that as χ increases, L increases and so $\Delta C/T\chi$ decreases.

5.   Particle Irradiation Effects

Davydov et al., reported [59] specific heat for irradiated and unirradiated 123-HTSC in 1988. Let us discuss a more recent report [60] of radiation damage to specific heat, C, and a.c. susceptibility, acs, of the n = 2 and 3 members of Tl2Ba2Ca(n-1)CunO(2n+4) superconducting pellets due to 40 MeV alpha-irradiation. The jump, ΔC, in the specific heat across the superconducting critical temperature, $T\chi$, is proportional to the superconducting volume fraction in the sample, which should be investigated in radiation damaged samples.

By making the sample thinner than the range of the irradiating alpha-beam, it was ensured that it passes out of the sample, causing pure radiation damage and no implantation. The 40 MeV alphas loose their energy through electronic excitations, important at such high energies, and elastic collisions with the lattice atoms. While looking for the mechanism of such radiation damage in oxide superconductors, it is often overlooked [61] that the process of electronic excitations can [62], unlike in the metallic superconductors [63], affect the covalent-like bonds in these oxides and thus damage their superconductivity. The specific heat, C = C(e) + C(l) + Co, of solids arises from the electronic (e) and lattice-vibration (l) parts and other contributions collectively represented by Co. Since the electronic as well as the lattice parts are directly affected by a high energy alpha irradiation in Tl-superconductors, and experimentally one measures the total specific heat, C, measurable changes in C can be anticipated in such irradiation experiments.

The work also addresses to the well-known problem of estimating $\Delta C/T\chi$ on account of the large phonon background, $C\lambda$, at temperatures of the order of 100 K for $T\chi$ and adopts a logical semi-empirical solution.

Tl-2212 and Tl-2223 pellets with about 50% of the theoretical density were shaped into 500 and 300 micron thick bars weighing about 55 and 66 mg, respectively, to avoid implantation as discussed earlier. Specific heat has been measured by a continuous heating adiabatic technique. Essentially the rate of temperature-rise of the sample, dT/dt, for its electrical heating, under nearly adiabatic condition, by a known heating power, P, is measured to determine the specific heat, $C(T) = P/(dT/dt)$. Using 1 Oe a.c.

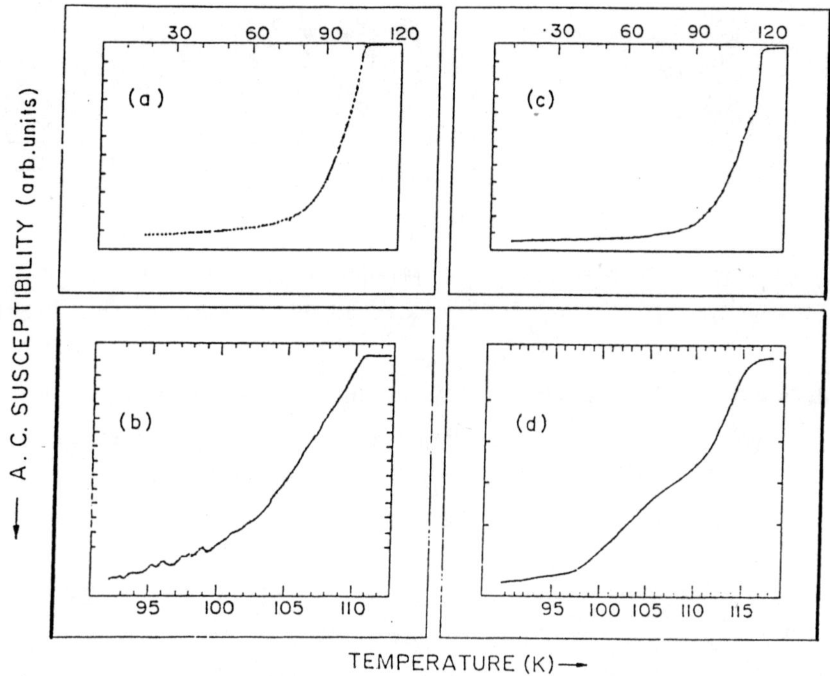

Fig. 1. Superconductivity onset in a.c. susceptibility for (a) unirradiated Tl-2212, (b) irradiated Tl-2212(c) unirradiated Tl-2223 & (d) irradiated Tl-2223 samples.

magnetic field, susceptibility has been measured (fig. 1) to estimate the on-set T_χ values. For room temperature alpha-irradiations in the cyclotron at the VEC Centre, Calcutta, an aluminum target holder with secondary electron recollection arrangement [64] was used. Earthing the insulated target-holder through a current integrator, beam-current as well as the fluence, 2×10^{16} $He^{++}.cm^2$, were accurately measured.

A. C. susceptibility data (fig. 1) shows that after irradiation the on-set T_c is practically unchanged to within about 1 K at about 110 K for Tl-2212 and at about 119 K for Tl-2223.

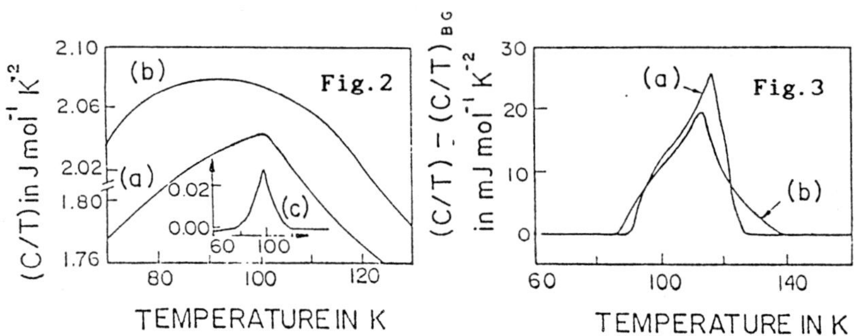

Fig. 2. Specific heat, C, vs. temperature T, data in C/T vs. T form for (a) unirradiated and (b) irradiated Tl-2212 pellets. Graph c in the inset is the unirradiated data minus its non-superconducting background (explained in the text) and represents C/T vs. T.

Fig. 3. Measured specific heat, C, minus its non-superconducting background, (C)BG, for (a) unirradiated and (b) irradiated Tl-2223 sample in [(C/T)-(C/T)BG] vs. T form.

Fig. 2 shows the measured (C/T) vs. T plots for Tl-2212. Compared to the clear kink at about 102 K or the unirradiated sample (graph a), later identified as the superconducting jump, there is no such structure in the irradiated sample (graph b). So there is no such structure in the irradiated sample (graph b). So there is no superconducting jump in the irradiated Tl-2212.

To make a jump in C/T vs. T data prominent, various authors subtract, from the measured specific heat, a phonon part, often estimated by fitting an un-explained combination [11] of Debye and Einstein functions. Such fitting functions have more fitting parameters but not yet [39] comprehensive physical model. Hence the present work [60] estimates (C/T)BG, the "background" appropriate for a non- superconducting but otherwise equivalent sample, by fitting a simple polynomial : $C/T = A_0 + B/T^2 + A_i T^i$, i = 1 to 5, to our data on both sides of the superconducting transition region. For example data below 80K and above 109K of the unirradiated Tl-2212 have been fitted to give $(C/T)_{BG}$ vs. T, which is subtrated from the experimental data (graph a) to get graph c in fig. 2 and hence conclude a jump : $(\Delta C/T\chi) = (33 + 3)$ mJ mol^{-1}K^{-2} at $T_c = 102$ K. This compares well with the values 35 + 10 and 25 to 29 mJ mol^{-1} K^{-2} reported [65, 66] earlier.

Similarly, "background" - corrected specific heat data for Tl-2223 is presented in fig. 3. It shows a jump of (28 ± 3) mJ mol^{-1}K^{-2} at 113 K for the irradiated sample. The unirradiated value is again comparable to the reports (39, 65) of (20 ± 10) and 29 mJ mol^{-1}K^{-2}.

The main result is that the superconducting fraction, as probed by the specific heat measurement, is reduced by the alpha irradiation, significantly in Tl-2223 and drastically to a unmeasurably low value in Tl-2212. However, the onset T_c, as measured by the a.c. susceptibility is practically unchanged by this irradiation. This tends to indicate that some traces of the original superconducting phases are left undamaged in both the samples to provide shielding currents and thus show almost unchanged onset T_c in the acs measurement.

More damage to Tl-2212 than to Tl-2223, can be explained from the difference that Tl-2212 has relatively more blocking/non-conducting layers, in which electronic excitations by the alpha-particles can cause [62] effective damage. Metal-like CuO_2 layers, relatively more in Tl-2223, are less affected by the electronic energy loss of the alpha-beam, as we have already indicated. This explains the observation and indirectly establishes the significant role of damage [62] via electronic energy loss in case of oxide superconductors.

6.   Conclusion and Summary

Basic aspects of specific heat in superconductors in general and the new features for the high temperature oxide superconductors have been discussed in the present review covering representative publications up to early 1995. Applications of specific heat based superconducting devices as in recent dark matter detection efforts [67, 68] will open up newer areas of further studies, not covered in this review.

The jump in specific heat in conventional superconductors takes place below 23K, temperatures at which the lattice contribution to the specific heat reduces practically to zero. This allowed a separation of the lattice and electronic contributions, attributing the specific heat jump at Tc to the change to superconducting state of the conduction electrons or charge carriers. Now that in the oxides Tc can be as high as 134K at ambient pressure, the lattice contribution is very important and mixed up with the electron or charge-carrier contribution. This complex situation cannot yet be described adequately in quantitative terms in any theory. Another manifestation of the high Tc, along with certain other aspects of the high Tc superconductors, is the much larger contribution of Gaussian and critical fluctuations. Specific heat measurements on more completely characterized samples including good single crystals [57] will guide the various theoretical approaches to really useful results. Further characterization of the nature of radiation damage is similarly important in understanding the specific heat of such damaged samples [59, 60].

The 68 references cited in this review form a part of the publications in this and related fields, and existing databases can be searched for an exhaustive bibliography.

Acknowledgement

We acknowledge that most of the figures are taken from published literature as mentioned in the figure captions. One of the authors (UD) thanks Dr. Bikash Sinha, Director VEC Centre, for academic support, and Prof. K. N. Shrivastava, University of Hyderabad, for encouragement.

References

1.   E. S. Rajagopal, Specific Heat at Low Temperatures (Heywood Books : London) 1966; for recent refs. on specific heat measurements : S. M. Pattalwar, Indian J. of Cryogenics **18**,   141 (1993)

2.   H. E. Fischer, S. K. Watson and D. C. Cahill, Comm. Cond. Matter Phys. **14**, 65 (1988).

3.   R. A. Fisher, J. E. Gordon and N. E. Phillips, J. Supercond.   **1**, 231 (1988).

4.   P. E. Sulewski et al., Phys. Rev. **B 35**, 5330 (1987).

5.   Z.Schlesinger et al., ibid **35**, 5334 (1987).

6.   J. R. Kirtley et al., ibid **35**, 8846 (1987).

7.   R. T. Collins et al., Phys. Rev. Lett., **59**, 704 (1987).

8.   L. Genzel, et al., Phys. Rev. **B 40**, 2170 (1989).

9.   S. Schlesinger et al., Phys. Rev. **B 41**, 11237 (1990).

10. F. Slakey et al., Phys. Rev **B 42**, 2643 (1990).

11. N. Okazaki, et al., Phys. Rev. **B 41**, 4296 (1990).

12. K. Kanoda et al., J. Phys. Soc. Jpn. **57**, 1554 (1988).

13. B. Batlog et al., Phys. Rev. **B 35**, 5340 (1987).

14. F. Seidler et al.,Physica **C 157**, 375 (1989).

15. W. Weber and L. F. Mattheiss, Phys. Rev. **B 37**, 599 (1988).

16. L. F. Matheiss and D. R. Harmann, Solid St. Comm. **63**, 395 (1987).

17. R. A. Fisher et al., Physica **C 153 - 155**, 1092 (1988).

18. E. Bonjour et al., Phys. Rev. **B 43**, 106 (1991).

19. W. Schnelle et al., Physica C **161**, 123 (1989).

20. D. M. Ginsberg et al., Physica C **153 - 155**, 1082 (1988).

21. M. J. Naughton et al., Phys. Rev. B **38**, 9280 (1988).

22. E. Braun et al., Z. Phys. B - Condensed Matter **84**, 333 (1991).

23. J. Mustre de Leon et al., Phys. Rev. B **44**, 2422 (1991).

24. J. Mustre de Leon et al., ibid **45**, 2447 (1992).

25. Lu Zhang, J. Z. Liu, and R. N. Shelton, Phys. Rev. B **45**, 4978 (1992).

26. L. N. Bulaevskii and O. V. Dolgov, Physica C **153 - 155**, 241 (1988).

27. Qiang Li et al., Physical Rev. B **46**, 3195 (1992).

28. J. R. Clem, Ann. Phys. (N.Y.) **40**, 268 (1966).

29. B. O. Wells et al., Phys. Rev. Lett. **65**, 3056 (1990).

30. D. J. Thouless, Ann. Phys. **10**, 553 (1960).

31. L. G. Aslamazov and A. I. Larkin, Sovt. Phys.- Solid St. **10**, 875 (1968).

32. S. K. Ma : 'Modern Theory of Critical Phenomena' (Benjamin, NY 1976) pp. 82 - 93.

33. S. E. Inderhees et al., Phys. Rev. Lett. **60**, 1178 (1988).

34. C. T. Rieck et al., Phys. Rev. B **39**, 278 (1989).

35. P. A. Lee and S. R. Shenoy, Phys. Rev. Lett. **28**, 1025 (1972).

36. R. Ikeda, T. Ohmi and T. Tsuneto, J. Phys. Soc. Jpn. **60**, 1051 (1991).

37. R. Ikeda and T. Tsuneto, ibid **60**, 1337 (1991).

38. R. E. Prange, Phys. Rev. B **1**, 2349 (1970).

39. D. Wohlleben et al., Proc. Int. Conf. Superond. (ISC) Jan 10-14 (1990) Bangalore (India), World Sc. Pub. Corp. Singapore.

40. M. Ishikawa et al., Physica C **153 - 155**, 1089 (1988).

41. B. D. Bhatacharyya, Proc. Int. Conf. Raman Spectr. Columbia USA ed. by J. R. Durig and Sullivan I. K (John Wiley, N.Y.) (1990).

42. B. D. Bhattacharyya, Ind. J. Phys. **66A**, 181 (1992).

43. K. Alex Müller, Z. Phys. B. - Condensed Matter **80**, 193 (1990).

44. P. Brüesch and W. Bührer, ibid **70**, 1 (1988).

45. A. R. Bishop et al., ibid **76**, 17 (1989).

46. S. D. Conradson and I. D. Raistrick, Science **243**, 1340 (1989); S. D. Conradson, I. D. Raistrick and A. R. Bishop, ibid **248**, 1394 (1990).

47. Xiong Shijie, Z. Phys. B. - Condensed Matter, **74**, 5 (1989).

48. Akira Oguri, J. Phys. Soc. Jpn. **57**, 2123 (1988).

49. Fu-Sui Liu et al., Physics Lett. **140**, 72 (1989).

50. D. H. Lowndes, D. P. Norton and J. D. Budai, - Phys. Rev. Lett. **65**, 1160 (1990).

51. A. S. Shcherbakov et al., JETP Lett. **49**, 121 (1989).

52. B. D. Bhattacharyya (unpublished) Presented as an Invited Talk at TIFR, Bombay, India (1989).

53. S. Massidda et al., Physica **C 152**, 251 (1988).

  H. Krakauer and W. E. Pickett, Phys. Rev. Lett. **60**, 1665 (1988).

  M. S. Hybertsen and L. F. Mattheiss, ibid **60**, 1661 (1988).

54. W. Schnelle, E. Braun, H. Broicher et al., (preprint) (1990).

55. S. Ruppel et al., (to be published).

  D. Wohllenben et al., (to be published).

56. J. Rosenblatt et al., Physics **B 152**, 95 (1988).

57. V. Breit et al., (preprint from KfK, Karlsruhe); R. Hauff et al., M^{2}S-HTSC IV, Grenoble, 5-9 July 1994.

58. D. Shi, et al., Supercond. Sci. Technol. 3, 457 (1990).

59. S. A. Davydov, A. E. Karkin, A. V. Mirmelshtein, I. F. Berger, V. I. Voronin, V. D. Parkhomenko, V. L. Kozhevnikov, S. M. Cheshnitskii and B. N. Goshchitskii, JEPT Lett. **47**, 236 (1988).

60. Udayan De, V. S. Pandit and Kayalan Mondal, Ind. J. Phys. **69A**, 37 (1995).

61. S. K. Bandyopadhyay, P. Barat, P. Sen, Udayan De, A. De, P. K. Mukhopadhyay, S. K. Kar and C. K. Majumder, Physica **C 228**, 109 (1994).

62. V. Hardy, D. Groult, J. Provost, M. Hervieu and B. Raveau, Physica **C 178**, 255 (1991).

63. Udayan De, R. Khanna, Y. Hariharan, V. S. Sastry, T. S. Radhakrishnan and D. Ghosh, Ind. J. Phys. **58 A**, 427 (1984).

64. Udayan De, V. S. Sastry, Y. Hariharan, D. Ghosh and T. S. Radhakrishann, in progress report (1981 - 1989), Variable Energy Cyclotron Centre, Dept. of Atomic Energy, Calcutta 700064, India, p. 186.

65. A. Junod, D. Eckert, G. Triscone, J. Mueller and V. Y. Lee, Physica **C 162-164**, 476 (1989).

66. T. Atake, H. Kawaji, M. Itoh, T. Nakamura and Y. Saito, Physica **C 162-164**, 488 (1989).

67. E. Fiorini, Nucl. Phys. **B 35**, 473 (1994).

68. S. Roychowdhuri, Udayan De and B. D. Bhattacharyya, Communicated. Also S. Roychowdhuri, Udayan De and B. D Bhattacharyya, Phys. Stat. Sol. (b), 1 May, 1995.

Mössbauer Effect of ^{119}Sn in $EuBa_2Cu_{2.98}Sn_{0.02}O_{7\text{-}\delta}$ Superconductor

Lydia S. Lingam and Keshav N. Shrivastava

School of Physics
University of Hyderabad
P.O. Central University
Hyderabad – 500 134, India

Abstract: The Mössbauer spectra ^{119}Sn in $EuBa_2Cu_{2.98}Sn_{0.02}O_{7-\delta}$ are re-examined in the normal state as well as in the superconducting state. The second-order Doppler shift (SOD) of the Mössbauer spectra in a superconductor. δ_{sc}. is calculated for the first time. The shift depends on the fourth power of the ratio of the gap energy of the superconductor to that of the matrix element of the electron-phonon interaction. At $T = T_c$ where the gap vanishes. the shift reduces to that in the normal state. The relative values of the SOD and the isomer shift for ^{119}Sn in $EuBa_2Cu_{2.98}Sn_{0.02}O_{7-\delta}$ have been found. The recoilless fraction of the ^{119}Sn in $EuBa_2Cu_{2.98}Sn_{0.02}O_{7-\delta}$ is compared with a previous theory. It is found that the ordinary BCS type gap agrees with the experimental measurements of the recoilless fraction only when the gap is modified to include a lattice distortion in the superconducting state.

1.   Introduction

Recently we have made considerable efforts to understand the Mössbauer spectra in superconductors. It has been found that correlations occur between the Mössbauer isomer shift and the susceptibility of the system [1]. We have reported [2] that the recoilless fraction is sensitive to the gap of the superconductor so that below the transition temperature the recoilless fraction is considerably reduced compared with the value in the normal state.

In this paper. we calculate the second-order Doppler shift of the Mössbauer spectra in a superconducting lattice and make efforts to estimate the SOD and the isomer shift seperately from the experimental measurements. The experimentally measured values of the recoilless fraction agree with the theory only for a special form of the gap. The extra term in the gap arises from the lattice distortions due to structural changes.

2. Second-order Doppler shift

We wish to calculate the shift in the Mössbauer spectra in going from the normal- to the superconducting state due to the change in the velocity of an atom owing to the gap in the dispersion relation of the superconductor. This shift is caused by the relativistic second-order Doppler effect because of the change in velocity of an atom in going from the normal- to the superconducting state.

The resonance energy of the γ-ray is reduced due to the second-order Doppler effect [3-6] by the amount,

$$\delta_{SOD} = \frac{1}{2} \frac{<v^2>}{c^2} E_\gamma \tag{1}$$

where v is the velocity of the atom which emits the γ-ray. E_γ is the γ-ray energy in absence of the second-order Doppler effect and c is the velocity of light. Since γ-ray looses energy, the gain in energy of the lattice is given by

$$\delta E = \delta(p^2/2m) = -\frac{p^2 \delta m}{2m^2} = -\frac{1}{2} \frac{v^2}{c^2} E_\gamma \tag{2}$$

where $\delta mc^2 = E_\gamma$ is the γ-ray energy. Thus an oscillator while emitting radiation looses its mass so that the frequency of the emitted photon is reduced by the amount δ_{SOD} accompanied by the gain in the energy δE of the lattice. For a Debye solid.

$$\delta_{SOD} = \frac{9}{16} \frac{k_B \theta_D}{mc} + \frac{9 k_B T}{2mc} \left(\frac{T}{\theta_D}\right)^3 \int_o^{\theta_D/T} x^3 (e^x - 1)^{-1} dx \tag{3}$$

in velocity units. where θ_D is the Debye cutoff temperature of the solid. For a Debye temperature of $\theta_D \simeq 500 K$, the mass of one ^{57}Fe atom as equal to 57 times the mass of the proton, $m_p = 1.6725 \times 10^{-24} g$, we find that $9 k_B \theta_D / 16 mc \simeq 0.15 mm/s$. for ^{119}Sn. m$= 119 \, m_p$ and $\theta_D = 320 K$ the same quantity is 0.046 mm/s. This accounts for about half of the centre shift. The remaining part of the shift is due to the isomer shift and due to the dynamics of the system.

In order to calculate the change in the second-order Doppler shift when the Debye lattice becomes superconducting, we consider the electron-phonon interaction.

The electrons in the conduction band interact with the phonons so that the Hamiltonian of the system is given by,

$$H = \sum_{k,\sigma} E_{k,\sigma} c_{k,\sigma}^\dagger c_{k,\sigma} + \sum_q \hbar\omega_q a_q^\dagger a_q + \sum_{k,q,\sigma} D_q c_{k+q,\sigma}^\dagger c_{k,\sigma} a_q + \sum_{k,q,\sigma} D_{-q} c_{k-q,\sigma}^\dagger c_{k,\sigma} a_q^\dagger \tag{4}$$

where the coupling constant of the electron-phonon interaction may be written as,

$$D_q = D_o(\hbar/2M\omega_q)^{1/2} \tag{5}$$

where $D_o = (\partial V/\partial r)_o$ is the distance derivative of the crystal potential evatuated at the equilibrium. Using the anticommutators for electrons and commutators for phonons, the electron-phonon interaction transformed [7] upto second order becomes,

$$H' = \frac{1}{2} \sum_{k,k',\sigma\sigma'} V_{k,k'} c_{k',\sigma'}^\dagger c_{-k',\sigma}^\dagger c_{-k,\sigma} c_{k,\sigma'} \tag{6}$$

where

$$V_{k,k'} = 2|D_{k-k'}|^2 \hbar\omega_{k-k'}[(E_k - E_{k'})^2 - (\hbar\omega_{k-k'})^2]^{-1} . \tag{7}$$

We introduce the gap as in the B.C.S. theory [8] as,

$$\Delta_k = \sum_{k'}{}' V_{k,k'} < 0|c_{k'}^\dagger c_{-k'}^\dagger|0 > \tag{8}$$

where k is associated with spin up while $-k$ is with spin down so that the gap corresponds to spin singlet state. Since the vacuum expectation value is taken, it is a c number. Substituting (8) in (6), we find,

$$H' = \frac{1}{2} \sum_{k,\sigma,\sigma'} \Delta_{k'} c_{-k,\sigma} c_{k,\sigma'} + \frac{1}{2} \sum_{k',\sigma,\sigma'} \Delta_k c_{k',\sigma'}^\dagger c_{-k',\sigma}^\dagger + \frac{1}{2V_{k,k'}} \Delta_k \Delta_{k'} . \tag{9}$$

taking all possible averages of the pairing terms. We take the attractive part of the potential for $E_k - E_{k'} \ll \hbar\omega_{k-k'}$ so that the average value of (9) is,

$$< H' >= 2\Delta^2/v_q \tag{10}$$

due to two different permutations of the last term of (9). Here,

$$v_q = -D_o^2/(M\omega_q^2) \tag{11}$$

is the attractive potential between electrons below the transition temperature. Substituting (11) in (10), we obtain,

$$< H' >= -\frac{2\Delta^2 M\omega_q^2}{D_o^2} . \tag{12}$$

This is the amount of attractive energy by which the energy of the system is
reduced in going from the normal- to the superconducting state. We use the
phonon dispersion relation as $\hbar\omega_q = \hbar qv$ where v is the velocity of sound so
that the above energy becomes,

$$< H' >= -\frac{2\Delta^2 M v^2 q^2}{D_o^2} . \tag{13}$$

The contribution of this energy to the velocity of an atom is given by,

$$\hbar^{-1}\partial < H' > /\partial q = -\frac{4\Delta^2 M v\omega}{\hbar D_o^2} . \tag{14}$$

The velocity of an atom in a superconductor, v_{SC}. thus becomes,

$$v_{SC} = v_N \left(1 - \frac{4\Delta^2 M\omega}{\hbar D_o^2} \right) \tag{15}$$

where v_N is the velocity in the normal state. The velocity of an atom in a
superconductor thus varies as the square of the gap energy.It is of interes
to find a numerical estimate of the change in the velocity in going from
the normal to the superconducting state whether it is observable. For thi
purpose numerical values of Δ and $D_q = D_o(\hbar/2M\omega_q)^{\frac{1}{2}}$ are needed. W
note that for $T_c \simeq 85$ K, $\Delta = 1.75 \ k_B T_c = 205.4\times10^{-16}$ erg. The phono
frequency may be estimated from the Raman and infrared measurement
[9.10]. Therefore, for the Bi containing superconductors [9] we assume, $\hbar\omega \simeq$
480 cm^{-1}, corresponding to Cu - O - Bi bonds and about 500 cm^{-1} for th
YBa$_2$Cu$_3$O$_{7-\delta}$. The order of magnitude of D_q may be estimated from th

attractive interaction, $D_q \simeq 320 \times 10^{-16}$ erg for $\hbar\omega \simeq 500\ cm^{-1}$ so that $2\Delta^2/(D_o^2\hbar/2M\omega) \simeq 0.82$. Thus the second term on the right hand side of (15) is about 82 per cent of the first term so that $v_{sc}/v_N = 0.18$. Therefore the velocity of an atom in a superconductor is predicted to be much less than in the normal state. This reduction in the velocity being proportional to T_c^2 is much larger for the high-temperature superconductors than for the low-temperature superconductors. In general, the temperature dependence of the gap of the superconductor is a numerically involved problem. However, near the transition temperature, $\frac{1}{2}T_c < T < T_c$, it may be described by the mean field theory, according to which it varies with temperature as,

$$\Delta = \Delta_o(1 - T/T_c)^{1/2} \tag{16}$$

so that the velocity of an atom in a superconductor becomes.

$$v_{SC} = v_N \left[1 - \frac{4M\omega\Delta_o^2}{\hbar D_o^2} \left(1 - \frac{T}{T_c} \right) \right] . \tag{17}$$

At $T = 0$. the velocity of an atom is given by $v_{SC}(o) = v_N[1 - 4M\omega\Delta_o^2/(\hbar D_o^2)]$ whereas at $T = T_c$. v_{SC} becomes equal to v_N. The shift of the Mössbauer line in a superconductor is thus found to be

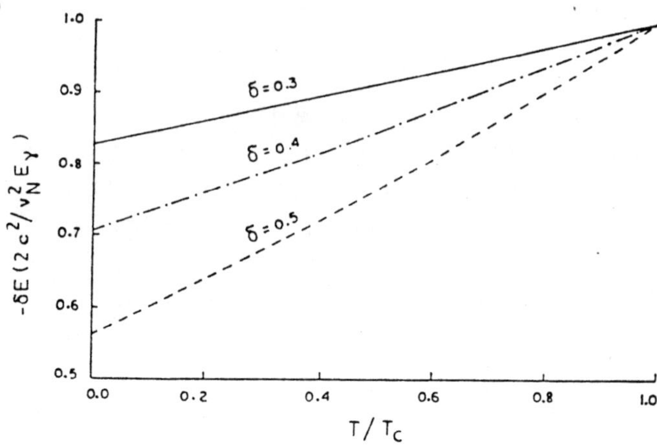

Fig.1: The second-order Doppler shift in the Mössbauer effect in a superconductor in units of $2c^2/(v_N^2 E_\gamma)$ as a function of temperature for three different values of the strength of the electron-phonon interaction and constant phonon frequency as for optical phonons from eq. (18).

$$\delta E_{SC} = -\frac{E_\gamma v_N^2}{2c^2}\left[1 - \delta_o^2(1 - T/T_c)\right]^2 , \qquad (18)$$

where

$$\delta_o = 2\Delta_o/[D_o(\hbar/M\omega)^{1/2}] \qquad (19)$$

The value of the parameter δ_o is of the order of the matrix element of
the electron-phonon interaction devided by the phonon frequency and hence
is always less than unity. The shift as a function of temperature describes
slightly parabolic behavior in the superconducting state owing to the fourth
power of gap energy $\delta E_{SC} \propto \Delta^4$.

For a reasonably strong electron-phonon interaction, we have plotted the
shift. δE_{SC} as a function of temperature in Fig. 1 for three different values of δ_o=0.3, 0.4 and 0.5. Recently we have reported [1] that part of the
isomer shift is proportional to the susceptibility. Therefore. the shift of the
Mössbauer line is determined by two terms, one proportional to the susceptibility and another proportional to the gap of the superconductor as given
by (18),

$$I.S. = a\chi(T) + b[1 - \delta_o^2(1 - T/T_c)]^2 . \qquad (20)$$

The magnitude of both the terms increases with increasing temperature except that the first term is negative while the second term is positive in superconductors. Such a small shift has been observed by Lin and Lin [11] in
the Mössbauer spectra of ^{57}Fe in $Bi_4Sr_3Ca_3Cu_{3.92}Fe_{0.08}O_{16.36}$ but the experimental data [12] of ^{57}Fe in $YBa_2(Cu_{0.98}Fe_{0.02})_4O_8$ is not adequate to
compare the theoretical predictions with measurements because of the structural distortions. Besides, Lin and Lin [11] did not report the susceptibility
of their iron containing samples. It was realized [12] that the deviation between the observed shift, δ_{obs} and the δ_{sod} (normal) for $T < T_c$ appears to be
a chemically induced effect due to onset of superconductivity. Our calcula-
tion suggests that the deviation is caused by the appearance of a gap in the
dispersion relation of the superconductor. This effect is much larger than the
phonon-induced isomer shift [13].

In $EuBa_2Cu_{2.98}Sn_{0.02}O_{7-\delta}$ there are two types of sites for the ^{119}Sn nuclei. A type in which the nearest neighbour oxygen sites are occupied and
the A' type in which the nearest neighbour oxygen sites are vacant [14]. A
the temperature of 30 K, the isomer shift of ^{119}Sn is 0.15(5) mm/s and a
A' site it is 0.28(5) mm/s. From the expression (3) for $\theta = 320K$, the firs
term is calculated to be about 0.046 mm/s and the second term is also c

about the same value. Thus about 0.09 mm/s may be assigned to the second order Doppler shift and the remaining to the isomer shift. Thus for A sites the isomer shift is about 0.06 mm/s and for A' sites it is 0.19 mm/s. Detailed measurements of the temperature dependence of the shift are not yet available for any further comparison between the theory and the experiments.

3. Recoilless fraction

The attractive between has been used [2] to find the phonon frequency and hence the recoilless fraction. According to this theory the logarithm of the recoilless fraction varies with temperature as.

$$\frac{-1}{3}k_\gamma^2 < u^2 >= -a_1 T^2 + b_1 T^3 (1 - T/T_c) + c \tag{21}$$

where k_γ is the wave vector of the γ-photon. $< u^2 >$ is the average value of the square of the atomic displacement and T_c is the transition temperature. The parameters, a_1, b_1 and c are constants of a given system. At $T = T_c$, the above value is given by $-a_1 T_c^2 + c$ whereas at T=0. the value becomes equal to c. The experimental measurements show that the value at T_c is higher than at T=0. Therefore the eq(21) does not agree with the experiments. The agreement between the theory and the experiment is obtained only when we add a term Δ_1 in the gap corresponding to the structural distortion. With the corrected gap. we find that the logarithm of the recoilless fraction can be written as

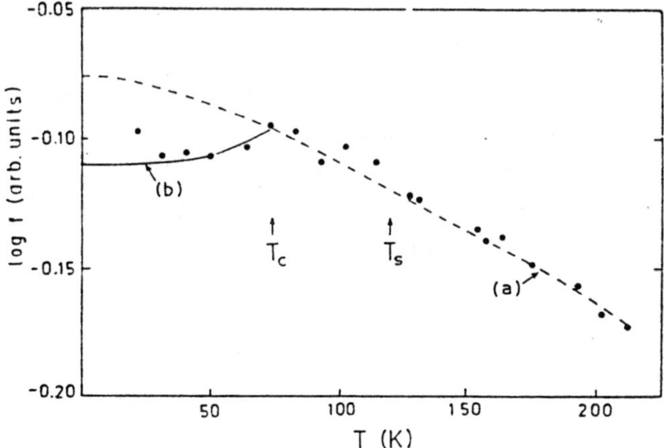

Fig.2: The logarithm of the recoilless fraction of ^{119}Sn in $EuBa_2Cu_{2.98}Sn_{0.02}O_{7-\delta}$ in the normal state is shown by (a) and in the superconducting state by (b).

The curves are calculated while the dots show the measured values. The normal value is calculated using the Debye cutoff temperature of 320 ± 7 K. There is structural distortion at $T_s \simeq 120K$.

$$\frac{-1}{3}k_\gamma^2 < u^2 >= -a_1 T^2 + b_1 T^3(1 - T/T_c) + b_2 T^3 + c \qquad (22)$$

In fig.2, we compare the experimental measurements [14] of the recoilless fraction of ^{119}Sn in $EuBa_2Cu_{2.98}Sn_{0.02}O_{7-\delta}$. For $T > T_c$ in the normal state the theory fits the data well for $\theta_D = 320K$. For $T < T_c$, the agreement between the calculated curve and the measured points is reasonable for $a_1 = 5 \times 10^{-7}K^{-2}, b_1 = 0.1 \times 10^{-7}K^{-3}, b_2 = 0.413 \times 10^{-7}K^{-3}, T_c = 74K$ and $c = -0.11$. The value of the transition temperature determined from the resistivity measurements for a slowly cooled sample is about 85 K. Therefore our value of 74 K found from the recoilless fraction is with ni reasonable agreement with value found from the resistivity in consideration of quenching effects. The wave vector of the γ-photon corresponding to the energy of 23.875 keV is $12.1 \times 10^8 cm^{-1}$. Therefore we predict the velocity of sound to be .

$$v = \left(\frac{k_B^2 k_\gamma^2}{12\rho\hbar a_1} \right)^{1/3} \simeq 9 \times 10^5 cms^{-1} \qquad (23)$$

for $\rho = 6gcm^{-3}$ this value of the velocity of sound is quite reasonable for a dense material.

4. Conclusions

In conclusion, we have calculated the second-order Doppler shift of the Mössbauer spectra from the velocity of an atom in a BCS superconductor. The calculated shift depends on the fourth power of the gap energy of the superconductor and hence in the mean field theory on the square of the temperature. For resonable values of the electron-phonon interaction the shift is a slightly parabolic function of temperature. We have also found that the measured values of the recoilless fraction of ^{119}Sn in $EuBa_2Cu_{2.98}Sn_{0.02}O_{7-}$ are in reasonable agreement with the theory.

I would like to thank Professor Punit Boolchand, of the University o Cincinnati for a useful communication and Dr. S.C. Bhargava of BARC fo discussions.

REFERENCES

1. Shrivastava KN, 1994, J. Phys. Condens. Matter **6**, 11151 .

2. Lingam LS and Shrivastava KN, 1995 J.Phys.;Condens.Matter**7**, L 231.

3. Greenwood NN and Gibb TC,1971, Mössbauer spectroscopy, Chapman and Hall Ltd., London.

4. Josephson BD, 1960, Phys. Rev. Lett. **4**, 331 .

5. Pound RV and Rebka GA,1960, Phys. Rev. Lett. **4**, 275 .

6. Shrivastava KN,1986, Hyperfine Interact. **24-26**, 817 .

7. Suguna A and Shrivastava KN,1980, Phys. Rev. **B22**, 2343 .

8. Bardeen J, Cooper LN and Schrieffer JR,1957, Phys. Rev. **108**, 1175 .

9. Cardona M, Thomsen C, Liu R, von Schnering HG, Hartweg M, Yan YF and Zhao ZX, 1988, Solid State Commun. **66**, 1225 .

10. Thomsen C, Cardona M, Gegenheimer B, Liu R and Simon A,1988, Phys. Rev. B **37**, 9860 .

11. Lin CM and Lin ST ,1993, J. Phys. Condens. Matter **5**, L247 .

12. Wu Y, Pradhan S and Boolchand P, 1991, Phys. Rev. Lett. **67**, 3184 .

13. Shrivastava KN, 1970.Phys. Rev. B1, 955 : ibid 1973 B7, 921 .

14. Boolchand P,Enzweiler RN, Zitkovsky I, Wells J, Bresser W, Mc Daniel D, Meng RL, Hor PH, Chu CW, and Huang CY, 1988, Phys. Rev .B **37**, 3766.

AC Susceptibility in Sintered Superconducting Doped (W/Mo) and Undoped (Bi,Pb)–2223

G. Narsinga Rao, P. Subhash and D. Suresh Babu

Materials Science Research Laboratory
Department of Physics
Osmania University
Hyderabad 500 007

ABSTRACT

We report measurements of temperature dependence of real (χ') and imaginary (χ'') parts of AC susceptibility of sintered doped (W/Mo) and undoped (Bi,Pb)-2223 samples in different AC magnetic fields. The $\chi''(T_p)$ value of undoped and W doped (Bi,Pb)-2223 samples is in good agreement with the expected value from Bean's model. In the case of Mo doped the value of $\chi''(T_p)$ is increased with increasing field. The values of $\chi'(T_p)$ for all three samples, are too large in comparison with the Bean's model result. Increase in the $\chi''(T_p)$ with AC field may be due to non-zero intergranular lower critical field and large values of $\chi'(T_p)$ may be due to large size of the grains.

INTRODUCTION

AC susceptibility measurements are often used to characterize superconductors. Since the discovery of CuO superconductors, there have been numerous reports on $\chi = \chi' + \chi''$, both theoretical and experimental. A recent review is due to Senoussi [1]. To summarize, it is generally agreed that in sintered samples one can observe both intra and intergranular effects at a few kelvin below T_c. χ'' exhibits a maximum at a temperature $T_p (<T_c)$, which reduces some what as the amplitude of the applied field, H_a, is increased. It is generally accepted that this peak in χ'' arises from losses due to intergranular coupling. Except at low fields (μH _ 10 T). χ' exhibits two characteristic transitions. A slow increase starts at T_c and is followed by a very rapid rise (at lower T_c) and reaches saturation. Presumably, the upper transition marks the onset of diamagnetism within the grains while at lower T the entire sample is involved. In the latter regime, volume shielding currents get setup in the pellet and must flow through the intergranular junctions. As proposed by Gömöry and Lobotka [2] it therefore becomes possible to treat the specimen as homogeneous, uniform superconductor. In particular they have involved Bean's [3] model, as well as other calculations [4,5], based on it, to interpret the observed values of χ' and χ'' at T_p for cylindrical samples and used the observations to determine the intergranular shielding currents, J_s. Gömöry and Lobotka [2] have found that for the cylindrical geometry and a homogeneous superconductor

$$\chi'(T) = -5/16 \qquad (1)$$

$$\chi''(T) = 0.212 \qquad (2)$$

$$J_s(T_p) = H_s/R \qquad (3)$$

They have also estimated the relation between probing field H_a and surface field H_s to be

$$H_s = H_a\,[(1-2D/3)/(1-D)] \qquad (4)$$

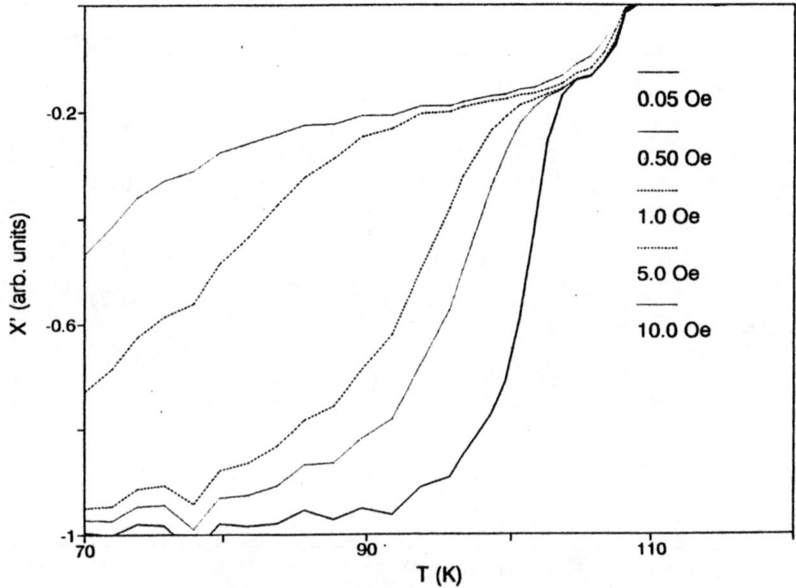

Fig 1(a). Temperature dependent real part of the AC susceptibility of sample $Bi_{1.7}Pb_{0.3}Sr_2Ca_2Cu_3O_y$ (S1)

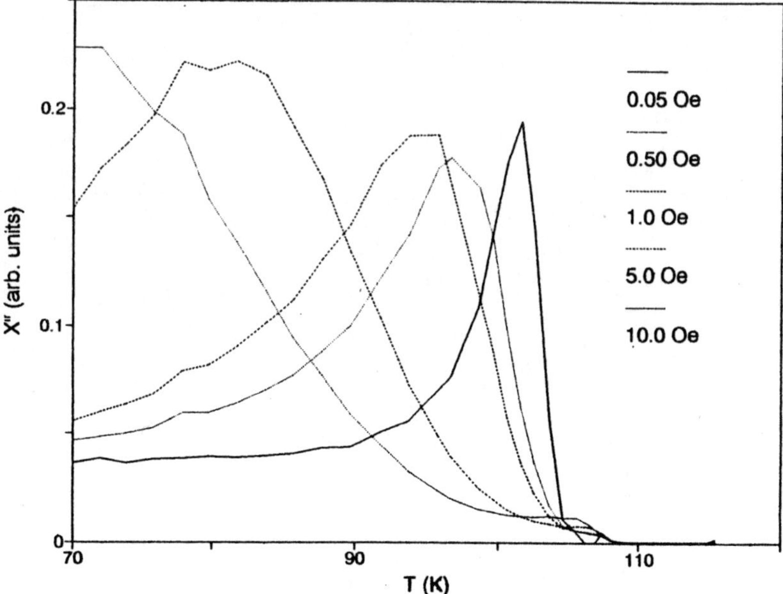

Fig 1(b). Temperature dependent imaginary part of the AC susceptibility of sample $Bi_{1.7}Pb_{0.3}Sr_2Ca_2Cu_3O_y$ (S1)

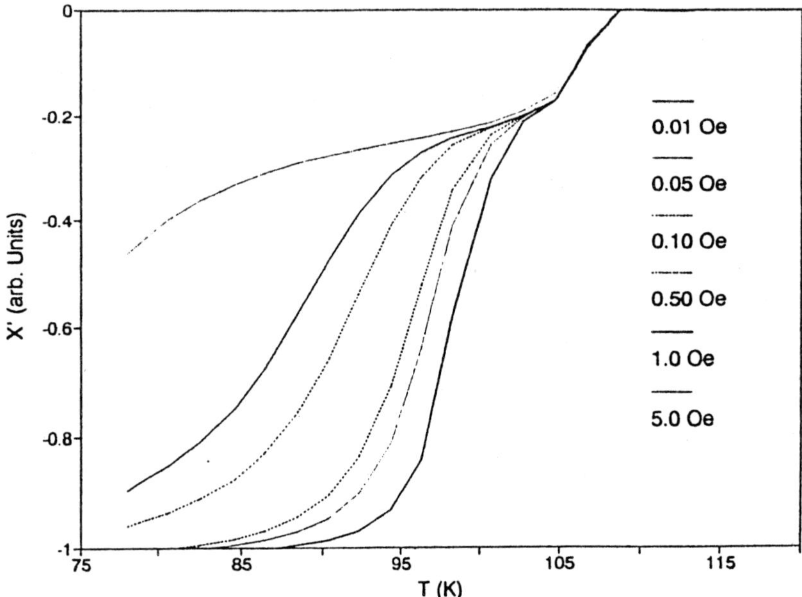

Fig 2(a). Temperature dependent real part of the AC susceptibility of sample $Bi_{1.6}Pb_{0.3}Mo_{0.1}Sr_2Ca_2Cu_3O_y$ (S2)

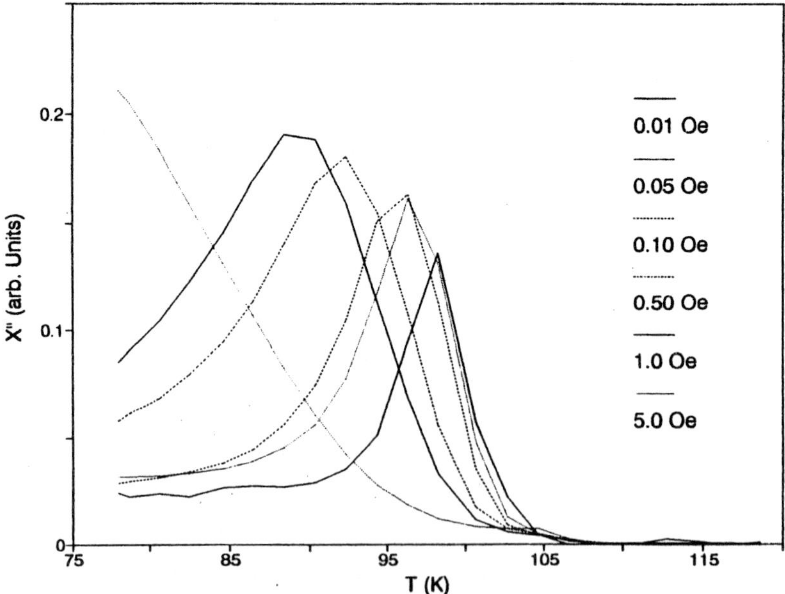

Fig 2(b). Temperature dependent imaginary part of the AC susceptibility of sample $Bi_{1.6}Pb_{0.3}Mo_{0.1}Sr_2Ca_2Cu_3O_y$ (S2)

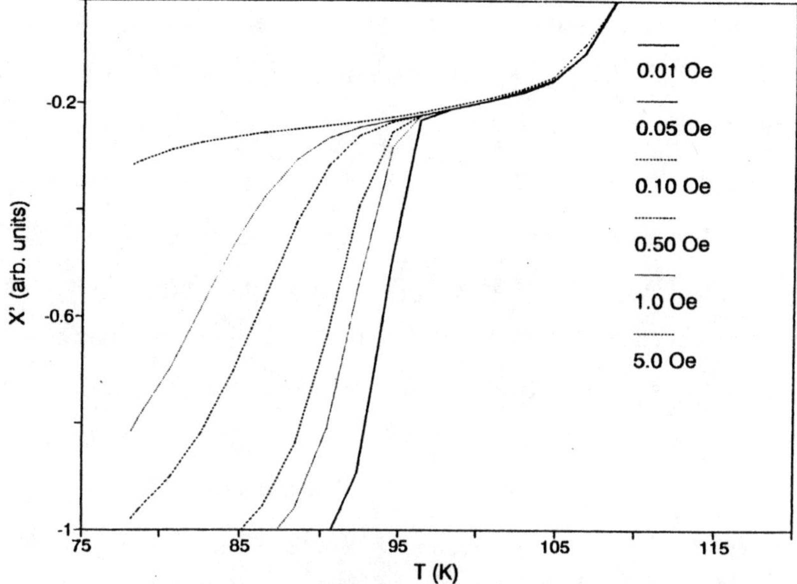

Fig 3(a). Temperature dependent real part of the AC susceptibility of sample $Bi_{1.6}Pb_{0.3}W_{0.1}Sr_2Ca_2Cu_3O_y$ (S3)

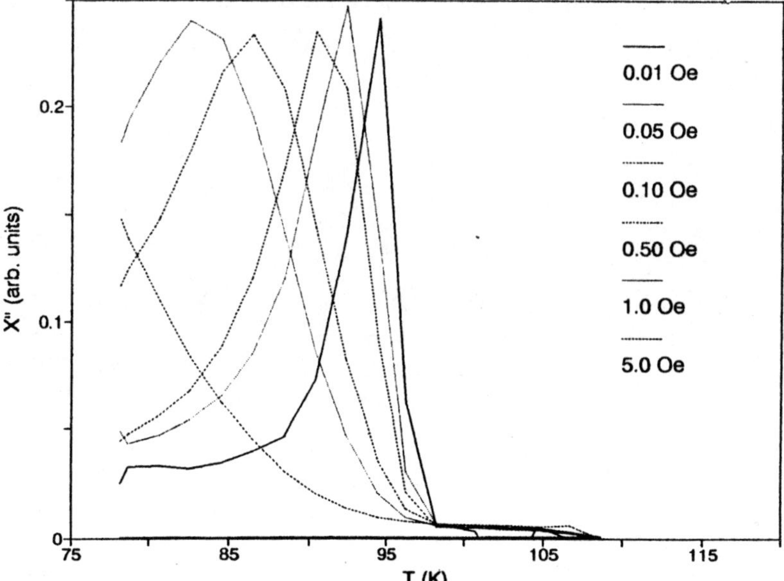

Fig 3(b). Temperature dependent imaginary part of the AC susceptibility of sample $Bi_{1.6}Pb_{0.3}W_{0.1}Sr_2Ca_2Cu_3O_y$ (S3)

where D is the demagnetizing factor for cylindrical samples [6]. In some samples it has been found [2] that $\chi''(T)$ increases with H_a, acquiring a value close to the value estimated from Bean's model at relatively high H_a values.

EXPERIMENTAL

Samples S1, S2 and S3 with nominal compositions $Bi_{1.7}Pb_{0.3}Sr_2Ca_2Cu_3O_y$, $Bi_{1.6}Pb_{0.3}Mo_{0.1}Sr_2Ca_2Cu_3O_y$ and $Bi_{1.6}Pb_{0.3}W_{0.1}Sr_2Ca_2Cu_3O_y$ respectively were prepared by two step solid state reaction method and these samples were characterized using XRD, SEM, AC susceptibility and EDAX measurements [7]. It was found that the doping of WO /MoO stabilizes the high T_c phase. AC susceptibility measurements were carried out on the Lakeshore AC susceptometer 7000 in AC magnetic fields ranging from 0.05 Oe to 10 Oe at fixed frequency of 666.7 Hz. Sizes of the cylindrical specimen used for these measurements are given in table (1) along with their T_c and D values. The applied field is along the length of the specimen. Absolute values of χ' and χ'' were determined based on an assumption that at temperatures substantially below T_c, where $d\chi'/dT = 0$, is $\chi' = -1$.

Sample	length (mm)	Diameter (mm)	Demagneti-zation	T_c (K)
$Bi_{1.7}Pb_{0.3}Sr_2Ca_2Cu_3O_y$	3.5	1	0.110	109
$Bi_{1.6}Pb_{0.3}Mo_{0.1}Sr_2Ca_2Cu_3O_y$	4.0	1	0.098	109
$Bi_{1.6}Pb_{0.3}W_{0.1}Sr_2Ca_2Cu_3O_y$	5.0	1	0.079	109

Table 1. Sizes of the cylindrical specimens used for AC susceptibility measurements along with their demagnetization factor and T_c values.

RESULTS AND DISCUSSIONS

Temperature dependence of AC susceptibility (real and imaginary) measured, for different H_a values, on samples S1, S2 and S3 are presented in fig.1, fig.2 and fig.3 respectively. The data required for checking equations (1) and (2) are listed in table 2, 3 and 4 for samples S1, S2 and S3 respectively.

For samples S1, S3 (table 2 and 4) $\chi''(T)$ is found nearly independent of field H_a. The average value of $\chi''(T_p)$ (=0.22 and 0.23 for samples S1 and S3 respectively) is in good agreement with the expected value from Bean's model (i.e. equation 2). In the

H (Oe)	$T_P(\chi'')$	$\chi''(T_m)$	$\chi'(T_m)$	J_s (A/cm^2)
0.01	101.6	0.19	0.43	0.167
0.5	96.7	0.18	0.49	8.33
1.0	95.8	0.19	0.38	16.7
5.0	81.8	0.22	0.44	83.3
10.0	68	0.23	0.54	167

Table. 2. Analysis of AC susceptibility curves of sample $Bi_{1.7}Pb_{0.3}Sr_2Ca_2Cu_3O_y$ (S1)

H (Oe)	$T_P(\chi'')$	$\chi''(T_m)$	$\chi'(T_m)$	J_s (A/cm^2
0.01	98	0.13	0.58	0.183
0.05	96.7	0.16	0.63	0.914
0.10	96	0.16	0.52	1.83
0.5	92	0.18	0.53	9.14
1.0	88	0.19	0.57	18.3
5.0	80	0.20	0.57	91.4
10.0	70	0.22	0.48	183

Table 3. Analysis of AC susceptibility curves of sample $Bi_{1.6}Pb_{0.3}Mo_{0.1}Sr_2Ca_2Cu_3O_y$ (S2)

H (Oe)	$T_p(\chi'')$	$\chi''(T_m)$	$\chi'(T_m)$	J_s (A/cm^2)
0.01	93.9	0.24	0.50	0.178
0.05	92	0.24	0.50	0.892
0.10	90	0.23	0.60	1.78
0.5	86	0.23	0.57	8.92
1.0	82.6	0.24	0.56	17.8
5.0	77	0.22	0.54	89.2
10.0	60.7	0.21	0.48	178

Table 4. Analysis of AC susceptibility curves of sample $Bi_{1.6}Pb_{0.3}W_{0.1}Sr_2Ca_2Cu_3O_y$ (S3)

case of sample S2 (table 3) $\chi''(T)$ is found to be 0.13 well below 0.21 (the Bean's model result, equation 2) for H = 0.01 Oe. $\chi''(T_p)$ is increased with the field acquiring a value (=0.22) at field H_a =10 Oe, equal to the bean's model result. Similar results were found in the literature [2] for Y-123 samples. Lofland et al [8] have ascribed, this increase in $\chi''(T_p)$ (the maximum loss) with field amplitude (H_a) to a non-zero intergranular lower critical field H_{c1}. They also found that (i) for small H_a, the maximum in χ'' occurs when the maximum field is slightly greater than that required for full penetration (ii) for increasing H_a , the maximum value of χ'' decreases monotonically (iii) for $H_a >> H_{c1}$, $\chi''(T_p)$ approaches the Bean's value (equation 3) and (iv) for H_{c1} =0, the peak pertains to full penetration.

The observed values of $\chi'(T_p)$, for all three samples, are too large in comparison with the Bean's model result (i.e. equation 1). Similar results were observed by Lofland et al [8] for Silver doped Y-123 wire. They explained that the large value of $\chi'(T_p)$ is arising from magnetic shielding due to large grains. With $r/\lambda(o)>>1$, (r = radius of the grains, $\lambda(o)$ = London penetration depth). One must include the magnetization produced by Meissner screening of the grains. This would increase $|\chi'|$, yet it would leave χ'' unchanged as can be seen by noting that by the continuity of H, $\delta H/\delta r$ remains unchanged and therefore the Bean currents remain unchanged.

If for simplicity, the grains are assumed to be spheres. One gets for the local magnetization of the grains,

$$M(r) = P (r/\lambda) H(r) \qquad (5)$$

where $\lambda(T)$ is the London penetration depth and

$$P(x) = 1-(3/x) \coth(x) + 3/x^2 \qquad (6)$$

as derived by London [9]. The average grain size in all three samples were found to be larger than 10 μm [7]. The average value of $\lambda(0)$ reported for (Bi,Pb)-2223 [8,10] is ~ 0.15 μm. Since $r/\lambda(0)$ is an order of 30 for these samples, $P(r/\lambda(T))$ is nearly constant (0.74 - 0.88) for $T<0.9T_c$ so that the contribution of the grains to $\chi'(T_p)$ is not small, hence these samples show the enhancement of $\chi'(T_p)$.

The shielding current density J_s is determined using equation 3 and 4 for samples S1, S2 and S3 and are given in table 2, 3 and 4 respectively. As expected, the magnitude of J_s is rather small in comparison with values deduced from measurements on single crystals or powders [11]. This is because the process limiting J_s involve intergranular coupling [12]. The temperature dependence of J_s of these samples is shown in fig.5. The sample S1 with higher values of J_s exhibits a temperature dependence closer to linear. It allows to conclude that the sample S1 has better coupling between the grains, while the sample S2 and S3 has weaker coupling between the grains.

    G. Narsinga Rao, P. Subhash and D. Suresh Babu

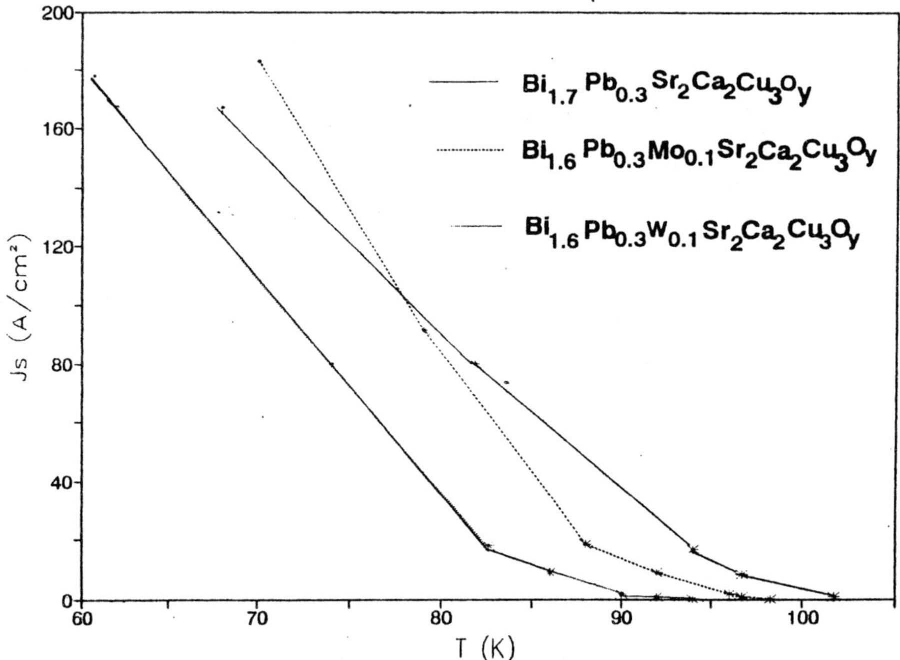

Fig. 4. Temperature dependent Screening current density Js.

ACKNOWLEDGEMENTS

The authors would like to thank Prof. M.Ganne, Institut des matriaux, de Nantes, France for providing the low temperature AC susceptibility data. We also thank the Department of Science and Technology, New Delhi, for financial support. One of the authors (PS) would like to thank CSIR, New Delhi for the award of SRF.

Imaginary AC Susceptibility in Superconducting Bi-Pb-Mo-Sr-Ca-Cu-O.

P. Subhash , G.N. Rao, and D. Suresh Babu

Materials Science Research Laboratory
Department of Physics
Osmania University
Hyderabad 500 007

Key words: AC susceptibility; Irreversibility line; Critical Current Density.

ABSTRACT

AC susceptibility measurements have been performed to determine the dependence of the intergranular component of the imaginary part of the susceptibility (χ'') of polycrystalline $Bi_{1.6}Pb_{0.3}Mo_{0.1}Sr_2Ca_2Cu_3O_y$ specimen on temperature and AC magnetic field (0.01 - 10 Oe). The intergranular loss peak temperature, T_p, shifts significantly to low temperatures with increasing field. The AC Penetration Field, H_p, AC Penetration Depth, λ_o, AC Loss Density, W and the Average Intergranular Critical Current Density, J_c were determined on the basis of Campbell model for Reversible Fluxoid Motion. In addition, the Irreversibility Line of the sample was determined.

INTRODUCTION

The measurement of AC susceptibility, $\chi = \chi' + i\chi''$, has been widely used to characterize high T_c superconductors, and has provided important information on flux dynamics in polycrystalline or ceramic materials [1,2]. These materials have been modeled as an array of Josephson-coupled strongly superconducting grains in which one distinguishes between intergrain and intragrain properties [2-5]. The susceptibility is then composed of intergrain and intragrain components where the real part, χ', represents the super current shielding and a nonzero imaginary part, χ'', is a manifestation of hysteresis losses and nonlinearities in the magnetization of the material. In particular, χ'' has been widely used to probe the nature of weak links, to understand intergranular and intragranular coupling, to investigate irreversibility lines, to estimate the average intergrain and intragrain critical current density and the pinning potential [5-7].

One can study the pinning behaviour in high T_c superconductors by investigating the irreversibility line [8-9], which marks the boundary between reversible (Zero energy loss) and irreversible (nonzero energy loss) behaviour in the field-temperature plane. The interpretation of this line has been the subject of intense study [10-16].

In this paper we report measurements of χ'' as a function of temperature and AC magnetic field for rectangular slab specimen of sintered $Bi_{1.6}Pb_{0.3}Mo_{0.1}Sr_2Ca_2Cu_3O_y$. From the field dependence of χ''_{max}, and peak temperature, T_p, the average critical current density was calculated. From the onset of intergrain loss, the irreversibility temperature, T_{irr} was determined.

EXPERIMENTAL

The sample with nominal composition $Bi_{1.6}Pb_{0.3}Mo_{0.1}Sr_2Ca_2Cu_3O_y$ was prepared by two step solid state reaction method and this sample was characterized using XRD, SEM, AC susceptibility and EDAX measurements [17]. It was found that the sample had a single 2223 superconducting phase along with $CaMoO_x$ type impurity. The diamagnetic onset at 108 K (T_c) was observed.

AC susceptibility measurements were carried out on the Lakeshore AC Susceptometer 7000 on a rectangular slab specimen (4 mm x 1.5mm x 1mm), in the AC magnetic fields ranging from 0.01 to 10 Oe at a fixed frequency of 666.7 Hz (the applied field is along the length of the specimen). Absolute values of χ' and χ'' were determined based on an assumption that at temperatures substantially below T_c, where $d\chi'/dT = 0$, is $\chi' = -1$.

RESULTS AND DISCUSSION

Figure 1 and 2 shows the typical temperature dependent AC susceptibility real (χ')and imaginary (χ'') respectively for the sample $Bi_{1.6}Pb_{0.3}Mo_{0.1}Sr_2Ca_2Cu_3O_y$ at different AC magnetic fields (H_a) and at a frequency 666.7 Hz. From the fig. 1, diamagnetic onset at 108 K was observed. It is also observed that an anomaly exist at 103 K for 0.01 Oe before reaching saturation. It can be seen from fig. 1 that the transition temperature at 108 K is hardly affected by AC fields, indicating this transition due to intragrain Bi (2223) structure. The temperature of the anomaly had decreased with an increase in AC field suggesting the anomaly is due to intergranular coupling.

As the amplitude of the applied field increases, the position of χ'' peak shifts to low temperatures, the height and the width of the χ'' peak increases (fig. 2). The shift of the χ'' peak due to the change in temperature can be explained by critical state model. In the critical state model, the AC losses are solely hysteretic due to the pinning of the flux lines in type II superconductors. The maximum loss always takes place when the magnetic flux line penetrates to the centre of the sample. As the driving field amplitude increases, larger screening currents are required to

     P. Subhash, G. Narsinga Rao and D. Suresh Babu

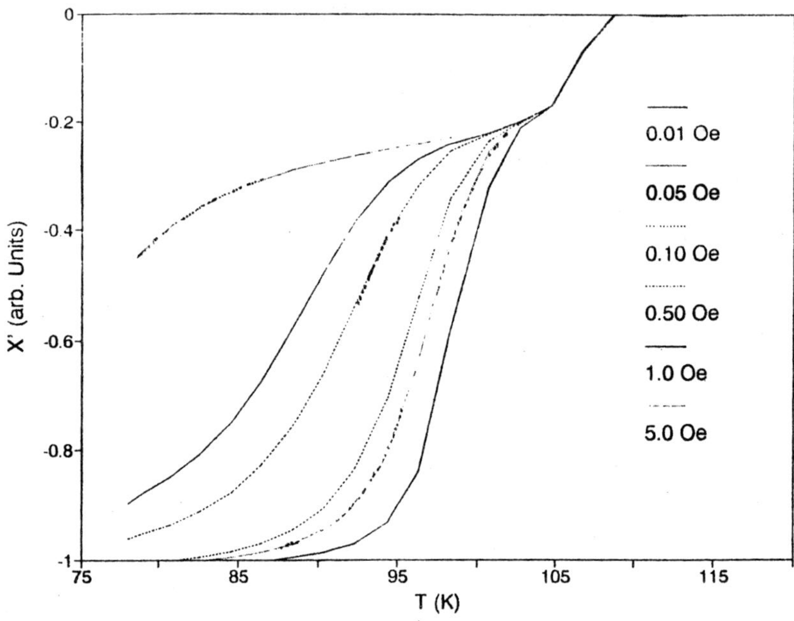

Fig 1. Temperature dependent real part of the AC susceptibility measured at different AC magnetic fields

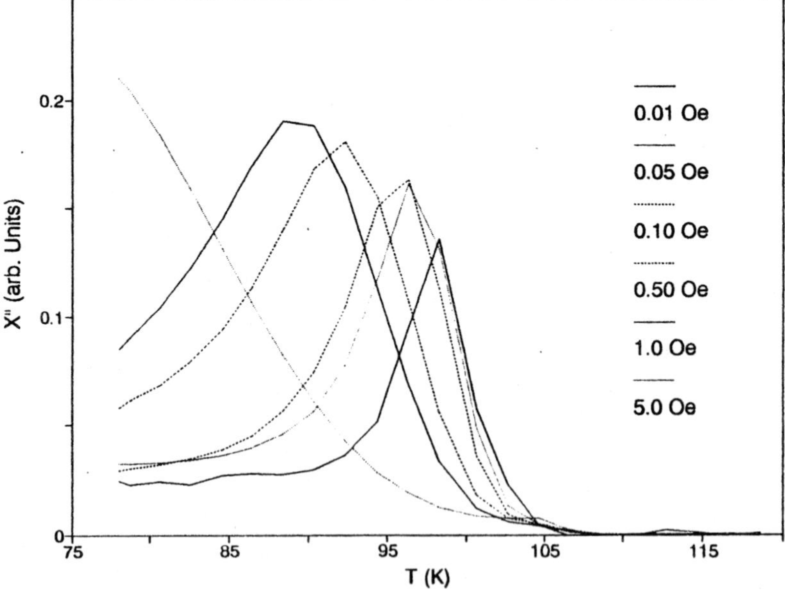

Fig 2. Temperature dependent imaginary part of the AC susceptibility measured at different AC magnetic fields

shield the applied AC magnetic field, as a result the position of X'' peak moves to lower temperatures. The increase in the height and the width of the X'' peak with increasing filed was explained by Matsushita et al [16] and is due to reversible fluxoid motion.

From the theory of Reversible Fluxoid Motion, according to Matsushita et al [16]. X'' can be expressed in terms of the AC loss density, W, as

$$X'' \quad W/\pi H_a^2$$

The fluxoid motion occurs within the distance of the AC Penetration Depth, λ_o, from the surface and their displacement is smaller than the interaction distance, represents half of the size of the pinning potential) d_i, due to small H_a. The reason why X'' or W are comparable in magnitude with the prediction by the critical state model in spite of the reversible fluxoid motion is that the numer of moving fluxoids and their displacement are much lartger than the predicted values and these factors enhance W [18]. For the test specimen AC maximum loss densities were calculated at different AC magnetic fields, H_a Fig. 3 shows the variation of W_{max} with T_p.

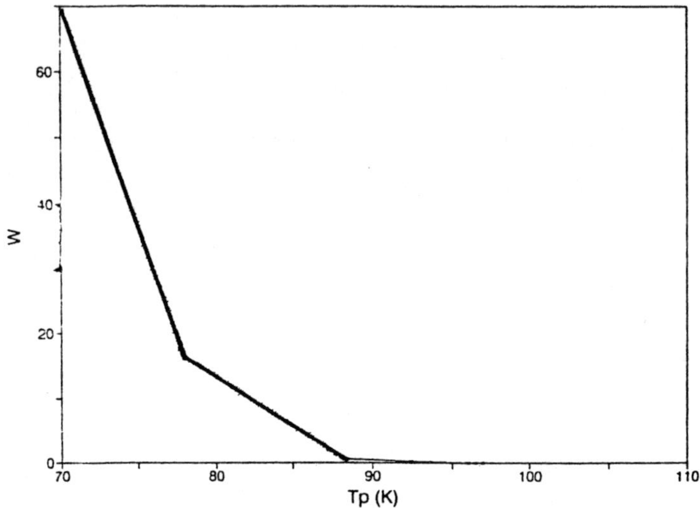

Fig 3. Variation of AC loss density (W) with X'-peak temperature Tp

        P. Subhash, G. Narsinga Rao and D. Suresh Babu

The reversible fluxoid motion is usually observed only when the AC field amplitude is very small, since the travelling distance of fluxoids must be limited for size of the samples larger than the AC Penetration Depth, λ_0, defined by Campbell [19]. According to this model [16]

$$\chi' = \frac{\mu_0 H_p}{[1+3(2\lambda_0/d)^2] H_p + H_a} \tag{1}$$

$$\chi'' = \frac{2\mu_0}{\pi} \frac{H_p H_a}{3[1+2(2\lambda_0/d)^2]^2 H_p^2 + H_a^2} \tag{2}$$

where d is the thickness of the superconducting slab, H_p is the Penetration field, H_a is the applied AC magnetic field and λ_0 is the AC Penetration Depth.

Equation (2) suggests that χ'' has the maximum value

$$\chi'' = \frac{\mu_0}{\pi 3^{1/2} [1+2(2\lambda_0/d)^2]} \tag{3}$$

at

$$H = 3^{1/2} [1+2(2\lambda_0/d)^2] H_p \equiv H_{ap} \tag{4}$$

Using equation (3) & (4), the AC Penetration Depth λ_0 and the Penetration Field, H_p were determined. From eqn (3) and eqn (4) we have

$$\chi'' = \frac{\mu_0 H_p}{\pi} = \frac{\mu_0 J_c d}{2\pi} \tag{5}$$

From equation (5) the average intergrain critical density for the whole circumferential path was determined. The figure 4 and 5 shows the T_p dependent average intergrain critical current densities, J_c and AC Penetration Depth λ_0, respectively. As the temperature approaches T_c, J_c is reduced to zero and λ_0 diverges.

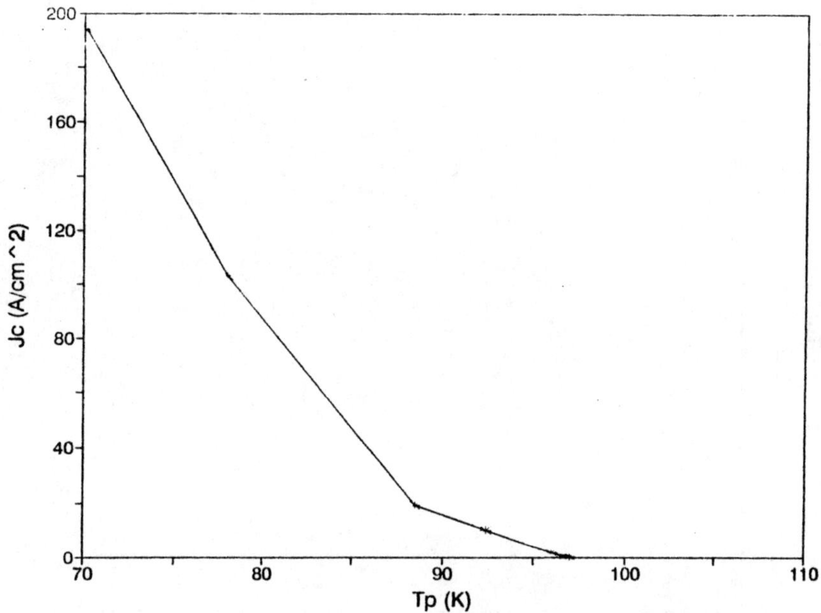

Fig 4. Variation of Average Intergranular Critical Current Density (Jc) with
the X"-peak temperature (Tp)

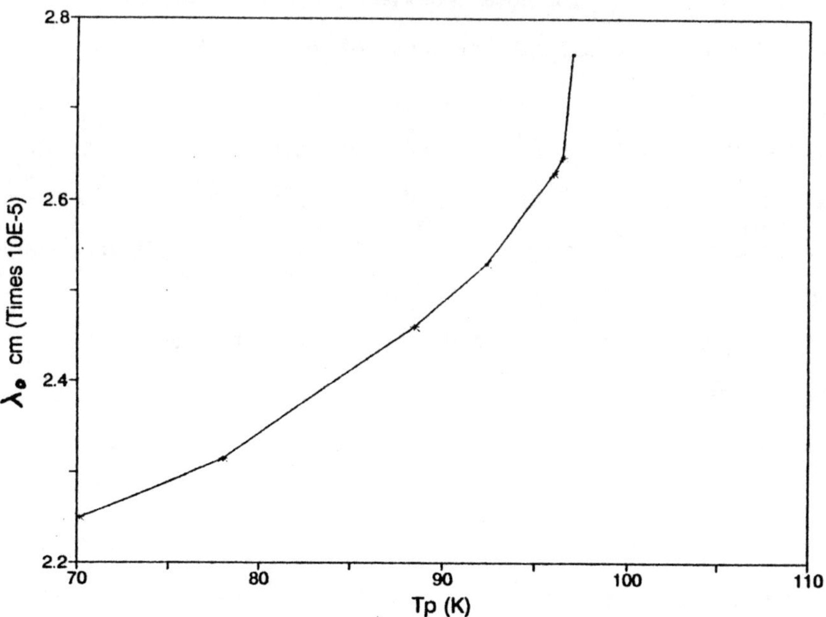

Fig 5. Variation of AC penetration depth (λ_o) with the X"-peak temperature (Tp)

     P. Subhash, G. Narsinga Rao and D. Suresh Babu

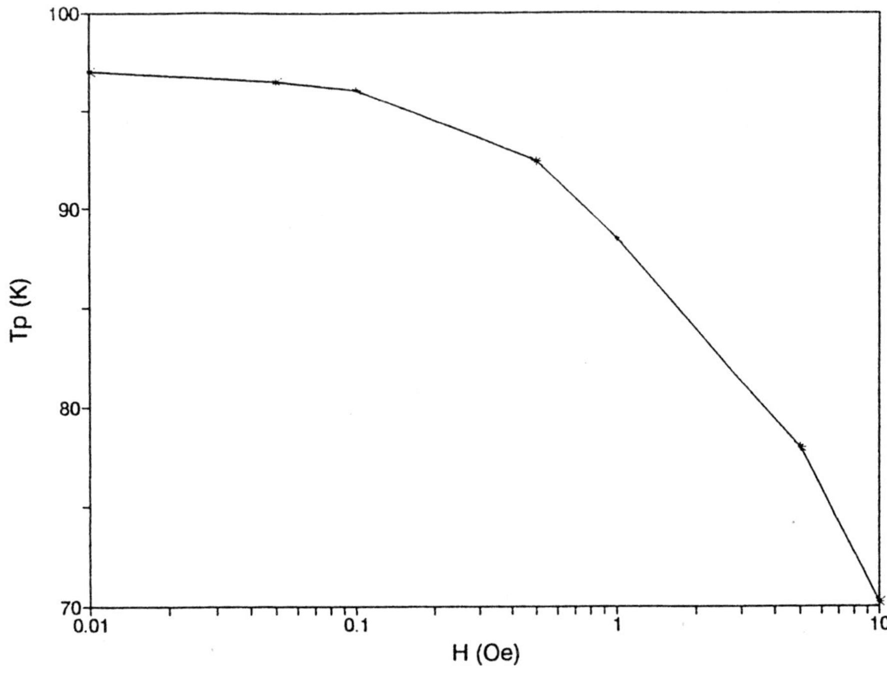

Fig 6. Field dependence of X"-peak temperature (Tp)

Fig. 6 shows the field dependence of peak temperature, T_p, for the sample $Bi_{1.6}Pb_{0.3}Mo_{0.1}Sr_2Ca_2Cu_3O_y$, determined at f = 666.7 Hz, For fields below 0.1 Oe, T_p is almost independent of field. For fields above 0.1 Oe there is dramatic shift of T_p to lower temperatures. This indicates that for the test poly-crystalline superconductor, the pinning of magnetic flux in the intergranular regions is very weak.

There has been practice in the past to determine the irreversibility line at the maximum of χ'', that is, to set the irreversibility temperature (T_{irr}) as the temperature at the χ'' peak (T_p) [8,13,14].This is unfortunate since not only does T_p depend on sample shape and frequency but also it occurs at the point where the magnetic flux fully penetrates the specimen, i.e., the hysteresis loop area is maximum, as pointed out by Müller [15]. This definition of T_{irr} places irreversibility well within the region of maximum hysteretic loss. Clearly in the reversible region, between T_{irr} and T_c, there must be zero loss since the pinning force density is effectively zero. Hence it is appropriate to define T_{irr} at the onset of hysteretic loss [15].

In our measurements the onset is easily determined since intragrain loss is absent at low fields [20]. Figure 7 show the irreversibility line (χ''-onset) for the sample $Bi_{1.6}Pb_{0.3}Mo_{0.1}Sr_2Ca_2Cu_3O_y$. Also shown for comparison are the lines determined using χ''-peak temperature, T_p. According to Savvides et al [20], this line is both sample-size and frequency dependent, with the reversible region decreasing with increasing frequency. Because the pinning potential decreases with increasing field, the irreversibility line shifts to lower temperatures with increasing AC field amplitude. The irreversibility line is also steeper compared to the line based on χ''-peak and is moved to higher temperatures so that the reversible region becomes quite narrow.

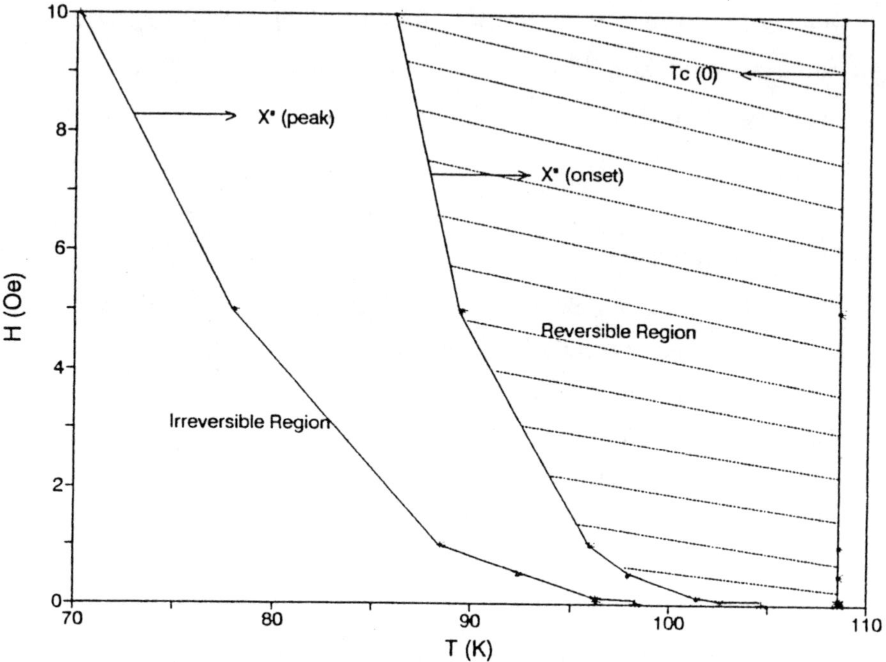

Fig 7. Irreversibility line for the polycrystalline $Bi_{1.6}Pb_{0.3}Mo_{0.1}Sr_2Ca_2Cu_3O_y$ sample

CONCLUSIONS

The temperature and AC magnetic field dependence of AC susceptibility of the superconducting sample $Bi_{1.6}Pb_{0.3}Mo_{0.1}Sr_2Ca_2Cu_3O_y$ were determined. On the basis of the Campbell model for Reversible Fluxoid Motion, the hysteresis loss (χ'') peak temperature dependent AC Penetration field (H_p), AC Penetration Depth (λ_o), AC loss density (W) and Average intergranular critical current density (J_c) were determined. It was found that the J_c reduces to zero and λ_o diverges as the temperature approaches T_c. The Irreversibility Line of the polycrystalline sample was determined using two different methods available in the literature and compared.

ACKNOWLEDGEMENTS

The authors would like to thank Prof.M.Ganne, Institut de Matriaux de Nantes (CNRS), Nantes, France, for providing the AC susceptibility data. The authors would also like to thank the Department of Science and Technology, New Delhi, for financial support. One of the authors (PS) thank CSIR, New Delhi, for the award of SRF

REFERENCES

1. R.B.Goldfarb, M.Lelental and C.A.Thompson, *Magnetic Susceptibility of superconductors and other spin systems*, eds. R.A.Hein, T.L.Francavilla and D.H.Liebenberg (Plenum, New York, 1992).

2. K.H.Müller, *Magnetic Susceptibility of superconductors and other spin systems*, eds. R.A.Hein, T.L.Francavilla and D.H.Liebenberg (Plenum, New York, 1992)

3. K.H.Müller, *Physica* **C 168** (1990) 585 ; ibid., **159** (1989) 717

4. K.H.Müller, S.J.Collocott, R.Driver and N.Savvides, *Supercond. Sci. Tech.*, 4 (1991) S325

5. R.Navaro, F.Lera, A.Badia, C.Rillo and J.Bartolome, *Physica* **C 183** (1991) 73

6. N.Savvides, A.Katsaros and S.X.Dou, *Physica* **C 179** (1991) 361

7. N.Savvides, S.X.Dou, K.H.Song, H.K.Liu and C.C.Sorell, In: *Adv. Supercond. III*, eds. K.Kajimura and H.Hayakawa (Springer, Berlin, 1991) 683

8. K.H.Müller, M.Nikolo, N.Savvides and R.Driver In: *Adv Supercond. III*, eds. K.Kajimura and H.Hayakawa (Springer, Berlin, 1991) 583

9. K.H.Müller, M.Nikolo and R.Driver, *Phys. Rev.* **B43** (1991) 7976

10. L.T.Sagdahal, T.Laegreid, K.Fossheim, M.Murakami, H.Fujimoto, S.Gotoh, K.Yamaguchi, Y.Yamauchi, N.Koshizuka and S.Tanaka, *Physica* **C172** (1991) 495

11. J.Vanacken, E.Osquiguil and Y.Bruynseraede, *Physica* **C 183** (1991) 163

12. I.Felner, B.Brosh, U.Yaron,Y.Yeshurun and E.Yacoby, *Physica* **C 173** (1991) 337

13. T.Matsushita, E.S.Otabe, T.Nakatani, B.Ni, T.Umemura, K.Egawa, M.Wakata and S.Utsunomiya, *Jpn. J. Appl. Phys.* **30** (1991) L1857

14. R.Navaro, F.Lera, C.Rillo and J.Bartolome, *Physica* **C167** (1990) 549

15. K.H.Müller, *Physica* **C 185-189** (1991) 1609

16. T.Matsushita, E.S.Otabe and B.Ni, *Physica* **C 182** (1991) 95

17. G.Narsinga Rao, P.Subhash and D.Suresh Babu, *Mat. Sci.& Engg.* **B 23**(1994) 405

18. T.Matsushita, N.Harada, K.Yamafuji and M.Noda, *Jpn.J.Appl.Phys*, **28** (1989) 356

19. A.M.Campbell, *J. Phys.* *C* **4** (1971) 3186

20. N.Savvides, A.Katsaros, C.Andrikidis and K.H.Müller, *Physica* **C197** (1992) 267

Penetration Depth in High–T_C Superconductors

Ajay Mohan Suvarna and C.S. Sunandana

School of Physics
University of Hyderabad
P.O. Central University
Hyderabad – 500 046

Abstract

In this paper we provide a brief review of penetration depth in high-T_c superconductors, an aspect that has a bearing on the mechanism of superconductivity in general and on the nature of pairing states in particular. We have also focussed on the decorative ESR method of determining penetration depth. We have described and discussed our experimental results on the temperature/concentration dependence of λ in Bi-Pb and Na-doped BSCCO superconductors in the light of Brandt's model for the triangular flux lattice.

Introduction:

The penetration of an external static magnetic field through the boundary surface of a superconductor is a typical example illustrating the general problem of penetration of an external effect through the surface of a solid. The general result is that the magnetic field and the associated current are reflected at the superconducting surface. The superconducting 'surface state' is controlled self consistently by the whole system which adjusts itself according to the external situation. The magnetic penetration depth which is small like the exponential field (which is exponentially dependent on distance) arises as the result of this adjustment. In fact penetration depth is the region where the magnetic contribution to the free energy density changes from its 'normal' or 'vacuum' value to that appropriate for the superconductor. An experimental determination of penetration depth as function of temperature, pressure, magnetic field and micro structural parameters is crucial to an understanding of the superconducting state in terms of the properties of the superconducting carriers (wave function, concentration, effective mass) including the excitation energies (i.e energy gap) thereby providing a sensitive test of the quantitative aspect of any theory of superconductivity.

Ever since F.and H.London [1] wrote down the Ohm's law for superconductors and identified a parameter that characterises the length scale at which magnetic flux penetrates into the superconductors, penetration depth has become one of the most important experimental-rather phenomenological parameters whose significance is basic to the coupling mechanism responsible for the phenomenon. The London penetration depth of dirty type II superconductors, in which the coherence length is not of the order of carrier mean free path. is given by

$$\lambda_L = \left(\frac{m^*c^2}{4\pi n_s e^2}\right)^{1/2} \tag{1}$$

where m^* is the effective electronic mass (in general a tensor) and n_s is the superconducting electron carrier density. Thus the penetration depth (λ_L) is directly proportional to the square root of effective mass tensor and inversely proportional to square root of n_s. It is also a manifestation of the 'stiffness' of the wave function of the superconducting state. The existence of λ_L is important in the discussion of thin film / small particle superconductors and in any discussion of the surface properties of the superconducting state. The temperature dependence of the penetration depth based on an interpretation of the London electrodynamics of a superconductor is given by [2]

$$\frac{\lambda(T)}{\lambda(0)} = -[\omega_e(t)]^{-1/2} \tag{2}$$

where $\lambda(0)$ is the penetration depth at absolute zero and $\omega_e(t)$ is the Gorter-Casimir two fluid order parameter, t being T/Tc and Tc is the superconducting transition temperature The functional form of $\omega_e(t)$ is given by

$$\omega_e(t) = 1 - t^4 \tag{3}$$

$$\lambda(T) = \lambda(0)[1 - t^4]^{1/2} \tag{4}$$

Thus the penetration depth in a nonlocal BCS superconductor with a nearly isotropic energy gap falls off to $\lambda(0)$ in a parabolic fashion (Fig.1) as do other quantities such as the critical magnetic field.

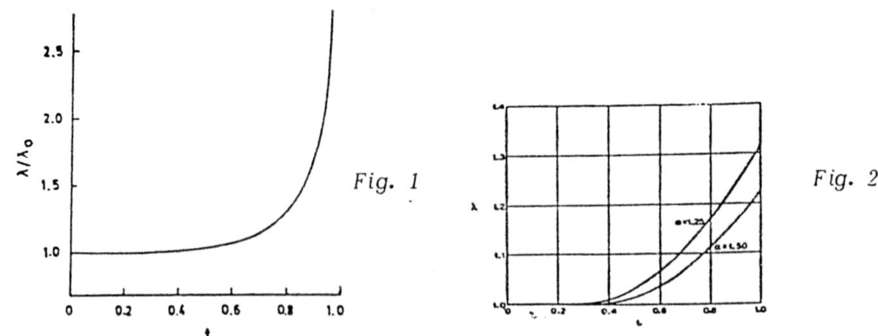

Fig. 1

Fig. 2

Fig.1 Reduced penetration depth λ/λ_o vs temperature 't', (T/T_c) in the Gorter-Casimir model.

Fig.2 Ratio of the penetration depth in the 'energy gap' model to that in the Gorter-Casimir model. both refer to the penetration depth at absolute zero.

A good fit of experimental data to equation (4) is usually taken to mean that the pairing in high T_c superconductors is of the conventional s-wave type with no nodal points or lines in the energy gap function. The general 'energy gap' model based on the exponential specific heat law is [3] gives the expression

$$\frac{\lambda(T)}{\lambda(0)} = [1 - \frac{(2\alpha + t)E(\alpha/t) - 2t\exp^{-\alpha/4}}{K'(o)E(\alpha)}]^{-1/2} \tag{5}$$

where $K'(0) = t^2K'(\omega_e)$. (Note that for the Gorter-Casimir model $K'(0)= 1/2$), $E(\alpha)$ is the standard exponential integral and α is a constant of the order of unity.) A more rapid approach of $\lambda(T)$ to $\lambda(o)$ than that predicted by the Gorter-Casimir model is a significant result of this quantitative model (Fig.2) which uses an arbitrary function $K(\omega)$ in the Helmholtz free energy expansion.

In sharp contrast to conventional (low T_c) bulk superconductors the ceramic high-T_c superconductors behave like a 3-dimensional array of Josephson junctions when subjected to an external magnetic field. The effective Josephson penetration depth λ_J beyond which screening currents prevent magnetic field flux from entering the superconductor is given by

$$\lambda_J = \frac{C\phi_o}{8\pi^2 aJ_{cJ}} \tag{6}$$

where $o_o = 2.07\times10^7 Gcm^2$, a is the spacing between nearest neighbors and $J_{cJ} =J_c /a^2$ the critical current density. This granular depth corresponds to the classical continuum value λ_L(eqn 1) if $aJ_{cJ}$is identified as the effective density of superconducting electrons. In type II superconductors below Tc, vortex lattices form to allow external field penetration. This results in inhomogeneities in the internal field causing a line broadening in the magnetic resonance experiment (ESR and NMR) or an enhanced spin-relaxation rate in muon spin rotation experiment (μSR). Assuming that the (Abrikosov) vortex lattice is a square array [11]. the mean square field inhomogeneity in the mixed state ($H_{c1} < H_{ext} < H_{c2}$) is

$$< \Delta H^2 >= H_{ext}\frac{\phi_o}{4\pi\lambda^2}[\frac{1 + 4\pi^2\lambda^2 H_{ext}}{\phi}] \tag{7}$$

where $H_{ext}=$ external magnetic field. $\phi =$ magnetic flux quantum $(hc/e^*) = hc/2e =$ $2.068 \times 10^{-8} Gcm^2$. and $\lambda =$ magnetic penetration depth. Thus a magnetic resonance experiment carried out on a type II superconductor as a function of temperature would yield $\lambda(T)$ which could be analysed and used to test models for superconductivity. If d is the spacing between vortices or the flux line lattice parameter then $d^2 = p/H_{ext}$ so that the above expression becomes

$$< \Delta H_{sq}^2 > = \frac{H_{ext}}{4\pi}(d/\lambda)^2[1 + (2\pi/d)^2]^{-1} \tag{8}$$

Corrections to equation (7) or (8) for a more realistic triangular flux lattice are shown to be negligible. However the exact expression for this case is [4]

$$< \Delta H_{tr}^2 > = H_{ext}[\frac{1}{2\pi\sqrt{3}}](\frac{\lambda}{\sqrt{3}/2d})^2[1 + 4\pi^2(\frac{\lambda}{\sqrt{3}/2d})^2] \tag{9}$$

The temperature dependent penetration depth is directly related to the temperature dependent energy gap through the local limit BCS result

$$(\frac{1}{\lambda(T)})^2 = (\frac{1}{\lambda(o)})^2[1 - 2\int_\Delta^\alpha(-\frac{\partial f}{\partial E})\frac{E}{(E^2 - \Delta^2)^2}dE] \tag{10}$$

where f is the Fermi function and Δ is the minimum value of the energy gap over the Fermi surface. Furthermore, for an isotropic superconductor

$$H_{c1} = (\frac{\phi_o}{4\pi\lambda^2})\ln k \tag{11}$$

so that

$$[1/\lambda(T)]^2 \propto H_{c1}(T) \tag{12}$$

The relation (11) affords an estimation of K. Measurements of relative penetration depth $\delta\lambda = [\lambda(T) - \lambda(0)]$ as a function of dc magnetic field [5] yield the value of H_{c1}, through a distinct change of slope:

$$\delta\lambda(T, H) = K(T)H^2 \tag{13}$$

where $K(T)$ is an experimentally determined quadratic coefficient related to the thermodynamic critical field in the Ginzberg-Landau model:

$$K(T) = \frac{3}{4}\frac{\lambda(T)}{H_{co}(T)^2} \tag{14}$$

A fit of experimental data to equation (14) would yield H_{co}.

The major question in high Tc superconductivity is whether or not there are nodes in the superconducting gap function. For any pairing state with a finite excitation energy the change in λ at low temperatures is

$$\Delta\lambda(T) \propto \exp(-\Delta/K_BT) \tag{15}$$

and Δ is the minimum value of the energy gap over the Fermi surface. For pairing state with line nodes on the Fermi surface

$$\Delta\lambda(T) \propto T^p \tag{16}$$

where $p = 1$ for the simplest form of the gap with d-wave symmetry. (In general, p depends on type of node in k-space). The limiting temperature dependence of $\lambda(T)$ as $T \to 0K$ decides whether the superconductivity is conventional or exotic.

Recent theories of the mixed state in layered superconductors have examined the effect of order parameter phase fluctuations on penetration depth near the transition temparature (T_C) and at temparatures $T \to 0$ It is also necessary to consider the contribution of these fluctuations in order to account for the linear temperature dependence of λ_{ab} and λ.

Roddick and Stroud [6] have described a simple model to describe the fluctuation effects in the classical limit and obtained a linear temparature dependence in $\lambda(T)$ which compares with experimentally observed magnitude. A variational calculation shows that this linear T-dependence persists even when quantum effects due to charging and dissipation are included in a simple model. Thus they conclude that a linear $\lambda(T)$ may not be a unique signature of a superconductor with line nodes at the Fermi surface i.e., the presence of a linear T dependence is not sufficient in itself. to detect an unconventional order parameter in the high-T_C superconductors.

Classical phase fluctuations in a nodeless order parameter

$$\psi(x) = \psi_o(x)\exp[i\theta(x)] \tag{17}$$

with the amplitude ($< \psi_o >$) assumed T-independent. could produce a $\Delta\lambda(T) = [\lambda(T) - \lambda(0)]$ proportional to T in an isotropic 3D material. The actual result is

$$\Delta\lambda(T) = k_B[8\pi\lambda^{1/2}(o)]/\xi_o\phi_o^2 \tag{18}$$

o_o being the flux quantum $hc/2e$. The proportionality constant roughly comes out as $1\mathring{A}/K$ with $\lambda(o)=2000\mathring{A}$ and $\xi_o = 10\mathring{A}$.

Using an anisotropic X-Y model and assuming the superconductor to be made up of many small grains whose size is comparable to ξ_o and coupled together by a Josephson-like interaction they have obtained the quantum mechanical result for λ, in the self-consistent phase phonon approximation. The temparature dependent penetration depth $\lambda_\alpha(\alpha = 1,2)$ is given by

$$\lambda_\alpha^2 = \frac{C}{4\pi K_\alpha}(\frac{h}{2e})\frac{a\phi_o}{2\pi} \tag{19}$$

where 'a' is the lattice constant ($= \xi_o$) and K_α is the parameter which minimizes the free energy of the superconductor. It is seen that the linear T dependence of λ is retained even when Coloumb interactions (both short and long range) and shunt dissipation are included. The useful results of this calculation are the following:

(i) the slopes of $K_1(T)$ and $K_2(T)$ yield $\Delta\lambda(T) = \lambda_{ab} - \lambda_{ab}(o)$.

(ii) the slopes $\partial\lambda_c/\partial T$ and $\partial\lambda_{ab}/\partial T$ depend on $\lambda_c(o)$ and $\lambda_{ab}(o)$.

(iii) the effect of thermodynamic phase fluctuations on $\lambda_{ab}(T)$ should be ruled out before the behaviour of $\lambda_{ab}(T)$ by itself is used to drain conclusions regarding the symmetry of the order parameter in high T_C superconductors. Starting from a Lawrence-Daviachmodel in the London limit. Coffey [7] has considered the effect of phase fluctuations on λ of a 3D superconductor in a Josephson-Coupled layered model to obtain the mean square phase fluctuation in intermediate parallel magnetic field. The results of his perturbation calculations are

$$\frac{\Delta\lambda(T)}{\lambda(0)} = \frac{K_B T/2}{\phi_o^2/(2\pi)^2\Lambda} ln2(1 - H/H_C)^{-1/2} \tag{20}$$

for the non-zero field case $(H \neq 0)$ and

$$\frac{\Delta\lambda(T)}{\lambda(0)} = \frac{\ln 2}{2}\frac{k_B T}{E_M} \; for \; H = 0, \tag{21}$$

where s = repeat distance along c-direction, $\Lambda = 2D$ screening length $= \lambda^2/s$, E_M magnetic energy $= óo^2/(2\pi)^2\Delta$ and the other symbols have their usual meanings. The coefficient of T in the second equation above is of the order of $10^{-4}/K$ for $\lambda(o) \approx 1500$ Åand $s = 12.\overset{\circ}{A}$. only an order less than the experimental value of 3.5 Å/K for $\Delta\lambda$ which again implies the non-negligible contribution of phase flutuations to the penetration depth.

A universal relationship between $1/\lambda^2$ (λ determined from μSR) and the transition temperature $T_c$for compounds with $T_c < 90K$ has been postulated [8,9] which would imply that for low carrier density compounds

$$T_c \propto \frac{n_s}{m^*} \tag{22}$$

The concept of penetration depth was introduced while accounting for the Meissner effect which involves a development of a non-trivial diamagnetic susceptibility for the superconductor. This idea has recently been used [10] in the determination of temperature dependence of cuprate superconductors. London's formula for the magnetic moment (m) of an isolated superconducting sphere of radius $'a'$ and a penetration depth λ is given by

$$m = -2\pi a^3[1 - \frac{3\pi}{a}coth(\frac{a}{\lambda}) + \frac{3\lambda^2}{a^2}] \tag{23}$$

If a net voltage $V_{ac}(\mu V)$ is measured in ac susceptibility apparatus for a superconductor of volume $V mm^3$.then the effective susceptibility $(\chi/\chi_o)_{eff}$ relative to a perfectly diamagnetic spherical grain of susceptibility χ_o is given by

$$\left(\frac{\chi}{\chi_0}\right) = \left(\frac{2}{3}\right)\frac{V_{ac}}{[V_s N - V ac F(1/3 - D)]} \tag{24}$$

where N is the calibration factor of the apparatus. D diamagnetisation factor which is zero for a perfectly diamagnetic superconductor and F is the filling factor.

For a polycrystalline sample this normal field susceptibility has to be integrated over the measured distribution of the grain sizes (using SEM) g(r). where r is the radius of th grain, so that

$$\left(\frac{\chi}{\chi_0}\right)_{eff} = \int \frac{1 - (3\lambda/r)coth(r/\lambda) + (3\lambda^2/r^2)r^3 g(r)dr}{\int r^3 g(r)dr} \tag{25}$$

Using Pb as standard,and employing magnetically aligned samples, both the anisotropy and temperature dependence of λ have been determined for YBa$_2$Cu$_3$O$_7$.

A high frequency dynamic susceptibility method involving the resonant absorption of microwaves through a superconducting sample decorated with a free radical has been developed for the determination of penetration depth [11]. This method relies on the increase in the line width of free radical ESR that results when the sample is cooled through the superconducting transition temperature. The observed peak-to-peak width of the Gaussian ESR, ΔH_{pp}, is related to the inhomogeneous broadening due to flux lattice contribution.

$$\Delta H_{pp} = \frac{1}{2}[\Delta H_s^2 + \Delta H_n^2]^{1/2} \tag{26}$$

where ΔH_n and ΔH_s are the contributions from the normal and superconducting components. ΔH_n persists through T_c so that its contribution has to be subtracted from the observed line width(ΔH_{pp}). Changes in ΔH_s are directly related to the penetration depth through the Brandt's formula derived for a triangular flux lattice .

ESR technique probes only the surface layer of the superconducting sample while μSR technique probes the entire bulk. Thus while the interpretation of the ESR results depends on the magnetostatics of the surface, [12] the μSR like NMR directly measures the decay of spin polarisation.

In this paper we describe and discuss our penetration depth measurements on undoped and doped BSCCO superconductors, carried out using the ESR of DPPH probe.

Experimental Measurements:

It is very important to use samples of uniform microstructure (particle size) in penetration depth measurements because any non-uniformity could lead to extraneous line width contributions that eventually affect the reliability of the results obtained. Thus we have chosen the sol-gel method for the synthesis of Bi-based superconducting ceramics both undoped (2212) and Pb/Na-doped (Pb-doped 2223 and Na-doped 2212). Samples with nominal composition,$Bi_2Sr_2Ca_1Cu_2O_y$ (sample A), $Bi_{1.8}Pb_{0.2}Sr_2Ca_2Cu_3O_y$ (sample B). $Bi_2Sr_2Ca_{1-x}Na_xCu_2O_y$ with x=0 (sample C), x=0.1 (sample D). x=0.2 (sample E) and x=0.3 (sample F) were taken. The general procedure is to dissolve stochiometric amounts of appropriate metal nitrates in water and add citric acid and ethylene glycol. This is followed by a condensation at $\approx$ 120°C to obtain a colloidal precipitate. Continued evaporation leads to a dark blue gel which is decomposed in a sand bath to obtain the precursor. Gradual ignition to 500° C. grinding and finally calcination at 800" C for 12 hr completes the decomposition and evaporation of all organic components. The powder thus obtained is cold-pressed into pellets and sintered over night at 800° C in air. Repeated anneals atleast three at 840° C for 24 hrs ,for undoped and Na-doped 2212 and at 870° C for 90 hr for Pb-doped 2223to obtain an optimal mono phasic material. [13]

The phase identity was established by X-ray powder diffraction. The lattice parameters of the synthesised phases are given in Table 1.

Table 1

Data on lattice parameters of Na-doped (free) 2212 and Pb-doped 2223

sample	composition	$a(\mathring{A})$	$b(\mathring{A})$	$c(\mathring{A})$
(A)	$Bi_2Sr_2Ca_1Cu_2O_y$	3.808	3.808	30.83
(B)	$Bi_{1.8}Pb_{0.2}Sr_2Ca_2Cu_3O_y$	5.413	5.413	37.13
	$Bi_2Sr_2Ca_{1-x}Na_xCu_2O_y$			
(C)	(x=0, T_c=92)	5.431	5.412	30.78
(D)	(x=0.1, T_c=81)	5.423	5.413	30.81
(E)	(x=0.2, T_c=83)	5.432	5.405	30.87
(F)	(x=0.3, T_c=86)	5.423	5.434	30.91

Superconducting transition temperatures (T_c) were determined by standard four-probe resistivity measurements on sintered pellets. T_c values are included in Table 1. The T_c values were confirmed by low-field microwave absorption studies, by plotting the derivative of the microwave absorption as a function of temperature.

Pure BSCCO samples do not show any ESR spectra around 300 mT field which preclude the presence of any isolated or clustered paramagnetic impurities in the samples. The superconducting pellets are decorated with the stable free radical DPPH (diphenyl picryl hydrazyl) in order to adsorb- physically adhere at the molecular level, the free radical to the superconductor surface, chunks of the pellet samples were immersed in an acetone solution of DPPH (25 mg DPPH dissolved in 30 ml acetone). The acetone solution turned clear purple indicating breakdown of the DPPH solid to molecular level. The very low boiling point of acetone ensures fast percolation and easy deposition of DPPH overlayer on the superconductor sample surface. The 'decorated samples' thus obtained were dried in air at ambient and subsequently in an oven maintained at 80° C. The ESR measurements were done on field-cooled samples using a JEOL-3X band spectrometer operating at 9.5 GHz. with dc magnetic field centered at 330 mT and modulated with an ac field of variable amplitude ($0\mu T$ - 0.5 mT) and fixed frequency of 100 KHz. The microwave power levels were kept low enough (1 mW) in order to minimise the line intensity saturation and line broadening effects. Sample temperatures were varied (from 77 K to 300 K) and controlled using the variable temperature accessory. The results obtained would be described and discussed in the next section.

Results and Discussion:

The ESR spectrum of the commonly used g-marker DPPH consists of an intense narrow symmetric Gaussian shaped resonance line at ambient temperature with isotropi (g = 2.0036). The narrowness ($\Delta H_{pp} \approx 0.3mT$) arises from the strong electron spin exchange. Down to 30 K, the line width increases slowly and monotonically, below whic it increases more rapidly [14]. Around 30 K DPPH undergoes an antiferromagnetic phase transition and the rapid increase in line width below 30 K might be an indication of this fact. Thus DPPH could be conveniently used as a probe through the superconducting phase transition which occurs in the region 80-110 K in Bi-superconductors besides Y,Tl and Hg superconductors, but not in the region $\approx 40K$ where the transition occurs in La-superconductors.

Fig.3 shows the typical ESR spectra of DPPH decorated Bi-superconductors (a) 2212 (b) Pb-doped 2223 and (c) Na-doped 2212 at room temperature ≈ 200 K $(T > T_c)$ and 77 K $(T < T_c)$.

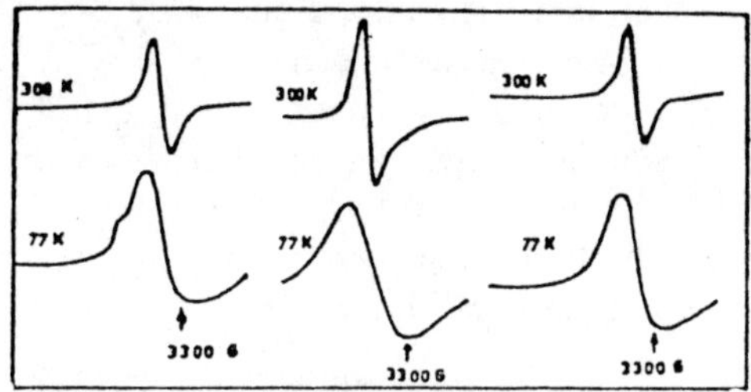

Fig.3 ESR spectra of DPPH- decorated pure 2212, Pb- doped 2223 and Na- doped 2212 at selected temperatures. Note the significant line broadening and assymetry below T_c.

One notes the effect of the superconducting transition on the features of the DPPH by way of a drastic increase in the peak-to-peak width and intensity below T_c, while the line shape is unchanged. The line broadening is an indication of the change in the exchange relaxation time -a manifestation of the coupling of the surface magnetic field to the magnetic moment of the DPPH probe.

Fig.4 shows the variation of the second moment as a function of temperature. Note that in the normal state , the second moment is fairly constant, characteristic of uniform magnetisation of the whole sample in its paramagnetic state and the absence of any flux line lattice.

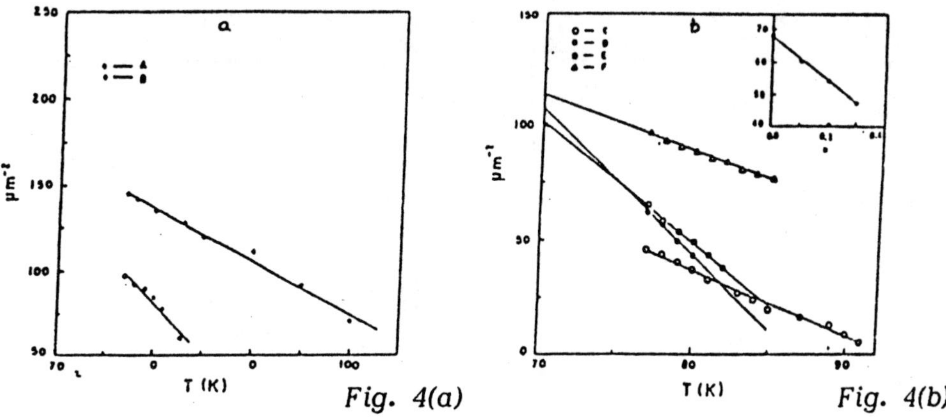

Fig. 4(a)   *Fig. 4(b)*

Fig.4 Temperature dependence of second moment for the samples (a) A, B (b) C, D, E. and F. Inset shows the concentration dependence of penetration depth.

As $T \to T_c$ from above diamagnetic shielding is gradually established, and magnetic flux is expelled from the sample regardless of what value of flux it had in the normal state. The effect of this change is an increased line width over and above the normal DPPH exchange line width which has to be extracted and analysed for the penetration depth $\lambda(T)$ to eventually determine λ_o. In fact this second moment is associated with n(B), the field density distribution near the sample surface that is n(B) is the probability that at a randomly chosen point inside the specimen. (the position of the unpaired spin in the probe molecule) the magnetic field has the value B.

To extract the value of the penetration depth at 0 K. λ, the second moment of the n(B) distribution is to be considered. What is the origin of the extra line width of the DPPH ESR observed below T_c ? It is the inhomogeneity of the flux line lattice which is a manifestation of the penetration of the magnetic flux. These flux lines are one tiny current vortices which repel each other and arrange in the form of a more or less ideal triangular lattice. The total observed line width

$$\Delta H_c(T) = V\sqrt{\Delta H_o^2(T) + \Delta H^2(T)} \tag{27}$$

where $< \Delta H^2 >$ is given by eqn (8). This average value, $< \Delta H^2 >$ has been calculated in the Gaussian approximation to yield

$$< \Delta H >^2 \approx 0.06091 \phi_o / \lambda^2 \tag{28}$$

where λ is given by equation (4) and ϕ_o is defined before. Plots of $1/\lambda^2$ vs T show no deviations from linearity near T_c which suggests that doping does not introduce any fluctuation effects implying that inter layer coupling is intact.

Fig.4 shows the plots of second moment vs temperature for undoped and Na- doped Bi-superconductors. The λ values extracted form equation (28) are fitted by the method of least squares and the plots are shown in Fig.5.

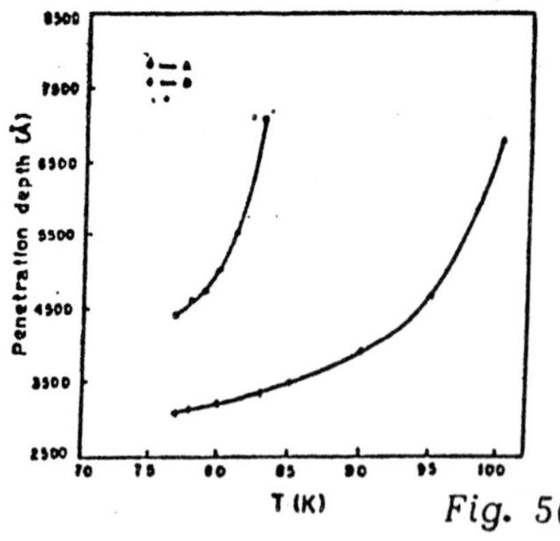

Fig. 5(a)

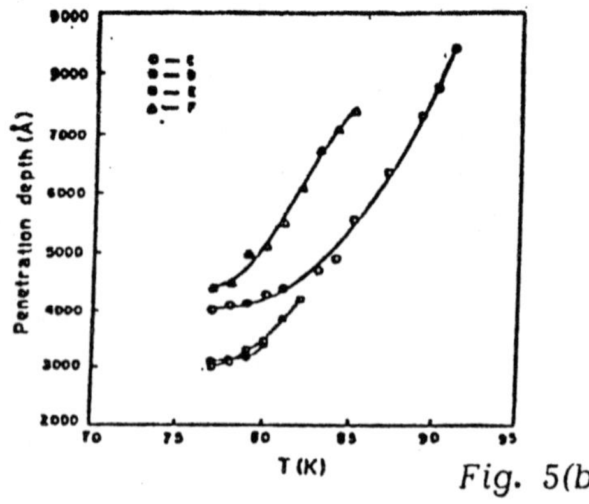

Fig. 5(b)

Fig.5 Temperature dependence of penetration depth for the samples (a) A. B (b) C. D. E. and F.

Table 2. compiles the penetration depth data on selected high-T_c superconductors obtained by various methods. In most of the case λ shows a temperature dependence consistent with the Gorter-Casimir formula. From an examination of this table the following points emerge.

1. λ is more anisotropic in Y-based superconductors and doping with Co and Zn at the Cu- site enhances anisotropy and fluctuation effects.

2. Among the several methods employed to determine λ , free radical ESR method gives values intermediate between magnetisation (lower) and other resonance (NMR. μSR) methods (higher).

3. The λ values obtained on magnetically aligned powders represents average values of λ_o lying between λ_c and λ_{ab}, with $\lambda_c < \lambda_{ab}$. The extreme anisotropy in λ indicate the effect of flux pinning and granularity of the vortex lattice.

4. The temperature dependence of $\lambda_{ab}(T)$ and $\lambda_c(T)$ that relate to the anisotropy of the energy gap need to be studied more extensively by particularly in doped samples. as doping would lead to anisotropy and fluctuation effects. Indeed deviation from linearity of $\lambda(T)$ Vs T near T_c could be used to determine the temperature range over which fluctuations could occur so that related experiments could be performed in that range.

Penetration depth studies on polycrystalline Bi-superconductors doped with Y for Ca and Pb for Bi using μSR have revealed that

1. 1/ λ_{ab}^2 of Y- doped compounds, $Bi_2Sr_2Ca_{1-x}Y_XCu_2O_{8+\delta}$ correlates with the nominal hole-concentration per Cu atoms.

2. $1/\lambda_{ab}^2$ is independent of probe concentration in Pb- doped samples implying that Pb^{++} does not increase the density of charge carriers, although it increases T_c.

Our ESR studies of λ on Na- doped Bi-2212 and magnetisation measurements of λ on Bi-2212 samples of various particle sites ranging between 0-0.3 μm reveal that the average penetration depth is magnetically sensitive to changes in the crystal structure and microstructure induced by doping and thermal treatment respectively, besides the face that doping and microstructure affects T_c

A universal linear relationship between $1/\lambda^2$ extracted from μSR and T_c seems to hold for compounds with $T_c < T_{cMax} = 90$ K which extrapolate to (La,Sr)-Cu-O , Bi-2212. Bi-2223. Tl-2212 and 2223 families [8,9]. This implies that for compounds with low superconducting carrier densities equation (22) is obeyed.

This correlation unlike the two-fluid model suggests a state in between BCS pairing and Bose-Einstein condensation.

In an interesting study of in-plane penetration depth on very thin film (50 nm) c-axis oriented $La_{2-x}Sr_xCuO_{4\pm\delta}$ (001) SrTiO$_3$ substituted. Jaccard et al [15] measured $\lambda_{ab}(T)$ using kinetic inductance, $L_k(T) = (\mu_o/d)\lambda_{ab}$, d= film thickness is analysed both in the critical region ($0.9 \leq T/T_c < 1$) and outside the region ($0.3 \leq T/T_c < 1$) while the surface resistance is analysed in the critical region surface resistance $R(T)$ (ac sheet impedance $Z = R + i\omega L_k$). 3D-XY model gives, for ($0.9 \leq T/T_c < 1$)

$$\lambda_{ab}(T) = \lambda_o(1 - T/T_o)^{-1/3} \tag{29}$$

Table 2

Penetration depth data on selected high-T_c superconductors.

composition	Type	λ_o(nm)	Method	Reference
$Pb_2Sr_2Y_{0.5}Ca_{0.5}Cu_3O_y$				
(T_c= 73 K)	single crystal	λ_c^o=290	free radical ESR	18
		λ_{ab}^o=440		
$YBa_2Cu_3O_y$				
T_c= 90 K	single crystal	=750	NMR	19
		> 700	μSR	20
		500-800	μSR	21
		700-900	μSR	22
		515	magnetisation	23
		625	ESR	24
$YBa_2Cu_3O_y$	magnetically alligned powder			
T_c= 92.9 K	pure	=140	AC susceptibility	25
		=1040		
T_c=94.4 K	1.5% Co	=202		
		=1160		
T_c= 80 K	4% Co	=200		
		=2000		
T_c= 41.5 K	5% Zn	=330		
		=1500		
$Bi_2Sr_2CaCu_2O_y$	poly crystalline	λ_o=240	ESR	13
$Bi_{1.8}Pb_{0.2}Sr_2Ca_2Cu_3O_y$	poly crystalline	=258	ESR	13
$Bi_2Sr_2Na_{1-x}Ca_xCu_2O_y$	poly crystalline		ESR	26
x=0, T_c=92 K		=310		
x=0.1, T_c=81 K		=278		
x=0.2, T_c=83 K		=258		
x=0.3, T_c=86 K		=215		
$Bi_4Ca_3Sr_3Cu_4O_{16}$				
T_{c1}=85	poly crystalline	185.5	magnetisation	27
T_{c2}=115		365		

while for the thermally activated flux flow behaviour observed in R near T_c

$$\ln R \approx \Delta U/T = \Delta U_o/T[1 - T/T_o]^{-1/3} \tag{30}$$

where ΔU is the vortex pinning activation energy for flux flow. $\lambda_{ab}(o)$ is minimum and ΔU, is maximum for the optimally doped sample. This behavior of $\lambda_{ab}(o)$ increases with carrier density in sharp contrast with London's prediction $1/\lambda_{ab}(o)^2 \sim n$.

Accurate ac susceptibility studies [16] on grain aligned $Y_1Ba_2Cu_3O_7$ powder samples of sizes 100 $\mu m < d < 20\mu m$ have shown a T^2 behavior of $\lambda(T)$ which has been attributed to an anisotropic gap or a d.gap in the presence of impurity induced disorder. This also shows that the onset of diamagnetism in polycrystalline high-T$_c$ superconductors occurs roughly at temperatures at which λ becomes of the order of the grain size. As $\lambda(T)$ diverges at T$_c$, this condition is met at lower temperature for small particles. In recent measurements of the dc magnetic field (H) dependence of λ_L in $Bi_2Sr_2CaCu_2O_y$ single crystals Maeda et al [17] have found a linear $\lambda_L(H)$ which they attribute to unconventional superconductivity.

Conclusions:

In this paper, we have provided a brief review of penetration depth in high T_c superconductors. a topic that has assumed tremendous importance because it has a bearing on the mechanism of superconductivity in general and on the nature of pairing states in particular. We have also focussed on the decorative ESR method of determining the penetration depth. This method is suitable for all forms of superconductors- thin films, polycrystalline and single crystal. We have described and discussed our experimental results on the temperature/concentration dependence of λ in Bi-Pb and Na doped BSCCO

superconductors in the light of Brandt's model for the triangular flux lattice. Future work in this area could concentrate on EPR imaging in an attempt for direct observation of the flux line lattice. The implication of the linear temperature dependence of λ in the region $(T < T_c)$ must remain controversial in view of the recent work on the role of the thermodynamic phase fluctuations in giving rise to a linear $\lambda(T)$. It would also be interesting to investigate the transition from Meissner to para-Meissner state [28-32].

Acknowledgements

One of us (AMS) thanks Council of Scientific and Industrial Research (CSIR)-New Delhi for financial assistance in the form of an SRF, and Prof. K. N. Shrivastava for valuable suggestions.

References

1. F. and H. London, Proc. Roy. Soc. **149**, 1, (1935).

2. J. R. Schrieffer, Theory of Superconductivity, Addison-Wesley, Reading, 1988, p.9.

3. H. W. Lewis, Phys Rev. **102**, 1508, (1956).

4. E. H. Brandt, Phys. Rev.**B 37**, 2349, (1988) and references therein.

5. S. Sridhar et al., Phys. Rev. Lett. **63**, 1873, (1989); D. H. Wu et al., Phys. Rev. **B 38**, 9311, (1988).

6. E. Roddick and D. Stroud, Phys. Rev. Lett.**74**, 1430 (1995) and references therein.

7. M. W. Coffey, Phys. Lett. **A 200**, 195, (1995) and references therein.

8. Y. J. Uemura, L. P. Le, G. M. Luke, Phys. Rev. Lett. **66**, 2665, (1991).

9. M. Weber, A. Amato, F. N. Gygax and A. Schenck, Phys. Rev. **B 48**, 13022, (1993).

10. A. Porch, J. R. Cooper, and N. Zheng, Physica C **214**, 350, (1993).

11. B. Rakvin et al., Phys. Rev. **B 41**, 769, (1990).

12. N. Bontemps, D. Davidov, P. Monod and R. Even, Phys.Rev **B 43**, 11512 (1991).

13. A. M. Suvarna and C. S. Sunandana, Solid State Commn. **95**, 11 (1995).

14. Yu. N. Shvachko et al., Physica C **197**, 27 (1992).

15. Y. Jaccard et al., Physica C **235-240**, 1811 (1994).

16. L. Guerrin et al., Physica C **235-240**, 1797 (1994).

17. A. Maeda et al., Phys.Rev.Lett **74**, 1202 (1950).

18. W. Korczak et al., Physica C **235-240**, 1803 (1994).

19. H. B. Brown et al., Physica C **57**, 297 (1991).

20. G. Aeppli et al., Phy.Rev **B 35**, 7129 (1987).

21. B. Pumpin, H.Keller and W. Kundi, Physica C **162**, 151 (1989).

22. M. Weber et al., Physica C **185-189**, 1547 (1991).

23. V. M. Zaransky et al., Physica C **162**, 1125 (1989).

24. M. Puri and Larry Kevan, Physica C **197**, 55 (1992).

25. A. Porch, J. R. Cooper et al., Physica C **214**, 350 (1993).

26. A. M. Suvarna and C. S. Sunandana, Physica C (in print).

27. M. G. Alexander, Phy.Rev **B 38**, 9194 (1988).

28. K. N. Shrivastava, Phys.Rep. (North Holland) **200**.51 (1991).

29. K. N. Shrivastava, Phys.Lett. **A 188**, 182 (1994).

30. K. N. Shrivastava, Solid state Commun, **90**, 589 (1994).

31. M. Sigrist and T. M. Rice, Rev.Mod.Phys **67**, 503 (1995).

32. D. J. Thompson, M. S. M. Minhaj, L. E. Wenger and J. T. Chen, Phys.Rev.Lett **75**, 529 (1995).

Microwave Measurements of
Granular Superconductors

A.G. Vedeshwar

Department of Physics and Astrophysics
University of Delhi
Delhi 110 007

1. Introduction
2. Theoretical background
   a. Surface resistance
   b. Josephson medium model
   c. Spin glass model
3. Experimental Techniques
4. Results and Analyses

1. INTRODUCTION

The microwave measurements of superconductors has been an important non-contact probe for investigating the surface imphedance since the days of classical superconductors [1,2]. One can easily explain quite qualitatively the RF and microwave losses in classical superconductors using two fluid model or Mattis-Bardeen theory [1-3]. These analyses made it possible to use superconductors as waveguides, cavity resonators, transmission lines etc. However, with the discovery of new generation of superconductors, the high Tc oxide superconductors, microwave measurements became an interesting useful tool of investigation and characterisation

[4-15]. It offered a wide scope to know about some of the materials paremeters like grain size, amount of non-superconducting phase, degree of porosity etc. [10]. It is very useful, perticularly, in the understanding of intergranular properties of granular superconductors. I have given only few very relevant references here just to give an account of the developement. The microwave measurements are done normally in three ways; the measurement of surface resistance at microwave or radio frequencies, measurement of the power absorbed as a function of applied dc field using unmodulated technique and the measurement of microwave absorption using modulation technique employing commercial EPR spectrometers. There are many interpretations based upon these types of measurements in the literature. The conventional and hard superconductors like lead, NbN, Nb_3Ge etc. were also reexamined to test the different formulations of interpretation [16,17]. I will briefly outline here the essential theoretical background of most important, routine interpretations of these measurements. I have also tried to review the state of the art till date emphasizing granular structure.

2. THERETICAL BACKGROUND

The microwave measurements of the new superconductors, in particular the ceramic or granular ones, are analysed using different models to account for the observed behavior in the superconducting state. One of the most obvious possibilities is the surface imphedance. Another competing process is the fluxon motion along the Joephson junction existing between various superconducting grains. This idea can be applied by modelling the granular structure as weakly coupled Josephson medium. However, the observed reversibility in susceptibility measurements of field cooled (FC) sample and the irreversibility in zero field cooled (ZFC) sample idicates a spin glass like behavior. Therefore, it is also quite relevant to model granular structure as a system of spin glass using the concept of frustration in the superconducting cluster formed through Josephson interaction. Some people believe that the loss is due to the flux movement in intergranular critical state. All the above models are used to analyse the appropriate microwave measurements of the new ceramic superconductors. However, a very few other kinds of interpretations are also available in the literature. They were unable to explain varios kinds of microwave experiments and subsequently could not be established. Therefore, I will give a brief account of the basic concepts of some of the important and useful models described above.

A. Surface Imphedance :

The concept of surface imphedance is important in understanding the behaviour of materials at high frequencies. The microwave surface imphedance can be calculated by taking the ratio of electric and magnetic fields at the sample surface,

$$Z \equiv \left(E_x / H_y \right)_{z=0} = E_x / \int_0^\infty j_x \, dz \qquad (1)$$

where j_x is the current density in the material. For a plane wave at normal incidence on a flat surface this is simply the imphedance of the layer of material of thickness equal to the skin depth. The complex imphedance of the plane surface of a conductor having a complex conductivity $\sigma = \sigma_1 - i\sigma_2$ is given by,

$$Z_s = R_s + iX_s \qquad (2)$$

where imaginary term represents the surface inductive reactance and R_s gives the surface resistance or losses per unit area per unit surface current density amplitude. The magnitudes of the real and imaginary parts of the imphedance are identical,

    A. G. Vedeshwar

$$R_S = X_S = 1 \,/\, \sigma\delta \qquad\qquad (\,3\,)$$

where δ is the skin depth, a measure of exponential penetration of plane electromagnetic waves into normal conductors. The skin depth can be expressed in terms of frequency (f) of electromagnetic waves, conductivity (σ) of the material and the magnetic permeability (μ) of the medium by,

$$\delta = (\, 1 \,/\, \pi\, f\, \mu\, \sigma\,)^{1/2} = (\, 2 \,/\, \omega\, \mu\, \sigma\,)^{1/2} \qquad (\,4\,)$$

where $\omega = 2\pi f$. Since the surface resistance of a conductor is related to the skin depth and surface conductivity, we can expect its dependence on frequency depending upon the electron mean free path l compared to the skin depth. The dependence of R_S on frquency for three possible situations can be **summarized** as following.

$$\begin{aligned}
R_S &\;\alpha\; \omega^{1/2} &\quad &\text{for}\;\; l < \delta \\
R_S &\;\alpha\; \omega^{1/3} &\quad &\text{for}\;\; l \sim \delta \qquad\qquad (\,5\,)\\
R_S &\;\alpha\; \omega^{\,2/3} &\quad &\text{for}\;\; l > \delta
\end{aligned}$$

The quantitative calculations of R_S for superconductors is not straight forward. One has to take into account the additional parameter λ, the London penetration depth, along with δ. The relationship between δ, λ and l is rather complex. δ can be expressed interms of λ, paired and normal electron densities (n_S and n_n respectively) and the relaxation time τ by simply substituting the two-fluid model expression for σ in place of σ of normal conductors and is given by,

$$\delta = \sqrt{2}\,\lambda \,/\, (\,\omega\,\tau\,(n_S \,/\, n_n\,) - i\,)^{1/2} \qquad (\,6\,)$$

It can easily be seen from above equation that when ω increases δ decreases. Therefore, for a superconductor the loss is lesser than the normal metal even for high microwave frequencies which is due to the reactive shielding current of electron pairs. The surface imphedance can be expressed in the two fluid model as,

$$Z_S = \left(\omega^2 \mu_O^2 \lambda^3 n_n \sigma_n / 2 n \right) + i \omega \mu_O \lambda \qquad (7)$$
$$= R_S + i\omega X_S$$

where σ_n is the normal state conductivity and $n = (n_S + n_n)$. It can be seen that in this model R_S varies linearly with ω^2. The variation of both the components are shown in figure 1. The temperature dependance of R_S in two- fluid model is given by [18],

$$R_S = \frac{A(\omega)\ t^4}{\left(1 - t^4 \right)^{3/2}} \qquad ; t = T / T_C \qquad (8)$$

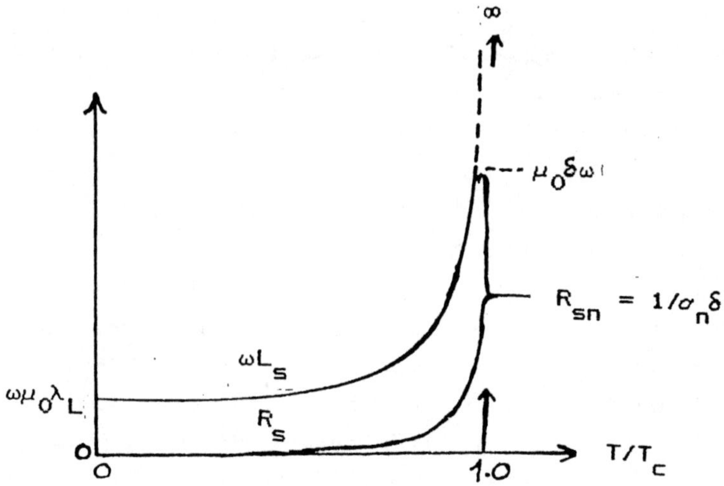

Figure 1. Temperature dependence of resistive & reactive components of Z_S of a superconductor.

However, one can expect that the incorporation of superconducting energy gap would improve the temperature dependance predicted by the two-fluid model. For $\omega << \Delta(T)/ \hbar$ and temperatures below 0.5Tc, the BCS theory of electromagnetic absorption in bulk superconductors leads to an approximate relation for surface resistance [19],

$$R_S = C \, \Delta / T \, (\omega / \Delta)^2 \ln (\Delta / \hbar\omega) \exp (- \Delta / k_B T) \qquad (9)$$

where C is a numerical constant and $\Delta(T)$ is the temperature dependet energy gap parameter of the superconductor. It can be seen that the above equation predicts a quadratic frequency dependence for R_S, that is, at a given temperature $R_S \, \alpha \, \omega^2$. Therefore, the measurement of frequency dependence of R_S of a superconductor is quite crucial.

B. Josephson Medium Model :

The key idea in applying Josephson junction model to the granular superconductor is because of the fact that the anisotropic stoichiometric superconducting grains are seperated by layers of non-stoichiometric non-superconducting interface material. The idividual superconducting grains first become superconducting as the temperature is lowered below Tc of the grains. All such grains form a percolative network through Josephson interaction between grains to give a bulk superconductivity. Therefore, the complete disappearance of resistance will take place at somewhat lower temperature T_{cj} than the Tc of the individual grains. This fact is reflected in the appearence of appreciable transition width ΔTc which is equal to Tc - T_{cj}. This idea generates many posibilities in the behaviour of granular superconductors arising from various material parameters like grain size, type of interfacing material, porosity etc. The knowledge of grain size and its homogeneity in the sample would simplefy the task of averaging the interaction between various grains to understand its superconducting properties. The type of interface material, that is, conducting, semiconducting or insulating will decide whether proximity effect or Josephson effect should be applied to calculate various electrical and magnetic quantities in the superconducting state. Based on these ideas various people have worked out the theory for different conditions [20-22].

I will briefly describe here the essential background of the model sfficiently enough to understand and analyse some of the microwave measurements of granular superconductors.

A typical granular high Tc superconductor in a pallet form can be modelled as consisting of superconducting grains arranged in three dimension. Well below Tc these grains will be coupled by Josephson interaction. The Josephson coupling energy E_j between any two grains can be expressed as,

$$E_j(T) = h\, I_c\, /\, 2e$$
$$= (\, h\, /\, 8e^2\, R_n\,)\, \Delta(T)\, \tanh\, (\, \Delta(T)\, /\, 2k_B T\,) \qquad (\,10\,)$$

where I_c is the tunnelling current between grains, R_n is the normal state resistance and $\Delta(T)$ is the BCS enrgy gap of the grains. The behaviour of E_j with temperature is determined by the type of junction, that is, proximity type or tunnelling type. The measure of the coupling strength in the sample can be determined by the ratio of macroscopic critical current density set by Josephson coupling $J_{cj} = I_c\, /\, a^2$ (a is the average grain size) to that set by the critical current density J_{cg} inside the grains. The typical values of J_{cj} and J_{cg} in YBCO is about $10^3 - 10^4$ A / cm^2 and 10^6 A / cm^2 respectively. However, it can be noted that J_{cj} depends on grain size aswell as tunnelling current between the grains.

In response to a magnetic field, a three-dimensional array of junctions sets up screening current analogous to that in a bulk sample to prevent the field penetration. By making use of the expression for current through a single Josephson junction and applying the appropriate laws of electrodynamics, one can calculate the effective penetration depth as,

$$\lambda_j = (\, c\, \phi_0\, /\, 8\, \Pi^2\, a\, J_{cj}\,)^{1/2} \qquad (\,11\,)$$

                    A. G. Vedeshwar

which is the characteristic exponential screening length for the applied magnetic field. Similarly, the coherence length for such an effective Josephson medium can be given by [22],

$$\xi_j = 2 a / \sqrt{3} \pi \sim 0.4 a \qquad (12)$$

for temperatures well below Tc, but nominally independent of temperature. The Ginzberg-Landau parameter κ_j can now be easily obtained using above equations and is given by,

$$\kappa_j = \lambda_j / \xi_j = (3 \phi_0 c / 32 J_{cj} a^3)^{1/2} \qquad (13)$$

The lower critical field for Josephson medium can be obtained by making use of the definition $H_{c1} = (\phi_0 / 4 \pi \lambda^2) \ln \kappa$ for a fluxon vortex to enter the medium and is given by,

$$H_{c1j} = (\phi_0 / 4 \pi \lambda_j^2) \ln (\sqrt{3} \pi \lambda_j / 2 a) \qquad (14)$$

Similarly, the upper critical field at which grains are decoupled to give a substantially reduced macroscopic critical current, H_{c2j} can be obtained in anology with the upper critical field given by the Ginzberg-Landau theory and can be expressed as,

$$H_{c2j} = 3 \pi \phi_0 / 8 a^2 \qquad (15)$$

We can take some typical representative values for some of the material parameters for YBCO to estimate variuos quantities given above. If we take $J_{cj} = 10^3$ A/cm^2 and a = 10^{-4} cm, we get λ_j = 5 μm, H_{c1j} = 0.5 Oe and H_{c2j} = 25 Oe. However, it should be noted that there may be distribution of grain size in the sample and in such cases average grain size should be considered. Also, all the above quantities are material sensitive through the grain size. We will see how these ideas are utilised in the interpretations of microwave absorption measurements and sample characterisation.

C. Spin Glass Model:

It was theoretically shown that there exists a spin-glass state in a randomly diluted granular superconductor consisting of three dimensional lattice of Josephson junctions [23-25] quite earlier than the discovery of high Tc superconductors. However, Muller et al [26] showed first the glassy behaviour in high Tc superconductors. They found reversibility in susceptibility of FC sample and irreversibility in ZFC sample. The two curves meet at a temperature $T_C^*(H)$ which is a function of the applied field. Above this temperature for a given field the reversibility exists and below a metastable state prevails just as in case of spin-glasses. Many of the microwave absorption measurements on high Tc superconductors were analysed in the light of this model. There has been fresh developement in theoretical aspects of the model after the discovery of high Tc superconductors [27-30].

In this model also we make use of the Josephson coupling between various grains in the sample. The grains are weakly coupled through the non-superconducting interface material by Josephson interaction. In the absence of magnetic field the coupling energy between the grains i and j can be expressed as [23],

$$E_{ij} = - J_{ij} \cos (\phi_i - \phi_j) \qquad (16)$$

where the exact form of J_{ij} depends on the type of interaction, that is, Josephson or proximity effect. In other words, it will be decided by the medium present between the grains which may be insulator or metallic. For Joephson interaction $J_{ij} = h\ I_{ij}\ /\ 2e$ where I_{ij} is the tunnelling current between the identical grains with same gap parameter $\Delta(T)$. It is already discussed in the previous section. If the coupling is through proximity effect, then

$$J_{ij} = C (1 - T\ /\ Tc)^2 \exp (- r_{ij}\ /\ \xi_n(T)) \qquad (17)$$

where C is a constant, r_{ij} is the seperation between the grains and $\xi_n(T)$ is the coherence length of the interfacing metal.

The entire array of grains can now be described by the Hamiltonian which takes the summation over all the pairs of grains and is given by,

$$H = -\sum_{<ij>} J_{ij} \cos (\phi_i - \phi_j) \qquad (18)$$

We can introduce the phase factor in the Hamiltonian to account for the behaviour of the array in a static applied magnetic field as the following,

$$H = -\sum_{<ij>} J_{ij} \cos (\phi_i - \phi_j - A_{ij}) \qquad (19)$$

where the phase factor $A_{ij} = (2 \pi / \phi_0) \int_i^j \vec{A} \cdot \vec{dl}$
It can be noted that the above formulation is over-simplefied by neglecting the complexities like the dependence of J_{ij} on the applied field, temperature, the grain size and the width of the junction etc. However, it is quite possible to compare these results with experiments by expressing them in experimental units. One such quantity is the ac susceptibility in the absence of the screening effect which is given by [24],

$$- \chi_{ac} = J n S^2 / \phi_0^2 \qquad (20)$$

where ϕ_0 is the flux quantum, J is the average coupling energy, S is the projected area of the loop perpendicular to the field and n is the number of such loops per unit volume. The field at which the first flux slip occurs, that is, where the ac and dc susceptibilities start to differ is given by,

$$H_c^* = \phi_0 / 2S \qquad (21)$$

The field at which the peak occurs in the modulated microwave absorption (MMA)measurements was identified as H_C^* [5].

The key idea of this model is that the superconducting grains are weakly coupled into closed loops. These loops support screening supercurrents in response to an external magnetic field. However, a large cluster can support many supercurrent carrying states of nearly equal energy. It is assumed that there are many such isolated clusters or loops. The diamagnetic response will be depending on the area of such loops, their orientation with the applied field and the number of loops per unit volume. The recent developements [25,27] predict the Meissner phase up to the field $H \leq H_C^{\ell}$, the superconducting glass phase for $H_C^{\ell} < H < H_C^{u}$ and Josephson spin glass phase for $H > H_C^{u}$. These newly introduced critical fields can be expressed interms of H_C^* as,

$$H_C^{\ell} = (3/4) H_C^* \quad \text{and} \quad H_C^{u} = 15 H_C^* \qquad (22)$$

3. EXPERIMENTAL TECHNIQUES :

The measurement of surface resistance of classical superconductors at microwave or radio frequencies was quite an important tool to verify the theory aswell as to test them for their potential applications at high frequencies. Still it is a popular characterisation technique even in high Tc superconductors. However, Microwave absorption measurements on new superconductors with or without modulation is quite a new technique in this area and has been interesting since the discovery of new superconductors. Therefore, I feel it quite necessary and appropriate to give some details about the measurement techniques and the concepts involved.

A. Measurement of Surface Resistance :

The surface recistance of the given sample is measured usually by measuring the quality factor Q of the resonant cavity employed with and without the sample. The fundamental approach involves the measurement of cavity Q as a function of temperature or field with and without the sample. The changes cuased by the sample can be used to determine the surface resistance R_S of the sample. The resonant cavities are normally made up of copper or niobium. They can be designed in cylinderical, rectangular or cubical shape as per the requirements. They can also be disigned to operate in TE_{011}, TE_{012}, TE_{112} etc. or any other desired mode. For a closed cavity either transverse electric (TE) or magnetic (TM) mode can exist. The term 'transverse' refers to the direction of wave propogation, as the direction of transmission of energy in waveguides. The subscripts abc refer to the number of half vave variations of the field along x, y and z directions of the cavity. Different modes resonate at different frequencies. Two such different cavities are shown in figure 2. The quality factor Q_O of any unloaded cavity is inversely proportional to R_S of the inner walls of the cavity with the proportionality constant depending upon the geometry of the cavity. In general, The surface resistance of the unloaded cavity can be expressed as,

$$R_{SO} = G \left(Z_O / Q_O \right) \qquad (23)$$

where Z_O is the characteristic imphedance of free space which is equal to 377 Ω and G is the cavity geometric factor. Therefore, the surface resistance of the sample can easily be determined by,

$$R_S = C \left(\frac{1 + \alpha}{Q_S} - \frac{\alpha}{Q_O} \right) \qquad (24)$$

where C and α are the geometry factors ($C = Q_O R_{SO}$) and Q_S is the quality factor of the cavity with sample. The cavity Q's can be measured using a network analyser. There are many techniques to measure Q's. Among them the bandwidth measurement technique and pulsed-rf methods are widely in use.

The details of these techniques can be found in any standard book on microwave electronics and it is beyond the scope of this article. The R_S can also be estimated by measuring the transmitted microwave power.

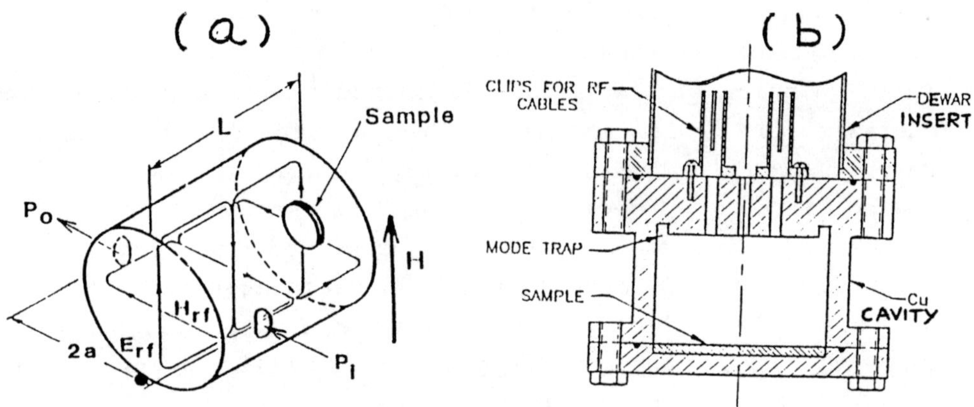

Figure 2. Sketch of the microwave resonant cavities usable for R_S measurements of a superconductor. (a) Cylinderical TE_{112} mode cavity. (b) General $TE_{011 \text{ or } 012}$ cavity.

B. Microwave Absorption Measurements :

Most of the microwave absorption measurements of new superconductors are carried out using commercial EPR spectrometers which employs field modulation technique for phase sensitive detection and gives the field differential absorption spectrum. There are also some measurements without using modulation technique in which conventional microwave waveguide set up with an external field was used. This is quite simple to set up. In this unmodulated absorption measurements either the transmitted or reflected microwave power is measured. The external magnetic field will be applied in such a way that the microwave magnetic field and applied field must be perpendicular to each other. The direction of the mrowave field can be known by the mode of operation (TE or TM) of the cavity as described earlier. The measure of the change in reflected microwave power gives the estimation of the absorption of microwave power by the sample as a function of applied dc field. The temperature dependence of the absorption which is very crucial for superconductors can be carried out by using a specially made double-walled vacuum sealed quartz dewar. Quartz is used to minimise the loss. A small quantity of the sample is introduced in a seperate thin quartz tube ($\sim$ 2-3 mm dia) which will be inserted into the dewar placed in the sample cavity. Liquid nitrogen will be poured into the dewar to cool the sample. However, an eloborate arrangement will be required to control the temperature and to go to liquid helium temperatures. It should be noted that the cavity should be retuned at different temperatures as the cavity goes out of balance due to the change in R_{SO} of the cavity.

The most popular method for measuring the microwave absorption is the use of commercially available EPR spectrometers. It is routinly employed to measure the microwave absorption at paramagnetic resonance in liquids or solids. This kind of resonant absorption of paramagnetic species is normally observed at higher fields of the order of kiloGuass. The microwave absorption of superconductors that we are dealing with is of the non-resonant type and is observed at very low field of the order of few Guass. The absorption or losses can, in general, be determined by measuring the reflection coefficient which is related to the dynamical susceptibility of the concerned material. The dynamical susceptibility per unit volume can be expressed as,

$$\chi = M / H = \chi' - i \chi'' \qquad (25)$$

where χ' represents the dispersion and is referred as in-phase component. χ'' represents the lossy part or aborption and is referred as out-of-phase component. The energy absorbed per cycle is proportional to χ''. The EPR spectrometers employ the modulation of the steady applied field by an audio frequency ($\sim$ 100 kHz) oscillating magnetic field for phase sensitive detection. Therefore, if the amplitude of the modulating field sweeps across the absorption line, there will be a corresponding modulation of the power reflected by the cavity. In this case we get the plot of $d\chi''/dH$ as a function of H. Some people simply call it as $d\Delta P/dH$ where ΔP is the power absorbed. It should be noted that the quantitative measurement of the absorption is very difficult and therefore usaually carried out in relative units. Majority of the spectrometers operate at microwave frequency of about 9.1 GHz. Some care must be taken to use EPR spectrometers for microwave measurements of superconductors. First of all one has to use a set of Helmholtz coils to nullify the remnant field of the electromagnets of the spectrometer. A small quantity of the sample is to be used to avoid the large losses and to help in balancing the bridge. Sample cutting tool should be non-magnetic becuase even very small contamination by magnetic substance will give a large absorption feature over a wide range of field. It should be remembered that the sensitivity of the spectrometer is extremely high and it will sense one part per million of impurity. A specially made dewar as described above can be made use of along with temperature controller for the temperature dependant measurements at low temperatures.

4. RESULTS AND ANALYSES :

After having enough background we can now be able to analyse and understand some of the results of microwave measurements of the granular superconductors. Eventhough there is a large number of data and types of results I will try to give only few representative data analysable using the above models.

A. Surface Resistance :

Most of the available data on R_S of high Tc superconductors were aimed to measure the frequency and temperature dependence becuase these can be understood in terms of BCS theory or two-fluid model. Indeed, almost all the measurements follow the predictions of these two models except very few exceptions. One of them is the difficulty of determining $\Delta(T)$ becauase the measured R_S has contriburtions from various origins. It can be noted that the measured R_S need not be purely due to the mechanisms of the above models. Therefore, the measured R_S can be witten as,

$$R_S(T) \quad = \quad R_{SC}(T) + R_{res} \qquad\qquad (26)$$

One can expect $R_S(0) = 0$ for a perfect superconductor which is not realised in practice. R_{res} represents the residual resistance which can arise from the origins like surface conditions, intergranular materiaals, defects etc. If R_S is caused by large normal conducting areas then it is expected to scale like $\omega^{1/2}$. If on the otherhand, the residual losses are predominantly originating in the intergranular region then the field penetration depth is determined mainly by the superconducting grains and R_{res} should be proportional to ω^2. In High Tc ceramic samples R_{res} dominates over R_S for T < Tc/2 Compared to conventional superconductors $R_S(T)$ should be very small below T < Tc/2 where R_{res} should dominate and near Tc R_S should dominate. Therefore, the frequency dependence of measured R_S at temperatures just below Tc and below Tc/2 should give the frequency dependence of R_S and R_{res} respectively. There is a general agreement on such a behaviour of R_S [31-35]. It is quite consistent with the above models and obeys $\omega^{1/2}$ dependence above Tc [31, 35] as described earlier.

The interesting part of R_S is the R_{res} which is found to be a function of temperature, measurement frequency and the applied dc field. However, there are no many reports in the literature highlighting these kinds of studies. Still, I have shown some representative behaviour of measured R_S as a function of temperature, frequency and applied magnetic field schematically in figure 3.

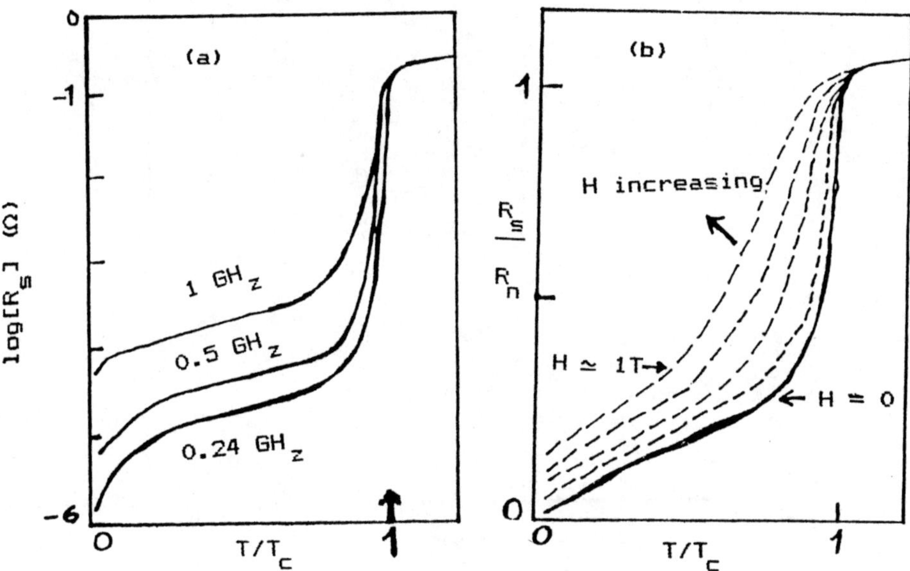

Figure 3. Variations of R_S with (a) temperature & frequency and (b) applied field.

We can further split our measured R_S in terms of various contributions as expressed below,

$$R_S(T) = R_{Sg}(T) + R_{Sj}(T) + R_{SO} \qquad (27)$$

where R_{Sg} and R_{Sj} represent the grain and weak link components of R_S respectively. R_{SO} is due to all other factors like surface roughness, defects etc. and is independant of temperature, frequency and applied field. All these contributions are shown as a function of temperature shcematically in figure 4. These ideas can be tested further by measuring R_S of the samples with different grain sizes in ZFC and FC experiments as a function of temperature. Figure 5

          A. G. Vedeshwar

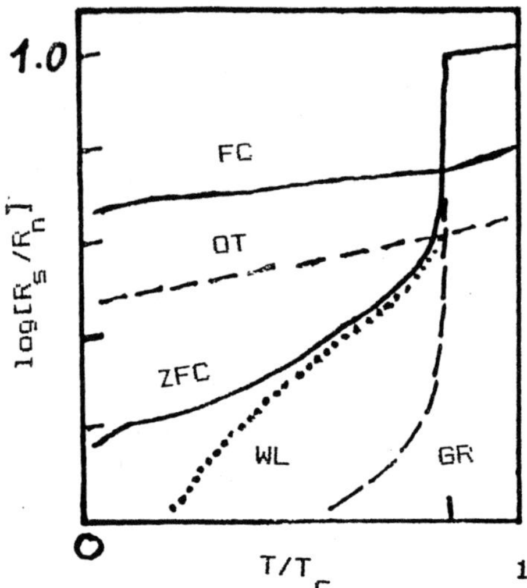

Figure 4. Schematic of the variations of different components of R_S with temperature as can be predicted by granular model.

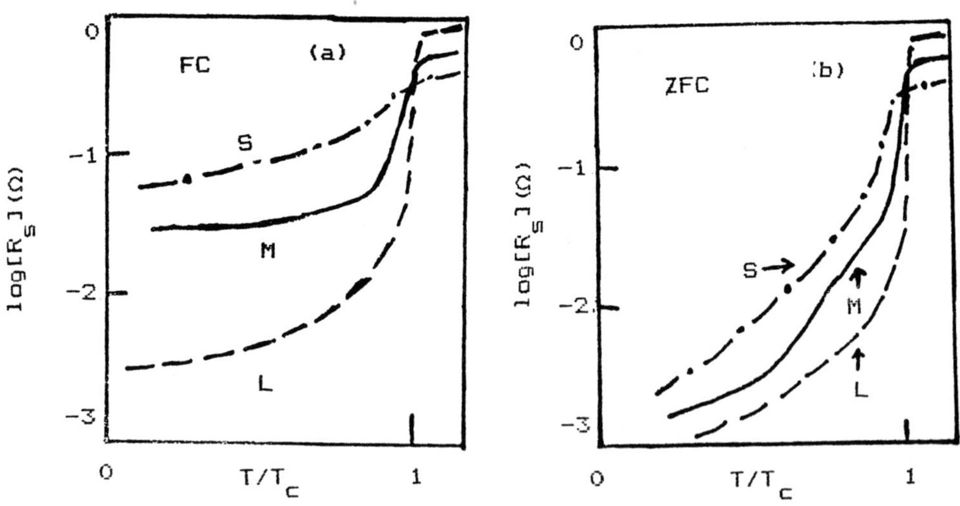

Figure 5. Variation of R_S with te mperature for three different grain sizes (qualitative) for (a) ZFC and (b) FC sample.

shows a schematic of the grain size dependence of R_S for three
defferent grain sizes as the function of temperature in both
ZFC and FC cases. There are some results falling on these
lines [34]. The measured R_S and its temperature dependence
should be able to estimate the superconducting volume fraction
in the sample. The expected dependence of R_S on
superconducting volume fraction is displayed in figure 6. The
label on each curve idicates the volume fraction of
superconducting phase in the sample. Therefore, one can have
the idea about the quality of the sample by just looking at
the steepness of the fall in R_S as depicted by the figure.
However, it can not be a quantitative measure, but can serve
as comparative.

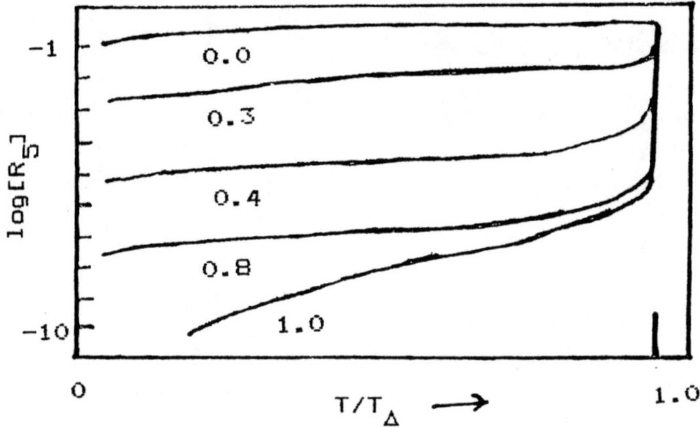

Figure 6. Dependence of R_S on the volume fraction of
superconducting phase in the sample.

B. Microwave Absorption Measurements :

The unmodulated microwave absorption measurements are
very few in the literature eventhough the method is quite
simple and inexpensive. However, it may not offer many finer
details of the absorption curve compared to the modulated
ones. In an unmodulated experiment there are only two fields

perpendicular to each other, the microwave field and the
applied field. The magnitude of the microwave field is too
small to drive the superconductor into the intergranular
critical state. Therefore, one can certainly rely on the
applied dc field at which any observed change in the
absorption spectrum. I have displayed two kinds of unmodulated
absorption measurements in figure 7. The first type involves
the field sweeping technique and the spectrum can be recorded
as shown in figure 7a for various amplitudes of field sweep.
Second technique is simply to record the power reflected in

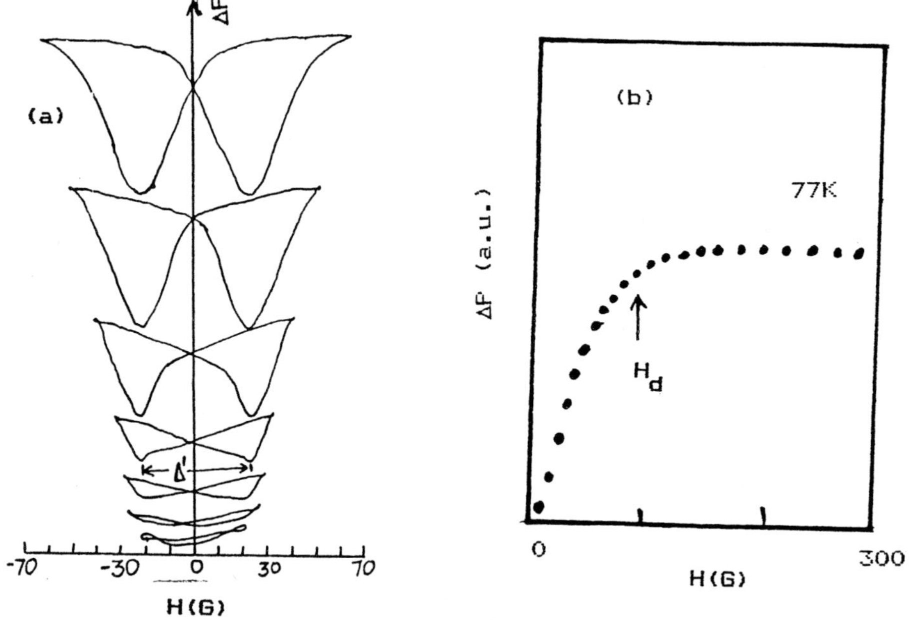

Figure 7. Unmodulated microwave absorption measured by two
different methods as explained in the text.

terms of dc voltage output from crystal detector at each
applied field as shown in figure 7b. The power absorbed is
mainly attributed due to the intergranular critical state.
There has been many attempts to explain these features
[36,37]. One can qualitatively explain these features by
incorporating Bean's critical state model for the

intergranular region at low fields. The result shows local power absorption is proportional to the local field [37]. The calculation of total power absorbed as the function of field reproduces the same features as shown in 7a. The main parameter of importance is the hysteresis splitting Δ' as defined in figure 7a. It can be used to estimate the macroscopic critical current density according to the relation $\Delta' = d\mu_0 J_C$ where d is the half of the sample thickness perpendicular to the applied field. This is a similar result to that of determining J_C from magnetization measurements. Δ' is a function of amplitude of the field sweep as can be seen from the above figure. It increases rapidly initially for small fields to saturate at a field of few Guass. The decreasing Δ' with increasing temperature as has been observed can be understood as due to decreasing J_C.

In the other kind of experiment shown in figure 7b the parameter of interest is the field at which the absorbed power gets saturated as indicated in the figure. This field was identified as the upper intergranular critical field or simply decoupling field at which grains are decoupled [38]. There are some reports on similar lines showing the dependence of absorption on grain size, frequency etc. [39-41]. However, the frequency dependence of Δ' is hard to understand as the above explanation does not predict any such behaviour. The powder data are still more difficult to digest.

The modulated power absorption measurements are quite diversified and plenty in the literature. Some of the parameters determined by these studies have already been established beyond doubt for sample characterization. I will mention only some important results here becuase the coverage of all aspects and results is quite impossible in this small artical. However, the reader can get more information and other details from the references cited here and many more in

the literature. This will give you an over all picture of the
phenomenon. The important thing to remember is to cool the
sample strictly in zero field in a ZFC measurement. It can be
noted that the earth's magnetic field is about 0.5 Guass which
is quite sufficient to drive the sample into intergranular
critical state. Real zero field situation can be achieved by
shielding the sample using a sheet of mu-metal during cooling.
A typical recorded modulated microwave absorption of Hafnium
doped YBCO using a Varian EPR spectrometer is shown in figure
8a while 8b shows Cu^{2+} paramagnetic resonance absorption which
may be due to impurity phase or enhanced valency of Cu [10].
The arrows indicate the direction of the scan. Horizontal bars
show the starting point and the broken curve shows the second
scan. The change observed in the second scan is attributed to
to the trapping of flux at intergranular region thereby

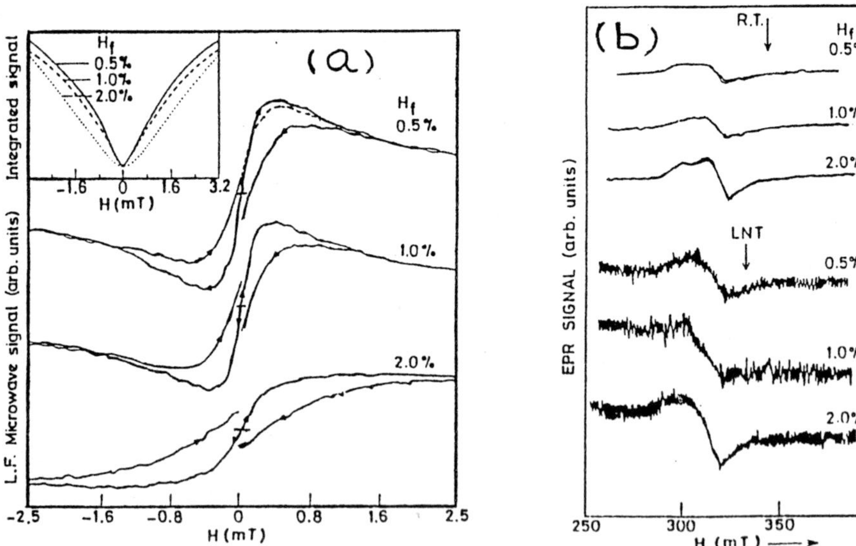

Figure 8. (a) MMA of Hf doped YBCO. The percentage of Hf
doping at Y site is indicated for each curve at extreme right.
(b) Cu^{2+} paramagnetic resonance absorption (see text.)

decreasing the dissipation. We will know more about it little
later. Since we know that signal is essentially $d\Delta P/dH$ we can

numerically integrate it to get ΔP as the function of H as shown in the inset for the purpose of comparison with unmodulated experiment. The agreement is not surperising because the spectra were recorded at very low amplitude of modulation and scanned upto few Guass of applied field. As a characterisation technique, it is quite evident from the figure that there are distinct changes in the spectra as a function of doping concentrations. The important parameter is the peak position. This can be identified as the lower intergranular critical field of the granular system. If we assume the grain size is not smaller than the junction length, we can make use of the relation for a single Josephson junction to evaluate the grain size and is given by,

$$H_O \;=\; \phi_O \,/\, 2\,L\,\lambda \qquad\qquad (28)$$

where λ is the London penetration depth of the grain. Here we have assumed the junction length L gives the measure of the grain size and the junction width is neglected compared to L. The average grain size observed by SEM of these samples tallies very well with the determined ones as above [10]. As can be seen from the figure, the area of the hysteresis loop increases with doping concentration. It clearly indicates the increase in J_{cj} which can quite clearly be understood as due to flux pinning.

The temperature evaluation of absorption spectrum along with the variation of some of its parameters with temperature is shown in figure 9 [11]. As explained in theory, the peak position is the measure of H_{c1j} and varies as expected in analogy with H_{c1} and is shown in 9b. The area of hysteresis loop ΔA decreases with increasing temperature becuase of decreasing J_{cj}. Therefore, it is quite consistent with the theoretical explanation as described earlier.

316   A. G. Vedeshwar

It is quite possible to determine both lower and upper
intergranular critical fields of the sample using MMA as
illustrated in figure 10 [12]. The determination of H_{c1j} as a
function of temperature is straightforward as swon in figure
10a. Now, if we integrate numerically the spectra of 10a, we
get figure 10b and the upper critical field H_{c2j} can be
determined as shown in the figure. The temperature dependence
of both H_{c1j} and H_{c2j} determined from figure 10 is shown in
figure 11. We can also calculate H_{c2j} using relation (15)

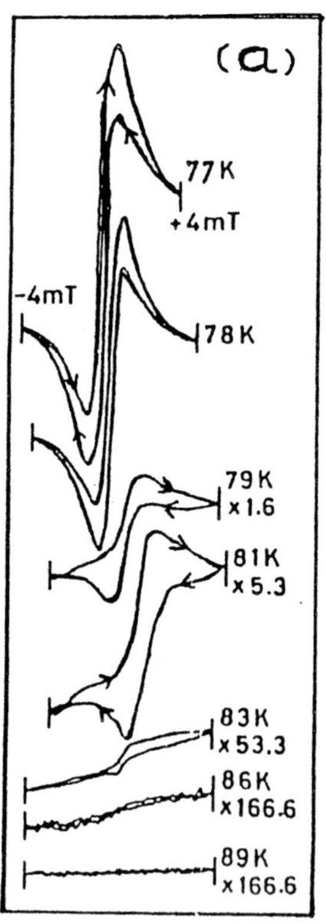

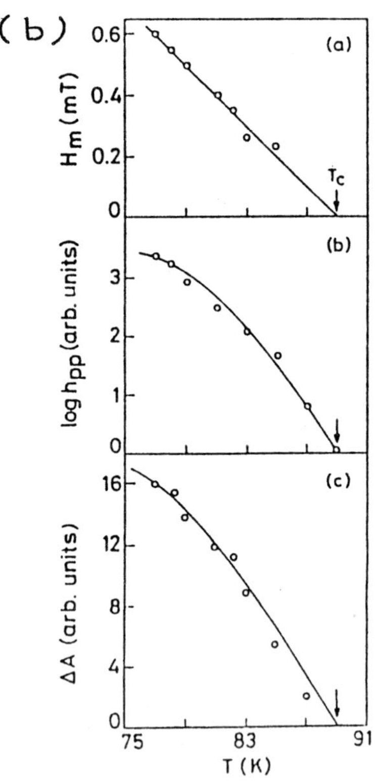

Figure 9. (a) Temperature dependence of MMA (b) Variation of
peak position H_m, peak to peak height h_{pp} and area under
hysteresis loop ΔA determined by (a) with temperature.

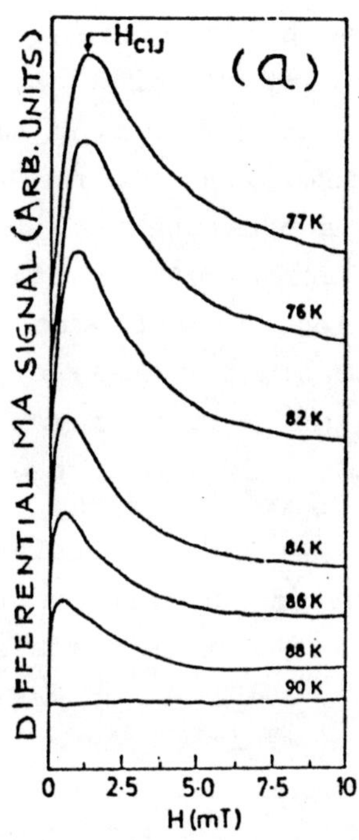
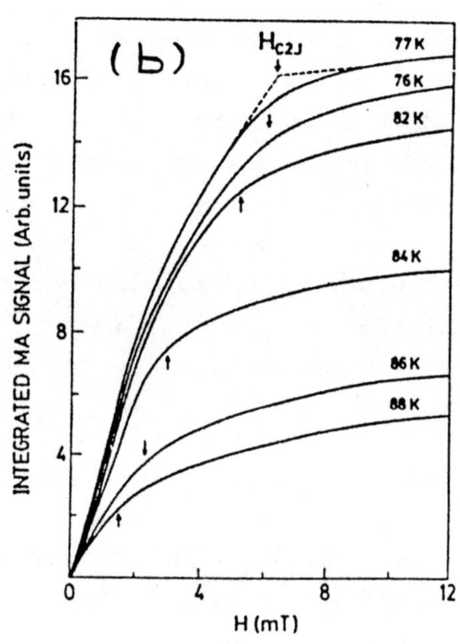

Figure 10. Determination of (a) H_{c1j} and (b) H_{c2j} for a YBCO sample as explained in the text.

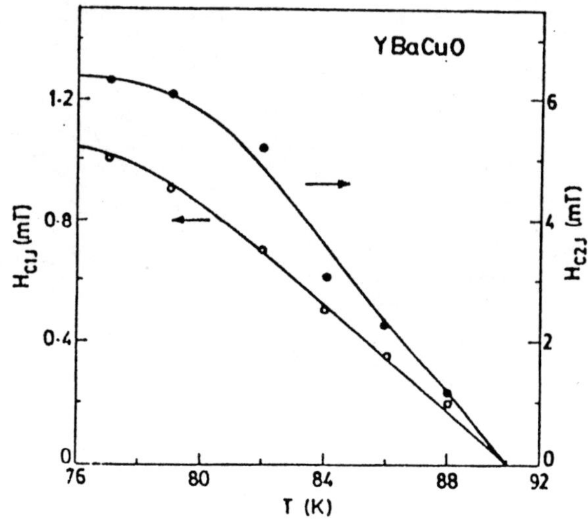

Figure 11. Variation of H_{c1j} and H_{c2j} with temperature as determined from figure 10.

provided grain size is known. If we believe the granular
system as consisting of cluster of superconducting grains as
discussed in spin-glass model, we can calculate the maximum
cluster size by identifying the peak position of figure 10a as
H_C^* and using the relation (21). If we substitute this cluster
diameter in place of grain size in relation (15) to calculate
H_{C2j}, the agreement with those determined from figure 10b is
very good. This fact indicates that the cluster diameter is an
appropriate parameter rather than the grain size if the grain
size is not homogeneous in the sample as observed in the SEM.

The use of cluster model in interpretting MMA
measurements can further be confirmed by FC and ZFC
experiments. There are various data like surface resistance,
ac susceptibility, magnetization, resistivity, microwave
absorption etc. for ZFC and FC cases. However, because of its

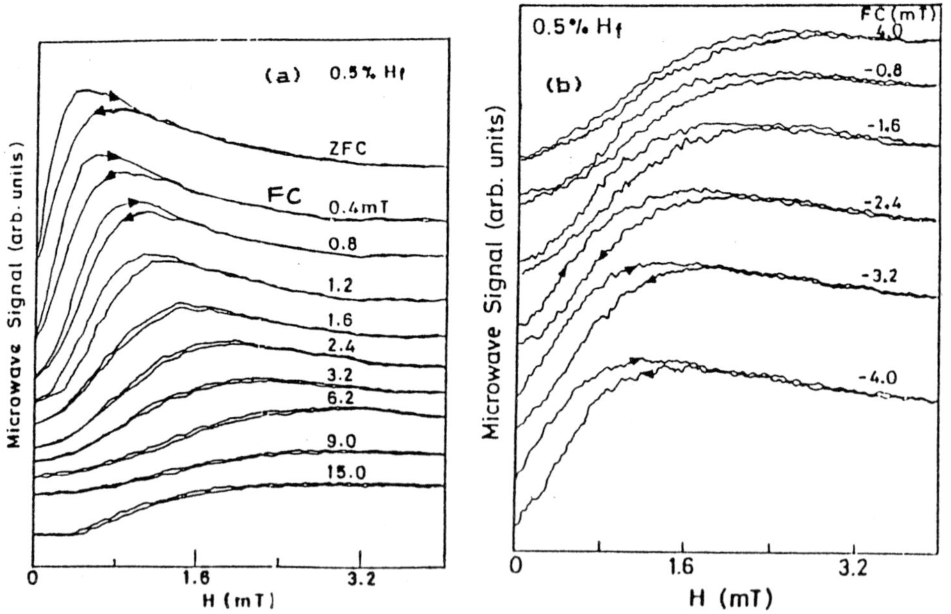

Figure 12. (a) MMA of ZFC & FC sample for various cooling
fields. (b) MMA recorded after applying various negative
fields to FC sample as explained in the text.

high sensitivity MMA is better for observing the difference between ZFC and FC cases especially at very low fields. In ZFC experiment sample is cooled below T_C in a zero field and then the field is increased to desired values. Depending upon the maximum field applied the measured absorption shows hysteresis as shown in figure 12a. Now, if we record once again it will follow the broken curve as shown. The hysteresis is known to be due to intergranular flux trapping. The interesting result emerges in FC case. If we cool the sample below T_C in a magnetic field (as labelled on each curve) and then record the spectra, we get the response as depicted in figure 12a. The main changes observed as a function of cooling field are the shift in peak position towards higher side and closing of hysteresis loop (decrease in ΔA). Decrease in ΔA can be understood as due to the initial trapped flux during FC and no further trapping of flux will take place for an applied field lesser than the cooling field. The shift in peak position can be easily explained invoking the cluster model. As discussed above, the peak position assumed to be H_C^* is inversely proportional to the area S of the cluster (eqn. 21). Therefore it is quite evident that the size of the cluster formed during FC goes on reducing as a function of cooling field becuase of the trapped flux between the grains. As we have seen the coupling strength depends on grain size via tunnelling current, only those grains with reduced coupling strength by trapped flux are excluded from the cluster thereby reducing the cluster size. So, the cluster size goes on shrinking with the increasing cooling field. This idea can furter be tested by removing the trapped flux from a FC sample by simply applying a negative field. As has been shown in figure 12b, the FC curve of the sample cooled at 40 Guass recovers towards ZFC case systamatically when the trapped flux is removed gradually. The number with negative sign labelled on each curve shows the negative field applied to rmove the trapped flux gradually. However, it can be noted that the

                           A. G. Vedeshwar

curve does not recover fully to ZFC case despite the removal
of all the trapped flux by applying a negative field equal to
the cooling field. This shows the presence of some trapped
flux which may be due to some other pinning mechanisms. All
the above arguments are summarised in figure 13 interms of
variations in cluster diameter d in various situations
described above [13]. Therefore, MMA measured carefully and
systematically offers lot of informations useful for both
characterization and understanding of basic physics involved.

The interesting features are obtained in the single
crystals also [42,43]. The MMA contains many closely spaced
spikes at a seperation of few milliGuass having overall
envelope similar to that of a granular sample. Since the

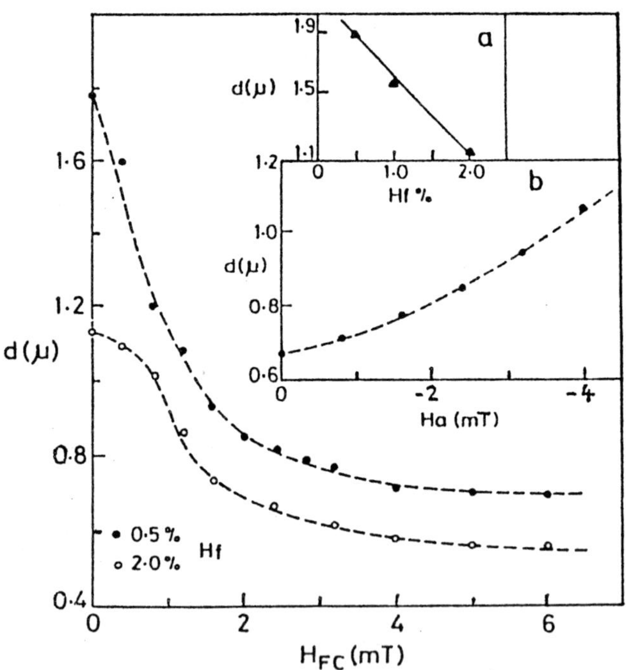

Figure 13. Variation of cluster diameter d with cooling field.
Inset (a) shows d as a function of doping and inset (b) shows
the recovery of cluster size after the removal of trapped flux
by applying negative field.

discovery of high Tc superconductors, microwave measurements gained momentum and is vastly expanding. New ideas and experimental techniques are put in to find various kinds of informations. Therefore, it is quite difficult to cover various diverging interpretations in this article of general nature. I have only tried to expose the tip of the iceberg of this fascinating technique to the reader. The effort was to give a down-to-earth picture of the whole scenario. There are still lot many things to resolve and understand various microwave data. Therefore, this area is quite active and offers lot of scope for further ivestigations.

REFERENCES

1. P. B. Miller, Phys. Rev. **118**, 928 (1960)

2. J. I. Gittleman and B. Rosenblum, Proc. IEEE **52**, 1138 (1964)

3. J. Halbritter, Z. Physik. **266**, 209 (1974)

4. S. V. Bhat, P. Ganguly, T. V. Ramakrishnan and C. N. R. Rao, J. Phys. **C 20**, 1559 (1987)

5. K. W. Blazey, K. A. Muller, J. G. Bednorz, W. Berlinger, G. Amoretti, E. Buluggiu, A. Vera and F. C. Matacotta, Phys. Rev. **B 36**, 7241 (1987)

6. K. Kachaturyan, E. R. Weber, P. Tejedor, A. M. Stacy and A. M. Portis, Phys. Rev. **B 36**, 8309 (1987)

7. K.N. Shrivastava, J. Phys. C 20, L 789 (1987)

8. A. Duclic, R. H. Crepau and J. H. Freed, Phys. Rev. **B 38**, 5002 (1988)

9. K. N. Shrivastava, Solid State Commun. **68**, 259 (1988)

10. A. G. Vedeshwar, Md.Shahbuddin, P.Chand, H.D.Bist, S.K. Agarwal, V.N.Murthy, C.V.N.Rao and A.V.Narlikar, Physica **C 158**, 385 (1989

11. A. G. Vedeshwar, Md.Shahbuddin, H.D.Bist, S.K.Agarwal,
    V.N.Murthy, C.V.N.Rao and A.V.Narlikar,
    Physics Letters **A 139**, 415 (1989)

12. A. G. Vedeshwar, Solid State Commun. **74**, 23 (1990)

13. K.N. Srivastava, M. Puri, J. Bear and L. Kevan

    J. Chem. Soc. Faraday Trans. <u>87</u>, 3893 (1991).

14. K. N. Srivastava, Phys. Repts. **200**, 51 (1991)

15. K. N. Srivastava, Solid State Commun. **85**, 227 (1993)

16. Yu. N. Shvachko et al, Solid state Commun. **69**, 611 (1989)

17. K.N. Srivastava, Bull. Mat . Sci. (India) <u>14</u>, 625 (1991).

18. T. Van Duzer and C. W. Turner, in " Principles of superco-
    nductive Devices and Circuits " Published by Edward Arnold
    London, 1981.

19. D. C. Mattis and J. Bardeen, Phys. Rev. **111**, 412 (1958)

20. J. R. Clem, Physica **C 153 - 155**, 50 (1988)

21. E. B. Sonin and A. K. Tagantev, Sov. Phys. JETP **68**,
    572 (1989)

22. M. Tinkham and C. J. Lobb, Solid State Physics **42**, p-91
    (Ed.:H. Ehrenreich and D. Turnbull, Academic Press,
    New York, 1989)

23. W. Y. Shih, C. Ebner and D. Stroud, Phys. Rev. **B30**,
    134 (1984)

24. C. Ebner and D. Stroud, Phys. Rev. **B31**, 165 (1985)

25. S.John and T.C.Lubensky, Phys. Rev. Lett. **55**, 1014 (1985)

26. K. A. Muller, M. Takashige and J. G. Bednorz,
    Phys. Rev. Lett. **58**, 1143 (1987)

27. I. Morganstern, K. A. Muller and J. G. Bednorz,
    Physica **B 152**, 85 (1988)

28. D. Stroud and C. Ebner, Physica **C 153 - 155**, 63 (1988)

29. V. L. Aksenov and S. A. Sargeenkov, Physica **C 156**, 18 and
    235 (1988)

30. T. Xia and D. Stroud, Phys. Rev. **B** 39, 4792 (1989)

31. J. R. Delayen and C. L. Bohn, Phys. Rev. **B 40**, 5151 (1989)

32. J. I. Gittleman and J. R. Matey, J. Appl. Phys. **65**, 688
    (1989)

33. Q. Li, K. W. Rigby, and M. S. Rzchowski, Phys. Rev. **B 39**, 6607 (1989)

34. M. Sato and T. Konaka, Supercond. Sci. Technol. **4**, 427 (1991)

35. C. L. Bohn, J. R. Delayen, U. Balachandran and M. T. Lanagan, Appl. Phys. Lett. **55**, 304 (1989)

36. A. M. Portis, K. W. Blazey, K. A. Muller and J. G.Bednorz, Europhys. Lett. **5**, 467 (1988)

37. A. M. Stewart, Supercond. Sci. Technol. **4**, 32 (1991)

38. R. Marcon, R. Fastampa, M. Giura and C. Matacotta, Phys. Rev. **B 39**, 2796 (1989)

39. S. Tyagi, A. Gould, G. Shaw, S. M. Bhagat and M. A. Manheimer, Phys. Lett. **A 136**, 499 (1989)

40. E. J. Pakulis abd T. Osada, Phys. Rev. **B 37**, 5840 (1988)

41. T. C. Choy and A. M. Stoneham, J.Phys.:Condens. Matter **C 2**, 939 (1990)

42. K. W. Blazey, A. M. Portis, K. A. Muller and F. H. Holtzberg, Europhys. Lett. **6**, 457 (1988)

43. K. N. Shrivastava, Solid State Commun. **68**, 1019 (1988)

Keshav Shrivastava did his B.Sc. degree at Agra University, M.Sc. at Allahabad University, Ph.D. at the Indian Institute of Technology, Kanpur, and D.Sc. at Calcutta University. He visited the University of California at Santa Barbara, University of Houston, University of Nottingham, State University of Utrecht, Royal Institute of Technology, Stockholm, International Centre for Theoretical Physics, Trieste, Technical University of Vienna, etc. He is Professor at the University of Hyderabad in India. He has discovered the energy shift arising from the spin-phonon interaction, flux quantized energy states in superconductors and made important contributions to the solid state. He is a fellow of the National Academy of Sciences, India and a fellow of the Institute of Physics, London, U.K..

Subject Index